SEVEN DECADES

Seven Decades

HOW WE EVOLVED TO LIVE LONGER

MICHAEL D. GURVEN

PRINCETON UNIVERSITY PRESS

PRINCETON & OXFORD

Published by Princeton University Press
41 William Street, Princeton, New Jersey 08540
99 Banbury Road, Oxford OX2 6JX

press.princeton.edu

GPSR Authorized Representative: Easy Access System Europe - Mustamäe tee 50, 10621 Tallinn, Estonia, gpsr.requests@easproject.com

Library of Congress Cataloging-in-Publication Data is available

ISBN 9780691231990
ISBN (e-book) 9780691232003

British Library Cataloging-in-Publication Data is available

Editorial: Alison Kalett, Hallie Schaeffer
Production Editorial: Elizabeth Byrd
Jacket: Chris Ferrante
Production: Jacqueline Poirier
Publicity: Matthew Taylor (US), Kate Farquhar-Thomson (UK)

Jacket Credit: diane39 / iStock

Printed and bound by CPI Group (UK) Ltd, Croydon, CR0 4YY

10 9 8 7 6 5 4 3 2 1

For Evie and Oliver

CONTENTS

SEVEN DECADES

Introduction

"OLD FOLKS are not afraid to talk. *Anic chij peyaquij mu'in.* [They know how to speak their minds]. And when they speak, the young people listen," explained Francisco, the chief of a Tsimane village located along the Maniqui River in the Bolivian Amazon, far from the nearest town.

He continued, "*Isho'muntyi* [old people] don't get lost in the forest, they know the short-cuts, they can find their way to get anywhere. . . . They lead [trips] into forbidden areas where the *jäjäba* forest guardians live."

The tropical moon was nearly full, and in the delicate shimmer of night, I could just barely make out his expression. He described other abilities and privileges held by older adults—the isho'muntyi with speckled hair who have lived many years.

His body quivered suddenly, and he began to chant, "*tsäquij tsäquij . . .* [it's dangerous, dangerous]. . . . To be on your own is dangerous. Children sometimes abandon their parents, they are stingy with them. Life can be hard." In the past, he explained, elders who were all alone would sometimes poison themselves by eating charcoal, and end their lives.

This conversation occurred almost five years after I had first visited Bolivia in 1999 and had the good fortune to live and work with the Tsimane (pronounced chee-MAH-nay) people. The Tsimane are Amazonian slash-and-burn "horticulturalists"—they are rain forest dwellers who make a living by growing a limited number of crops in small fields, crops such as sweet manioc root, corn, and rice. Like most other Amazonian groups, they also hunt, fish, and gather wild fruits and honey.

When I first came to Bolivia, I wanted to better understand all the intricate ways people work together to eke out a living. Not just the technical aspects, but the social and cultural means of binding fates—how do these help promote a long, healthy life? How do people like Francisco even reach old age in the harsh *infierno verde* (the "green hell," as it was called by outsiders who waded through the Bolivian jungles in the early twentieth century)?

As a graduate student in the late 1990s, I had studied how cooperation and sharing help buffer hunter-gatherers against food shortfalls and other common risks. I was privileged to study among the Ache of Paraguay for my doctoral dissertation.[1] Living among the Ache first opened my eyes to how survival in extreme environments was as much, if not more, about trusting others and having others trust you than about figuring out how to start a fire with wet logs or avoiding snake bites.

But in Bolivia, I was hoping to untangle how departures from a full-time foraging lifestyle might change survival strategies. How might a less nomadic lifestyle, where people farm and live in semipermanent settlements, affect how people survive the day-to-day and how they cope with different hardships and disasters? I wanted to know how these experiences might leave their imprint in the body, and how the body ages under such conditions. In the popular imagination, life in the wild is tsäquij, just as the Tsimane sometimes sing—requiring vigorous strength, close ties with others, and the know-how that comes from long-standing cultural traditions. Life, in the timeless and cynical words of the philosopher Hobbes, is "nasty, brutish and short." It's a young person's world. And so maybe there is no real aging under such conditions, just lives cut short close to one's prime; elders are just the lucky few to survive life's literal slings and arrows.

The above would seem to support the dominant narrative among many people—both lay public and scholars alike—that human life expectancy increased dramatically over the past two centuries but was low over much of human prehistory. And so, among our preindustrial ancestors, reaching a late-age life stage was a combination of lucky happenstance and honorable defiance against the elements. As the story goes, the dawn of mass sanitation, public health awareness, and antibiotics gave rise to the improved conditions that allowed growth, development, health, and aging to be transformed in fundamental ways. Certainly there are some sweeping differences: the height of an adult man in the early nineteenth century was between 5 feet, 3 inches (Russia) and 5 feet, 7 inches (UK), and he lived to about 25 to 35 years of age, on average, whereas, by the end of the twentieth century, average height had reached about 5 feet, 10 inches in both countries, and men lived to about 60 years in Russia and 74 years in the UK. Though these facts about the changing human conditions over the past two centuries are true, we can still flip the traditional narrative on its head.

The dominant narrative ignores a basic biological fact: our body was built to last roughly seven decades, and this longevity has been a fact not only for recent centuries, but for many millennia. Our massive gains in life expectancy over the past two centuries reflect a larger percentage of people over time living to older ages, rather than any dramatic extension of longevity. Extended life has long been a part of our human experience. It is an immutable characteristic of our species. This human life course, like that of any species, is a coordinated system, whose moving parts that seem to age at different rates make sense only once we realize that we evolved to produce and convert energy into babies. But we achieved this in a particular way, given the new physical and social environments our hominin ancestors inhabited and helped create.

Modern amenities may further prolong healthy human lifespan, but major improvements are still in the realm of science fiction. Nevertheless, more people are reaching their longevity potential than ever before. This is a boon for the majority of us likely to celebrate many more birthdays. It's also a potential crisis, with headlines decrying the many implications of our rapidly graying population—on work, retirement, health care, housing, and caregiving. We're at a pivotal time to rethink aging, to imbue elderhood with purpose, and to entertain new opportunities to harness the experience, wisdom, and know-how of older adults.

———

In this book, I argue that our potential for a seven-decade lifespan first evolved at least 50,000 years ago, way back during the time of our hunting-and-gathering ancestors. Our seven-decade lifespan didn't emerge after the Agricultural Revolution some 12,000 years ago, and it certainly didn't emerge after the Enlightenment of the eighteenth century or even after the Industrial Revolution. The evidence points to emergence in the early days of our anatomically modern species, *Homo sapiens*. But why? To better understand the mosaic of conditions that shaped the life history of our ancestors over past millennia, we can look to contemporary hunter-gatherers and other small-scale subsistence populations. These few remaining populations can give us a glimpse of how our long life came to be, and why. We'll come to see that a long lifespan is more than just a quirky feature of us hairless apes. It defines who we are and enables all that we generally take for granted—from our slow-growing children, our over-stuffed brains and the rich mythologies that fill them, and our multigenerational social grooming.

This capacity for long life is part of a bundle of traits that makes us unique among mammals and even among our primate cousins. This human "life history" includes the capacity to live not just a long life, but to live decades beyond menopause—an outlier in the natural world, where making babies to pass genes to future generations is the main metric of evolutionary fitness. For most animals, the body dwindles and declines at about the same rate as the diminishing ability to bear young. The decoupling of our ability to procreate from every other aspect of our physical body is a zoological anomaly—one that has only increased further in the twenty-first century. Globally, 24% of the human population, or 1.9 billion people, are now living at "postreproductive" ages (that is, over age 50)—a demographic feat unheard of in the history of our species. Even in 1950, the proportion of "over 50s" was 16%, or 0.4 billion, and by 2100, the UN projects it to be an astonishing 40%, or 4.4 billion people. By 2050, it's predicted that one in six of the global population will be age 65 and over. The number of people aged 80 and over will triple in the next three decades. United Nations projections label these trends a global "longevity revolution."[2]

The key to understanding aging and lifespan for any species is to grasp its natural history—the enduring exposures and selection pressures that shaped the way species evolved. For humans, this means appreciating aging in the context of a lifestyle of hunting and gathering. As a species, we were full-time hunter-gatherers for most of our evolutionary history (which many still practiced until five hundred years ago). And for millennia we also included domesticated plants and animals as part of our ever-changing livelihoods as farmers and pastoralists.

This historical fact may seem uncontroversial, even obvious. But the biomedical and pharmaceutical industry, hell-bent on extending lifespan by slowing, halting, or even reversing aging, operates as if natural history matters little. Moreover, some researchers who focus on aging believe that study of our past and of high-mortality populations in the present only obscures understanding of how aging operates. The leading belief that dominates the longevity industry is that human aging can be studied and understood only in low-mortality populations like the United States, Japan, and Sweden, living under the best of conditions. Such a view is myopic and cannot provide much insight into the flexibility of human physiology.

What is relatively fixed as part of our species-typical biology versus modifiable through changes in lifestyle, diet, and our social interactions? Can aging be further slowed if mortality rates at late ages continue to decline according to current trends?

To address these and other gnawing questions, we'll first consider what old age is "for." To paraphrase a worn but wonderful quote: nothing about our body makes sense except in the light of evolution.[3] I harness evolutionary thinking to help reveal *why* we can live long lives and age the way we do. On our journey, we'll explore old age among hunter-gatherers and the critical but sometimes forgotten role of older adults, or elders as I respectfully call them (not elderly), in our own societies. We'll examine what older adults actually do in foraging and other subsistence populations, and how they live and thrive. This approach is relatively new territory because the study of aging and lifespan itself is a modern luxury. Most of what we know about aging and longevity comes from the past half century, and from Global North countries like the United Kingdom, United States, Sweden, and Germany.

The health-conscious public is usually captivated by considering potential lessons to be learned by stripping away the many layers of our contemporary uber-"civilized," ultraprocessed, and protected existence.

This sadly often reduces to platitudes about the simple joys of outdoor living among romanticized "noble savages." Or wild appreciation for the modern comforts that distance us from a less romantic view of what past life must have been like. Balancing the rich ethnographic record with my own good fortune to have lived with several small-scale subsistence populations, I hope to walk the fine line between selective grazing of "ancient" lives and overgeneralizing from too few examples. By stepping outside what is most familiar, we can gain fresh insight about our golden years. That insight can help us rethink how to gain purpose and meaning from our elderhood, and how to spend those years a little healthier and less lonely.

Practitioners of "geroscience" and other fields focused on molecular and cellular aging are overly confident that they can make what's merely possible become a reality. Gains may be achieved by pushing back with technological tools against vitality-sapping entropy—with a forward arrow of progress toward ever-longer lives.

Yet the study of nonindustrial populations offers a privileged opportunity for examining the conditions that have shaped not just our bodies and biology, but also our emotions and our minds. The quest to understand the whys and wherefores of old age across human populations helps orient us, not by showing how far we've come but by illuminating what might be possible about how to live, and to live well. Understanding how we lived and thrived in the past can help inform current initiatives to maximize human "healthspan," and help us better confront the current crises brought about by the "silver tsunami" of older lives. Such insights are sorely needed as we face the many challenges of global aging in the twenty-first century.

Organization of the Book

I've divided this book into three parts, operating on the idea that all good things come in threes (*omne trium perfectum*).

In part I, I address what we actually know about human lifespan and aging over the course of human history. Across time and space, many people have indeed survived well into their sixties. This evidence supports my claim that the human body was built to last about seven decades under preindustrial conditions. Hunter-gatherers and other groups whose lifestyles are our best guess of what life might have been like prior to intensive agriculture and the Industrial Revolution all contain older adults as group members. Their life expectancies may be low, but a signal of long life potential manifests nonetheless. Our massive gains in life expectancy over the past two centuries reflect a larger percentage of people over time living to older ages, rather than any dramatic extension of longevity. Our whirlwind tour will take us from Australopith ancestors 4–6 million years ago through the origins of anatomically modern *Homo sapiens*. We'll go from prehistory to the earliest written indications of what old age may have been like, including examples from ancient Greece and Egypt.

We all hold different views of what aging is and is not, and my argument will shed light on this complex terrain by revealing why aging exists in the first place. To best situate "old age" as part of the natural history of our species, we'll cartwheel through many disciplines, especially evolutionary biology, demography, anthropology, and psychology. As I move through the demographics and theory on aging and longevity, I'll take some small detours along our journey, but I promise they'll be fun and interesting.

In part II, I lay out a proposal for how human longevity came to be in the first place. From the perspective of evolutionary biology, living beyond one's reproductive years is like a superpower. It's a puzzle rarely observed in nature, except in the case of a few mammals. To help reveal why postreproductive life itself is usually rare, I introduce the concept of the declining "force of selection" with age. Longevity and other unique traits make us stand out among our large-bodied primate cousins: our children grow slowly and require years of nurturing and care; our fat-rich energy-demanding brains help foster a lifetime of learning. It takes up to three decades to become proficient enough in the varied skills needed to make a living as a hunter-gatherer, but success yields surplus. Cooperation and sharing of bonanzas within and among generations are not just quirks of being a generous primate—they are fundamental to making our human life history possible. We'll play with some contentious ideas in this arena, and debate whether we should thank women or men for our evolved longevity. We'll also contend with whether late life even has a "function," or whether it is instead just a by-product of greater robustness in adulthood.

The multigenerational organization of extended families with divisions of labor go hand in hand with cognitive and social abilities primed to promote efficiency and fairness in small groups: attention to others' abilities, work effort, level of need, and reputation. These nurtured sensibilities help boost the gains from helping and being helped. With this theoretical gloss rooted in the social economics of hunter-gatherers, we'll consider the many ways that elders in traditional societies help others—through food production, childcare, pedagogy, ritual expertise, storytelling, leadership, and conflict mediation, and as breathing, walking repositories of ecological knowledge. These roles provide benefits to kith and kin and help ensure that transfers of food and other goods and services flow along the generational ladder, from older generations to young, and back up the ladder as old age security.

Are proposed services provided by elders in small-scale societies of hunter-gatherers, farmers, and pastoralists—like education, leadership, and babysitting—just different flavors of entertainment filling idle time, or might these activities be important enough to help ensure that enough older adults are around to provide them? In the absence of these elders, would society collapse?

Cooperation is a critical human enterprise, and elders may be immersed in the daily goings-on in small-scale societies, but this doesn't mean that older adults are always treated well. To what degree are older adults held in high esteem, and what might be the tangible value of respect on individual lives? What do elders actually *do* in their "retirement"? Is there even a retirement to speak of? Why, as Francisco said, would suicide ever be considered a desirable exit, either for the sad neglected elder, or for the relatives left behind?

We'll cover the full range of treatment and regard, from gerontocracy to geronticide. The notion of "interdependence," the extent to which older adults supply irreplaceable goods or services, helps explain where older adults stand in this regard. Being of use in an interdependent social landscape reliably predicts how much others look after older adults. Once utility is low enough, our own incentives to stay alive decrease, and others' vested interests in keeping us alive may plummet as well. Most cases of geronticide and neglect occur when families and communities deem older adults no longer useful. Given conflicts of interest within extended family networks, the opportunity for elder abuse, neglect, and exploitation is the dark underbelly of being prosocial.

In part III, we bring our new appreciation of ancestral human demography and of how lifespan fits within our evolved life history to help think about healthy "aging in the wild." Understanding "healthy aging" is big business, a worthy goal to maximize physical and emotional well-being for as long as possible. Is it realistic to maintain vigor and delay physical decline until the last of our years? The human body may have been built to last seven decades under more traditional, nonindustrial conditions, though in much of the developed world today, it appears to be eight or nine decades. *What has changed that now shifts our warranty period by over a decade?*

And how do we make sense of the patterned way in which things fall apart with age? You would think that natural selection should lead to a simultaneous rise and fall of all body functions and processes. And it would certainly be easier on us if everything fell apart abruptly and simultaneously only at the end of our days, much like Oliver Wendell Holmes's wonderful one-hoss shay—built to last intact until "it went to pieces all at once." Is there a logical method to the madness of physical change over the life course? Our knees tend to go out before our hamstrings, and our kidneys have a harder time working efficiently at later ages than our liver. These kinds of observations over the course of my career have made me wonder whether there is any part of our body that stays the same, or even improves with age. Though the body is a system of many systems, we'll cut through the morass and make sense of age-related changes in a few key domains: muscular strength and endurance, our immune system(s), and several types of cognition affecting decision-making, memory, and wisdom.

From there, we'll tackle key questions, like these: Are the chronic diseases of aging—the modern scourges of the Western world, like heart disease, diabetes, cancer, and dementia—inevitable consequences of living past seven decades? Have they always been with us? Again, we look to contemporary hunter-gatherers and other subsistence societies to see whether they endure the same diseases that we've come to fear and dread in our old age. We take head-on whether these "diseases of civilization" are instead by-products of our evolved physiology, mismatched with modern lifestyles and the current environment.

With so much focus on physical aging and disease, it's easy to lose sight of the importance of mental health. But understanding how people maintain purpose, value, and connection despite declining physical health is essential to our discussion here. We'll assess well-being among older adults in small-scale societies. We'll learn that midlife crisis followed by late-age contentment is not universal, despite its commonly observed occurrence in the West. We also gauge the role that retirement has on our psychological well-being. Retirement is a modern luxury that, in its current form, contributes to less-than-ideal quality of life.

As social creatures, it may come as little surprise to learn that chronic feelings of loneliness may have the same effect on mortality as cigarette smoking and alcoholism. Social isolation, and the so-called loneliness epidemic, are now major public health concerns in the industrialized world. We're more likely to be living alone and to have lower community engagement ("bowling alone," as popularized by the political scientist Robert Putnam) than in decades past. If people are healthier at late ages than they were in the past, and more financially secure, then why does our psychological well-being remain so pegged to our connectedness with others? Throughout the book, we look beyond the familiar landscape of predator-free contemporary urban lives and discover the social world of older adults in small-scale societies.

What to Expect

Debates rage about the future of human lifespan. Strident optimists believe lifespan will continue to increase unabated, whereas cautious optimists expect only small increases on the horizon. Pessimists believe we have already reached the maximum limit afforded by our biology, and thus tissue and organ regeneration may be the only way to immortalize our soma. Amid venture capital pitches for mTOR inhibitors, senolytic drugs, CRISPR gene-splicing organs, and parabiotic vampiric blood-sharing with the teenager down the street, the approach I take here is more humble. I confess that you won't find any radical new way to think about cellular aging in these pages. No miracle methods for slowing down aging. No prehistoric mind tricks to stave off dementia. There is no shortage of good books on those topics, nor is there likely to be in the near future. We need those books, too, because tackling aging will require different types of expertise.[4]

So, what does it mean to consider aging and lifespan from a broader evolutionary perspective, as I do in this book? While living a long, satisfying life is a nice achievement, evolution does not care about how long our lives are or how satisfied we are, if those don't serve to help maximize our reproductive fitness. Over thousands of generations, the mindless sieve of selection helped shape our bodies, our brains, and our variegated human nature. But the realization that natural selection has reproduction as its guiding light doesn't mean there aren't important lessons for urban dwellers in modern nation-states. And vice versa for nonindustrialized populations now experiencing rapid changes to their lives and livelihoods. By the end of the book, I will show how we all can best gain from the benefits of modernization without incurring all the tagalong costs.

This seven-decade lifespan is the anchor to which we will keep returning—whether talking about the filtering capacity of our kidneys, the fatness of our retirement savings, or the motivations we have for making new friends. Once we recognize that older age is built into our species-typical design—and is not just chocolate sprinkles on the ice cream of life—we are in a good position to reconsider our relationship with the past, and to gain insight to help reimagine our own life plan. Current longevity is not unusual but is, rather, an extension of a pattern that was already present. This simple fact is still unappreciated. Even with more people living to seven decades today than ever before in human history, ageism is alive and thriving, with elders often viewed as problems, burdens, or threats. In contrast to that gloomy view, the thread throughout this book is how long healthy life is just the means to a functional end.

As President John Kennedy said in a now-famous 1963 speech a few short months before his death, we need to add life to years, not just years to life. More than what we eat, or how much we move our bodies, this involves our finding meaning and purpose. Helping the elderly become elders. By positioning older age as a normal part of our evolved life course, I hope to normalize later life and imbue it with the importance it has long held.

Part I

1

Defining the Age of Aging

Old age is always fifteen years older than I am.

—UNKNOWN

A VARIATION on a common philosophical musing: If a Tsimane woman living in the remote forest has a birthday, but doesn't know her age, does she still feel one year older? How much does perception matter to the nuts and bolts of biological aging? How old would you think yourself to be if you suddenly forgot your age?

The core of aging (or at least of the word) is age, but many languages spoken by foragers and farmers living outside of the Global North don't have number concepts beyond seven or eight and, in some cases, beyond three. The Tsimane people I work with traditionally say *dai* to refer to any number larger than about seven, and the Ache people of Paraguay have no words for numbers larger than two (*breko*, same word for spouse). When asking many Tsimane adults their age, I'm usually met with blank stares, or a polite laugh, and in some cases comically impossible claims—like when one villager seated next to his youngest grandchild while whittling a toy knife from balsa wood told me that he had just turned 10 years old. He insisted that this was correct, even flashed his ten fingers for emphasis. I placed him at about 57 years old.

It's hard for us to imagine not knowing, much less not caring, about how many times the Earth has spun around the sun in our lives. And for those who don't mark time systematically, questions about how aging feels, or what is considered old, can seem absurd. It might even make us worry less if we didn't know our ages. But time moves forward regardless of whether or how we tick the days away. How, then, might we define old age? Even in our modern, industrialized society, we approach aging from a relativistic perspective. Ask teenagers what age is "old" and they'll tell you 40 is ancient; 40-year-olds go with 60; 60-year-olds say 80. We're accustomed to thinking that old age begins when you're eligible for movie discounts (usually age 60 or 62) or an AARP membership (age 50+). In another year, I'll be eligible for discounts at more than 500 US museums, galleries, and planetariums.[1] Age may just be a number, but *old* age seems to defy any universal designation.

A more natural definition of old age might focus less on a target number of years, but instead ascribe meaning to the fundamental changes in the life course. Maybe you think of retirement as the onset of old age. In the US, the age of retirement is typically 65 years, set in the 1930s by the Social Security Administration and based on business pension plans in place at the time. If retirement marks the transition from "productive" to unproductive, viewing old age as postretirement equates old with being unproductive, a life of leisure and little responsibility. Official retirement ages range from 58 (Turkey) to 67 (Denmark, Norway, Greece), with governments planning to increase this age in many countries, due in part to concerns that greater survival may lead to too many dependents and not enough productive workers.[2] Decisions about changing this age are not taken lightly—the wild protests in Paris in March 2023 after President Macron proposed raising the retirement age from 62 to 64 highlights the difficulty of pleasing everyone with a one-size-fits-all policy.[3]

For many, becoming a grandparent might instead mark the transition to old age. In the US, the average age of becoming a grandparent is about 50, and is increasing, based on current trends of people having children at older ages.[4] But hunter-gatherers, who on average first become parents at age 17–19 years, are typically grandparents by their mid-30s. Many Tsimane are great-grandparents by the time most Americans are grandparents.

Language is often used to help define aspects of age. Tsimane recognize babies (*joijno*), children (*miquis'munsi*, literally "little people"), teenagers (*nanas*), adults (*son'*, *pen*), and older adults (*isho'*).[5] The Asmat of Papua New Guinea consider around mid-40s to be "old," with menopause being an important aging marker for women.[6] The Ju/'hoansi (commonly known as the !Kung) of Botswana and Namibia see old adulthood as the period when childbearing has ended and the work of childrearing is slowing down. In the Ju/'hoansi language, the respectful suffix *n!a* meaning "big" or "old" is added to names of people in their 40s or 50s who are no longer raising children.[7] The suffix "=/da!i," meaning "nearly dead," is used for the very old, mostly over the age of 70, even if they're still vigorous.[8] Life, it seems, proceeds in discrete functional stages.

Let's zoom out and take a more systematic approach to asking what defines the transition to "old age." Thankfully, one such study was conducted in the 1970s, when scholars were first catching on that cultures might vary not only substantially with regard to views of aging, but also in patterned ways affected by a group's history, its means of subsistence, and other features of its culture and ecology. Among a representative sample of fifty-seven small-scale nonliterate societies from around the world, the majority have explicit definitions of old age. Almost half of the definitions compiled in these studies involve some abrupt change in an individual's social role, including their work patterns, whether their kids are grown and/or married, and menopause for women. When other preferred definitions were mentioned, such as those involving physical limitations, senility, and visible physical changes (for example, white hair, loss of vision), these are almost always accompanied by some reference to a changing social role.[9]

Studies like this are limited by what ethnographers, explorers, and others paid attention to and thought worthy of writing down. While anthropologists are often drawn to older adults because of their rich knowledge and vivid memories of past traditions (plus they're usually easier to find), the actual experiences and perceptions of older adults are rarely the object of study.

In the West, we say someone has reached old age when they have accumulated some number of birthdays. But in small-scale societies, old age in many cultures is generally seen by their members to start when active participation or contribution to their families and community falters. Although not often named using specific terms, a conceptual distinction is usually made between the "old but capable" and the "decrepit and dependent."

In this chapter, I encourage a shared understanding of what we mean when we say "old" and "aged." Number of rotations around the sun is an odd but unbiased way to clock ourselves. We'll see how other information is needed to best interpret the larger significance of our years. The good news is that the latest evidence suggests we may be biologically younger than our chronological years today than was the case in centuries past. While more of us are reaching our longevity potential than ever before, the bad news is that future longevity gains seem unlikely without radical innovations. In any case, we need to learn some tools of the trade to best evaluate lifespan from the time of our species origin and onward, much less where we might be headed.

From Individuals to Populations to Individuals Again

In popular discourse, aging also refers to "population aging"—the notion that the demographic profiles of communities and countries are shifting toward more older people—both in absolute numbers and in total share of the population. By now, we're familiar with the stats usually used to incite panic or alarm in the industrialized world: 17% of the US population (56 million people) were age 65 or over in 2022. It was only 4% back in 1900 (or just 3 million people). But the trend is not limited to high-income countries of the West. It's global. Two-thirds of the world's older people live in low- to middle-income nations, where this fraction is increasing at a rate faster than in the West. In more than two hundred countries, the numbers and proportion of adults over age 60 are expected to increase over the next several decades.[10] There isn't a country on the planet not expecting to witness a boom in elders in the coming decades.

With the global surge in the numbers of seniors, new awkward terms have been coined to divide "old age" into meaningful categories that might help capture the heterogeneity in lived lives, social participation, and health needs. The first attempt for an international definition by the World Health Organization makes age 65 the entrance to elderhood. "Young-old" often refers to those ages 65–74, "middle-old" from 75 to 84, and "oldest-old" from 85 onward. In the industrialized world, young-old refers to those who are still reasonably active and independent, the so-called Third Age or golden years of adulthood, positioned between retirement and the onset of age-related physical and/or cognitive limitation.[11] The Fourth Age reflects the period of decrement, dependence, and decline leading to death. In this latter category, women usually outnumber men. Reliance on medical care and services is high during this stage.

Like individuals, populations can age because of higher survivorship at all ages. As the age structure of the population shifts toward more older individuals, the average age of people in an older population increases. But populations also age for other reasons. Most notably, populations can age rapidly because of declines in fertility. And fertility has declined substantially in most countries of the world over the past century. A mere five decades ago, the average Earthling had five children. Now she has just 2.5. What difference can that make to a population's age?

Consider Finland. In 1878, the life expectancy in Finland was 39 years of age, and the average Finnish woman had 4.7 kids over her reproductive life. If we imagine that mortality hypothetically didn't change since then, but fertility dropped to the current low level of 1.4 kids, the share of Finnish adults age 65+ would increase almost fourfold, from 2.0% to 7.3%.[12] In this illustration, population aging occurs not because people are living longer, but because the bottom half of the age pyramid narrows over time because of fewer births, which leads to an older population.

Life Expectancy

Now that I've clarified how aging can apply to both individuals and populations of individuals, I introduce the most widely used metric of lifespan: "life expectancy." Although life expectancy gives us a nice average age at which someone born could expect to die, it's really a deceptive mathematical composite that fails to capture what adult lifespan looks like in practice. Nor is it terribly useful for addressing whether we're living longer or healthier lives now than we have in the past.

Here's why. Life expectancy at birth is the average number of years a newborn can expect to live (abbreviated e_0 for the expected lifespan for someone just born). It's a population measure, but it is used to estimate how long any one individual from that population could live, in the absence of other information. These metrics can be tabulated for each birth cohort, reflecting all people born in, say, 1950. Or for "synthetic cohorts," which represent the mortality rates experienced by people of all ages during a single slice of time, typically a year. So, in 2019, for example, e_0 for the world population was estimated to be about 72 years old. Among countries that year, e_0 ranged from 51 in Lesotho to 84 in Japan. That global average e_0 of 72 didn't exist in any country of the world prior to 1950![13]

Because life expectancy is an average, it takes into account the probability of living or dying at all ages. And so early deaths will have a bigger effect on e_0 than those occurring at late ages. Where infant and child mortality rates are high, e_0 is on the lower end of the spectrum. In countries with low infant and child mortality, e_0 skews higher because people largely die at older ages.

For most of human history, death was usually associated with youth—most deaths were among the young, not the old. A population with an e_0 of 30 years doesn't mean that everyone will tend to live until age 30, and then abruptly die. And it certainly doesn't mean that no one in that population will ever see past 30. As we'll see in chapter 3, the e_0 of 30 is not the most typical lifespan observed. At the extreme, nobody dies at the average life expectancy, e_0. Consider a world where a third of the population perishes by age 5 and the remaining two-thirds live to age 65; the e_0 will be in the mid-40s, though in this hypothetical example, no one actually dies in their 40s—and the majority of deaths are in the 60s.

Beyond Life Expectancy at Birth

Demographic wizards zero in on the notion of the typical lifespan in several ways. One way is to simply look at typical life expectancy once people have survived past a certain age, like 15 or 40. This makes a good deal of sense if the reasons for death early in life are unrelated to what

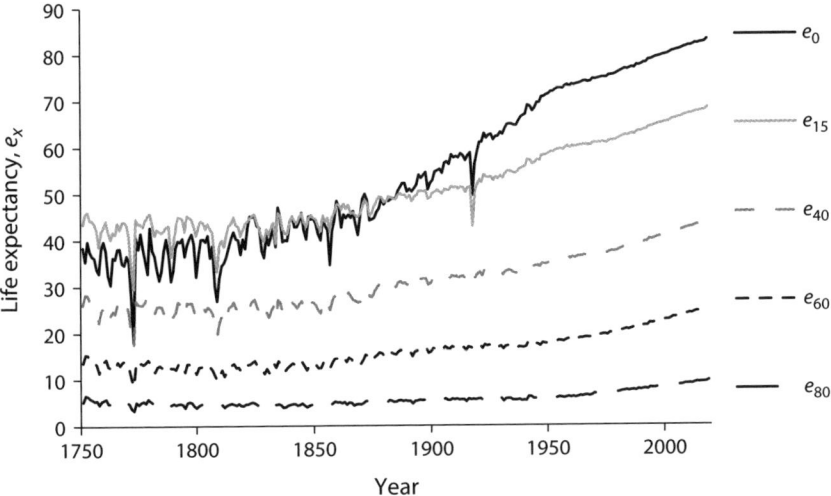

FIGURE 1.1. **Life expectancies in Sweden from 1751 to 2019.** These show average remaining years at birth (e_0) and at ages 15 (e_{15}), 40 (e_{40}), 60 (e_{60}), 80 (e_{80}). The absolute gains in extra life over time after age 60 were minimal until the mid-twentieth century. *Data source:* Human Mortality Database.

kills us at later ages. Children don't typically suffer from the heart attacks or strokes that afflict older adults.

Because it has excellent vital records going back centuries, and until the 1970s had the highest life expectancy in the world, Sweden offers important insights into how adult life expectancies have changed over time.[14] In fact, the country provides some of the longest high-quality demographic data out there.

Figure 1.1 shows the life expectancy at birth (e_0), and at different ages across a 269-year period of Swedish history, from 1751 to 2019. Sweden is a prosperous uber-democratic superpower now, but back in the mid-eighteenth century, Swedes were agrarian farmers, as was the case with many other people throughout the world. In the figure, you can see that e_0 increases by forty-five years over this time period (from 38 to 82 y), an increase of about two months of extra life gained per year. This is a massive achievement, as Steven Johnson, author of *Extra Life: A Short History of Living Longer*, called "one of the greatest in the history of our species"[15]—the result not so much of complex surgeries and pharmaceuticals, but rather of the falling rates of infectious diseases, and better access to clean water and improved sanitation.

In 1751, those who reached age 20 could expect to live an additional four more decades, surviving until age 60 (that is, life expectancy at age 20, or $e_{20} = 40$). By 2019, they could expect to live over six more decades, reaching an impressive 83 years. Over the course of several centuries, that amounts to an extra thirty-two days of adult life added per year. Note that by age 80, remaining life expectancy (e_{80}) ranges from an additional five to just under ten years over the centuries. In relative terms, this is still a big deal—a doubling of e_{80}. But the absolute gain is small when compared to the absolute gains at earlier ages. The gain in e_{80} over this period of huge e_0 increase is only an extra six days per year. And most of these gains occurred only since 1950.

Lifespan a la Mode

Another population measure gives a different, more intuitive, impression of how long we typically live. This requires considering not the mean or the median, but the mode—the most common age at which people die. If we examine the numbers of people dying at all ages across a population, we can observe some clear regularities.

Figure 1.2 illustrates a typical "age-at-death distribution," here for Finnish women in the late nineteenth century. The mode is easily identifiable as the peak of the distribution. It occurs earliest in life because, at that time, most deaths were concentrated in the delicate years of infancy and early childhood. But deaths reach a minimum after the age of about 10 y or so, then creep up steadily during adulthood. After reaching another peak at some later age, the number of lives left to count decreases. This peak represents the characteristic modal length of life in late adulthood that looks like a towering mountain of death— let's call it Death Mountain.

The German statistician Wilhelm Lexis (his full name is too long to fit in this book)[16] was the first to suggest that this "modal age of death" might reveal a more meaningful and useful sense of life's natural length— once you peel away the marks of early-life demise and other premature

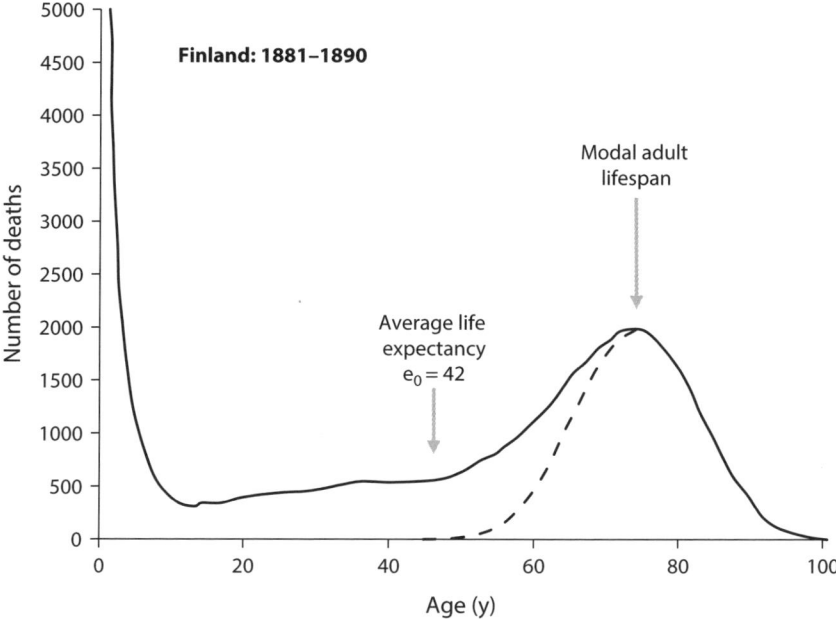

FIGURE 1.2. Modal length of life, Finland women 1881–1890. Dashed line represents the bell-curve-like spread of senescent deaths centered at the mode. *Data source:* Kannisto 2001.

deaths. What's left when you strip these away is a bell curve of "normal" deaths.[17] The mean of that bell curve represents the modal age of adult death. Notice this mode is greater than life expectancy, e_0, because all the deaths that occur earlier in life drag the mean down. When the number of those deaths are sufficiently high, as occurred throughout much of human history, e_0 will fall within the age range of premature deaths—even though most deaths occur either early in life, or around the adult mode. Though less popular a measure, the modal age of adult death seems to tell us something more informative about longevity than an average measure based on expectations at birth. (Lexis himself lived to age 77, close to the mode shown in figure 1.2.)

Recall that in Sweden, the life expectancy e_0 increased by more than forty years in the period between the nineteenth and twenty-first centuries. Up until the end of the eighteenth century, one out of every three or four babies born never lived long enough to celebrate their fifth birthday. Thereafter, the rate lowered slightly to one in four or five through the nineteenth century, and then lowered still more, down to one in twenty by the mid-twentieth century.

But what if we look at the changes in the modal age of death in Sweden over the same time period? In figure 1.3, we observe that the mountain of Swedish lifespans shifts to later ages over time. The peak is at age 64, in the mid-eighteenth century; 70 by the beginning of the nineteenth century; and late 70s by the time Lexis published his work on the mode. The modal peak hits the 80s during the twentieth century, and by 2019, the peak reaches a whopping 90.

Notice that, as the peak shifts to older ages over time, Death Mountain narrows and becomes steeper, as more deaths concentrate around the mode. If we liberally consider premature deaths as all those cut short prior to age 60, then two out of three deaths in Sweden were premature in the eighteenth century. About half were premature in the nineteenth century, down to less than one in five by the mid-twentieth century. By 2019, only 5% of deaths occurred before age 60.

This fairly rapid reduction in premature deaths is a remarkable feat unprecedented throughout history. The timing and details vary from place to place, but such a pattern is typical of the industrialized world. With the massive reduction in early life mortality, life expectancy at birth and modal age of death are closer now than they've ever been before.

Two points are clear and prominent: the typical adult lifespan has increased by up to two decades, and many more people are likely to attain that typical lifespan. As such, inequality in lifespans has decreased substantially over time.

Survival Mode

The phenomenon in Sweden, in which Death Mountain gradually shifts rightward over time, is also seen in every other country with reliable demographic data. Figure 1.4 shows the modal age of death across many countries spanning a couple centuries. The modal lifespan hovered between 65 and 75 years prior to the mid-nineteenth century. It was in the lower end of this range during periods of famine or epidemics, such as

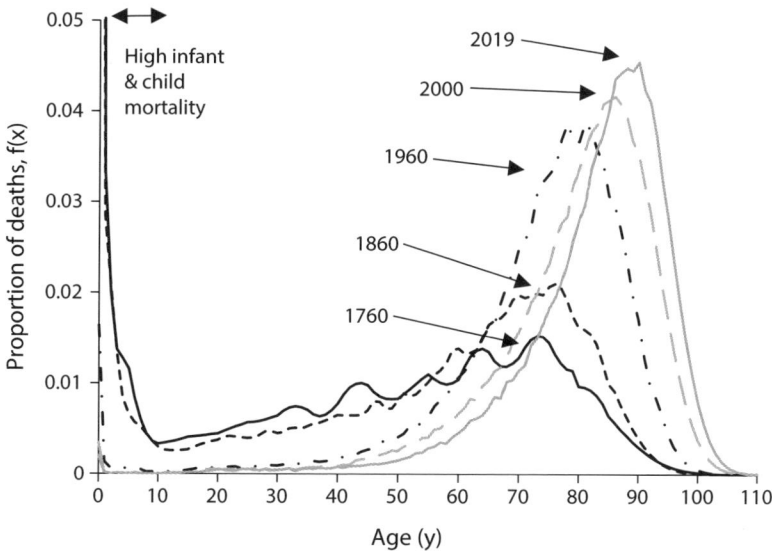

FIGURE 1.3. **Modal ages of death for Sweden (1760–2019).** The modal adult lifespan increased by two decades, from ~70 years in 1760 (when life expectancy $e_0 = 40$ y) to ~90 years in 2019 (when $e_0 = 83$ y). *Data source:* Human Mortality Database.

Iceland in the 1840s. Then, from the mid-nineteenth to mid-twentieth centuries, modal adult lifespan inched up to between 70 and 80. Since the mid-twentieth century, that mode climbed closer to nine decades. The increase in the mode, especially after the mid-twentieth century, shows the sustained improvements in adult survival.[18] As modal lifespan increased, the proportion of deaths occurring at those exact ages increased as well. The mountain narrows as it shifts to later ages. This is true everywhere it has been examined: in England, Italy, Japan, and other Eurasian countries.[19]

Another intriguing aspect of this phenomenon is that the average life expectancy once a person has reached the modal age of death has *decreased* over time. In other words, fewer people are living very far past the modal age of death. While the mode continues to march forward, the fall off that cliff becomes steeper and steeper. All of this suggests that we may be coming closer to hitting the proverbial wall of death, a limit to how long we might live.

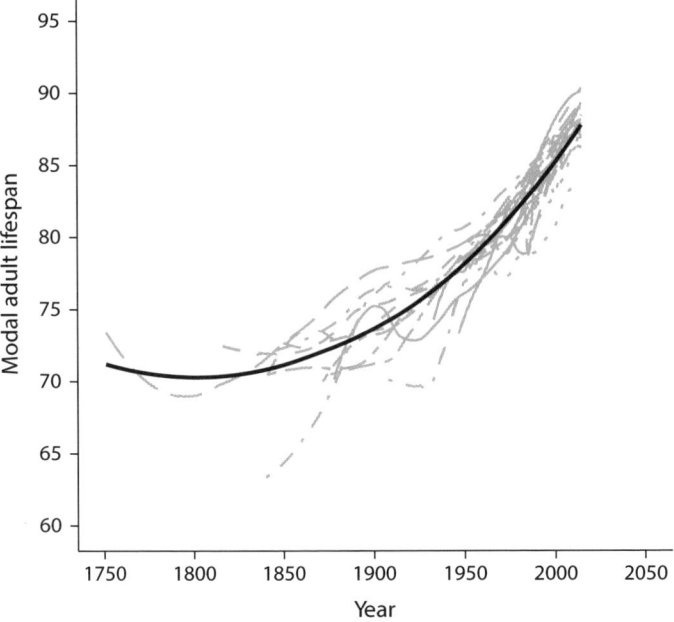

FIGURE 1.4. **Modal lifespans (ages of death) over time: Twenty-one countries.** Modal lifespans have increased everywhere there is longitudinal data. The overall pattern (in bold) shows a seven-decade modal lifespan prior to 1900. *Note:* 1915–1919 and 1940–1944 periods excluded in some countries because of flu pandemic and World War II, respectively. *Data source:* Human Mortality Database.

Coming of Age

So far, in our brief survey of changing patterns of survival over the past few centuries, we've looked only at chronological age. But is a 65-year-old today healthier than a 65-year-old in, say, 1870? Which would spend more time in a hospital waiting room? What's important is not just the quantity of life lived, but its quality.

Given differences in health conditions, does it really make sense to rely on a single chronological age as the classic marker of "old"? To get around this obstacle, the economist Warren Sanderson and the demographer Sergei Scherbov came up with a simple approach—to measure backward from death, rather than forward from birth. Their "prospective measure of aging" defines the onset of being "old" as the age at which there are just fifteen more years left to live, on average.[20] Using the notation we defined previously, "old" age would occur when $e_{old} = 15$. This metric assumes that 60-year-olds, say, with fifteen years left of life remaining are in better condition than 60-year-olds with only four years left.

Returning to our example of Sweden, the prospective measure of aging would suggest that "old age" began at age 58 y back in 1751, dipped a little prior to the closing of the eighteenth century, then got pushed higher and higher, reaching 65 y by the mid-twentieth century (the thick black line in figure 1.5). It doesn't stop there, however. By 2019, "old age" begins at age 72 y.

We can use the same approach to also ask what's the "new" 40, relative to the 40-year-olds from mid-eighteenth century Sweden. By this measure, a 59-year-old in 2019 is the equivalent of a 40-year-old from 1751—both ages have a mean of twenty-six years of life remaining (see bottom line in figure 1.5). Figure 1.5 shows a range of new adult ages from 40 to 65 over time compared to the baseline condition in 1751. In each case, the new ages today are more than fifteen years "younger" than those of the mid-eighteenth century. Just like we saw before, when looking at e_0 and modal age of death, the majority of this "rejuvenation" has occurred since the mid-twentieth century.

If we use a more recent time period, say, 1950, as our baseline of comparison, then we can see that a 40-year-old in 2019 is the equivalent to a 30-year-old in 1950. Similarly, 50 is the new 40, and 74 is the new 65. A whopping difference of a decade!

If the number of years of life left remaining is a good proxy for physical condition, then we could stand to lower the volume of the alarm bells ringing over rapidly rising population aging. Consider, for example, retiring at the chronological age of 65. Intense debate hinges on whether retirement ages need to change—to keep pensions solvent and to stay out of the red. Using the new metric we just learned, a Swede aged 65 in 1751 had a prospective age hovering around their chronological age of 65 for about a century. Then there was a slow climb

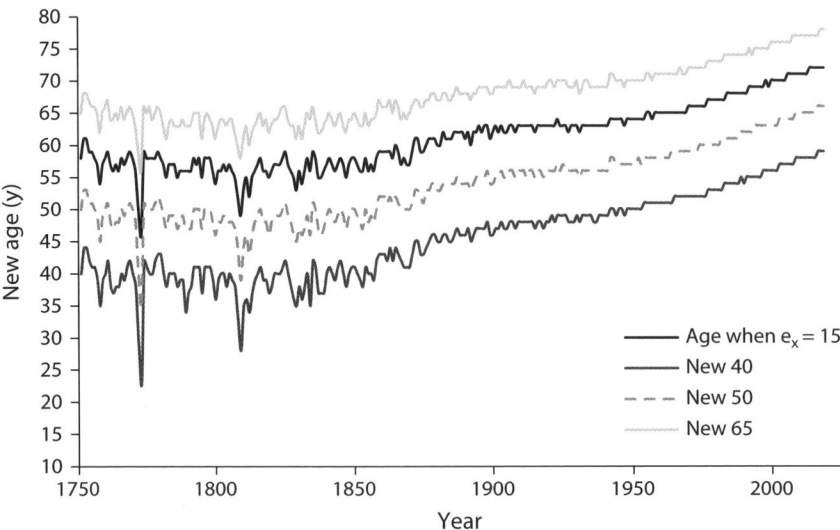

FIGURE 1.5. **Prospective measures of "old age" and other "new" ages relative to chronological age in 1751 Sweden.** For example, 54 was the new 50 in 1899, but 63 was the new 50 in 1999. The age where remaining life expectancy is just fifteen years (in bold) increased from 55 to 60 years prior to the twentieth century, to late 60s by 2019. *Data source:* Human Mortality Database.

such that, by 1950, the new 65 was about 70. Then a steeper climb toward 78 in 2019. Imagine if retirement age was based on these new interpretations of age 65.

In this new framing, the proportion of postretirement-aged adults increases more modestly over time than if we relied on the current system based on chronological age. Figure 1.6 compares the proportion of Sweden's census population that is aged 65+ with the proportion who are at or above the "new" age equivalent of 65 based on the mid-eighteenth century baseline. The number of 65+ year-olds is somewhat steady until the 1870s, then more than triples to about 20% by 2019. Yet if we consider that the health of 65-year-olds improved over time, then the proportion of adults deemed older by the prospective measure increases around the turn of the twentieth century, from 5% to a maximum of only about 9% today. While the actual number of seniors aged 65+ is about double that, this lower percentage gives a rosier perspective on

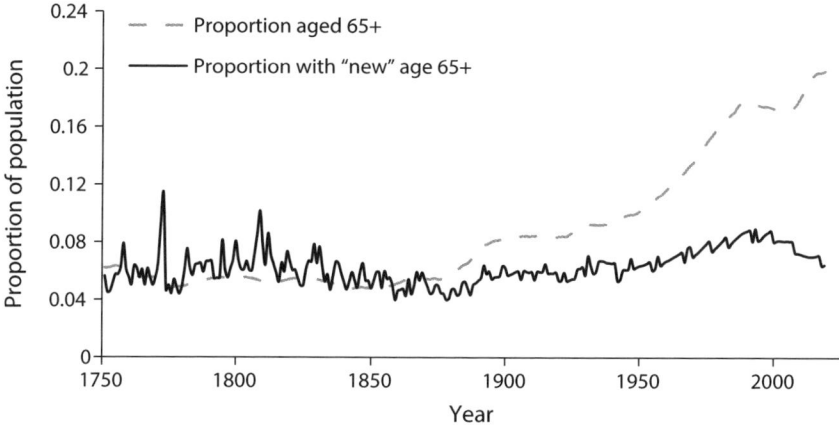

FIGURE 1.6. **Proportion of older adults in Sweden, 1751–2019.** While the proportion of people with chronological age 65+ has tripled over the past two centuries, the proportion of people with "new" age 65 has remained relatively constant over time.

the degree of potential "burden" usually ascribed when defining "old" based only on chronological age.

Another measure of population age that draws concern among politicians, scientists, and voters is the dependency ratio, or the ratio of the number of seniors age 65 and over, to working-age adults, typically ages 20–64. This is an important measure because it affects everything from the strength of the economy to public and private costs of health care. If we also replace chronological age 65 here with the prospective age equivalent, the familiar horror story about rapidly rising dependency and depleted social security coffers doesn't look so bad. Take Japan, for example. Japan has one of the highest fractions of its population as seniors. In 1955, Japan's old-age dependency ratio was 10%. By 2005, it was 32%, and the UN projects this ratio to jump to a whopping 75% by the year 2045. If we instead use prospective ages to define "old," however, this ratio then drops to 16% in 2005, and 27% by 2045.[21] Why the discrepancy? Because fewer people count as "old" in the numerator of the new ratio, and more chronologically "old" people would still count as working.

Longevity Record Breakers?

Instead of thinking about averages or modes, what about our longevous heroes breaking world records? Don't they reveal the longevity potential lurking in all of us? And if the record varies over time and population, might that reveal something about changes in lifespan? I'd be remiss if I didn't address the supercentenarians, usually defined as those having reached the glorious old age of 110 years. Amid this rose-tinted glory is the realization that nobody living past 100 lives independently, and less than half are cognitively intact.

The French supercentenarian Jeanne Louise Calment holds the current record for being the longest-lived human to have ever roamed the globe (at least post-Biblically), at 122 years and 164 days. Eating two pounds of chocolate a week, and smoking until she was 117, this doyenne of humanity never lost her sense of humor, stating, at 110, "I've never had but one wrinkle, and I'm sitting on it."[22]

There are no confirmed centenarians (yet) among subsistence-level populations that I'm aware of. Only twenty-one of the 387 Tsimane adults who died over the past two decades lived to reach at least their eightieth birthday, and only five to their ninetieth. Two of the oldest Tsimane I got to know personally were living testimonials of resilience. The first was a jovial woman with blonde hair who was 76 years old when I met her in 2003 (Tsimane women often go blonde at very late ages). She helped found her community, and was often observed making *shocdye'* beer, telling stories, laughing with family (and ordering people around). In her small community alone, she was surrounded by seven adult children and twenty-seven grandchildren. After being widowed for the second time at 82, her health and mobility declined, and she died within eighteen months at the age of 84. Another man who was 79 in 2005 when I first met him, and who was deaf and mute, later died at age 87. He is an astounding example of persistence, given his lack of hearing and of ability to verbalize his thoughts. While he was unable to hunt, he was a tireless farmer and fisherman. He, his family, and those who knew him well communicated using hand gestures.

The oldest Ju/'hoansi studied by Nancy Howell in the 1960s was a woman estimated to be 81, who died in 1972 at 86. According to Kim Hill, the oldest Ache ever recorded died in 1999 at age 89, after getting lost walking between two distant camps during a rainstorm. From conversations with friends who have worked with other similar groups, none can attest to having recorded or even heard of someone living a full century. While it is fascinating to hear the stories, perspectives, and favorite recipes of record-breaking long-lifers, for most of human history those lucky record-holders might have been anomalies. Maybe they had everything going right for them in order to outlive their peers. The same logic applies to our primate cousins. In 2017, the oldest chimpanzee at the time, Little Mama, died of suspected kidney failure at an estimated 79 years, having lived her last few decades as a retiree in Florida.[23] This far exceeds even the most long-lived chimpanzees in the wild.

Extreme outliers like Jeanne Calment and Little Mama crop up because of the large size of the base population from which they came. How many humans had to have been born over recorded human history to give us Jeanne Calment? In a population with a similar mortality profile as Sweden in 1751, you'd need to witness at least a hundred lives to find one person who made it to age 90. To find a genuine centenarian (ignoring for the moment any difficulty in confirming that birthdate), you would need to census at least 4,348 people just to find one, and more than 100,000 to find someone celebrating their 105th birthday.

Issues of accurate birth dates aside, the emergence of centenarians indeed precedes the Industrial Revolution, and by several thousand years. The world would have borne witness to its first centenarian once the world population surpassed about 100 million, say around 2500 BC.[24] In other words, some share of the rising maximal lifespan confirmed over the past century is due to the rapid rise in population size. Each new birth is another potential centenarian. You can even make headlines by claiming that the world's first 150-year-old has already been born.[25] In any case, differences in maximal lifespan reported over time are much less dramatic than the changes in life expectancy or modes we discussed above. Looking at trends in supercentenarians, the longevity superstars who live to at least 110 years, can tell us whether our most

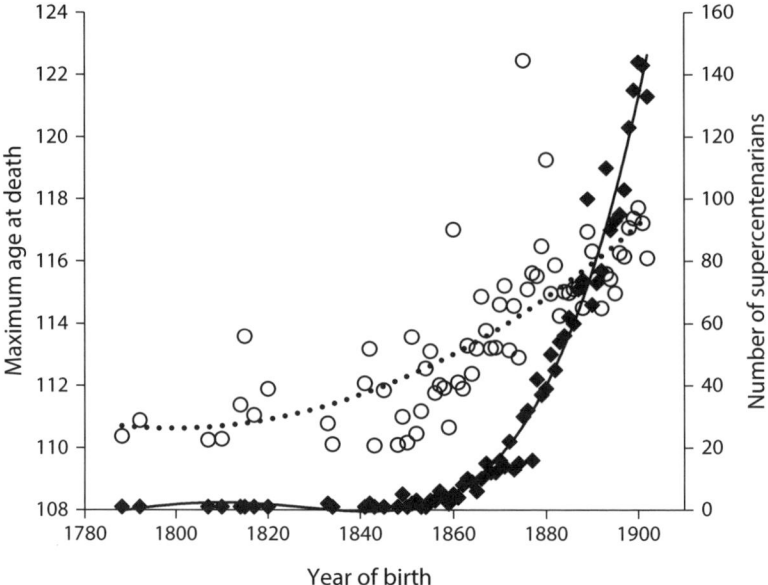

FIGURE 1.7. **Supercentenarians (survivors past age 110 y).** Number of supercentenarians in the Gerontology Research Group registry (black diamonds), and the maximum age lived by birth cohort (open circles). While the number of supercentenarians has increased over time, the longest lifespan recorded has increased by about six years. *Source:* Accessed June 24, 2021, from Gerontology Research Group website.

successful Methuselahs are living longer.[26] Across over a hundred-year period, the maximum recorded lifespan increased from just under 111 to over 117 (figure 1.7). But the world population over the time period when these folks surpassed their hundredth birthdays grew from about 1.6 billion to more than 6 billion people. A better way to think about "maximal" age is to look at changes in the average age of death among supercentenarians. It turns out that the average age of death barely budged over the past century.[27] So while the maximum age has increased with the growth of the global population, most supercentenarians are still dying within a couple years of attaining their super status. In fact, most supercentenarians die before their 114th birthday.

Instead of relying on just the record breakers, or even those who lived to age 110, demographers often focus on the age at which only a small robust few are still alive; for example, the age at which only 5% of a birth cohort is still alive. For our Swedish example, this age ranges from 83 in the mid-eighteenth century to 86 in the beginning of the nineteenth century. It then jumps to 90 by the end of World War II, inching higher to about 97 years today. While this increase is impressive, it's a change of just over a decade in almost three centuries—a far less impressive trend than the increases in life expectancy or even modal age of death over the same period—which is a theme we will return to: while longevity has increased over the course of human history, the much bigger achievement is that many more of us are now likely to reach it.

When a Year Is Not a Year

Is there a point at which extra years gained will no longer be desirable? While we welcome extra years added to our lives, we hope that longer life will be fulfilling, that we can lead active and capable lives, and not gain extra years marked by disability and hardship. In the tabulations we've seen so far, a year of life gained at 40 is treated the same as a year of life gained at age 80. Currently, 79% of all years lived with disability in the industrialized world come from chronic diseases, like cardiovascular disease, cancer, and chronic respiratory conditions.[28] Much of the suffering and eventual death from such conditions manifests among adults over age 70.

And so with increases beyond eight decades, is the extra life gained a time of healthy enjoyment, or is it, as Shakespeare evokes (in the voice of melancholy Jaques) about the seventh, final stage of man in *As You Like It*, "second childishness and mere oblivion, sans teeth, sans eyes, sans teeth, sans everything"? This vital question is what churns debates over whether the optimism about increases in longevity should be favored or feared. More years are good, if they are healthy years. Few may wish extra years spent in a hospital bed, tangled in tubes and unable to partake in the usual activities that give life meaning and purpose. The ideal situation invokes what the physician James Fries called the "compression of morbidity," whereby prevention, treatment, and "healthy aging" pushes disability to the very end of life. With morbidity compression, adult life, including extra years gained from healthy living or interventions, would be spent free of disability.[29] This idea, which Fries came up with in the late 1980s, contrasted sharply with the dire fear that further increases in lifespan would result in a pandemic of disability. Without delays in physical decline and reductions in morbidity, extra gained years are just "manufactured survival."[30]

In order to assess the quality of any extra years gained and determine the implications on policy, planning, and our own prospects, we need different measures that combine the probability of living or dying at each age with an assessment of how good our health is at that age. Healthy or "disability-free" life expectancy counts the expected number of healthy, able-bodied years one can expect to live. This requires knowing how many people of each age are likely to suffer from a number of limiting, disabling conditions. Some subjectivity here makes these measures a bit more difficult to standardize and compare over time or across populations. We'll be hearing more about these kinds of measures in the public arena as we attempt to prioritize both quantity and quality of life. Gains in lifespan without concomitant gains in "healthspan" are already leading to new types of inequality, a lifespan-healthspan gap.

Healthspan as the New Lifespan

What these "healthspan" measures show us so far is mixed with regard to morbidity compression in the industrialized world. Across twenty-five European Union countries, an adult reaching age 50 can expect to live an additional three decades, with almost two of these decades spent relatively free of disability. That's about a decade gap between lifespan and healthspan. Even in Europe, however, the variability in healthy life expectancy among countries is large: ranging from a low of nine to ten years of healthy life upon reaching age 50 in Estonia to twenty years in Denmark.[31] So far, there isn't strong evidence of large morbidity compression over time. Instead, increases in life expectancy continue to come with extra healthy and unhealthy years.

Since 1970, US life expectancy at birth has improved by about nine years for men and five years for women. By age 65, adult life expectancy improved as well: five more years for men and 3.5 years for women. But only half of these boosts are spent disability-free. In fact, the proportion of years remaining that are disability-free have not changed in women or men since 1970.[32] In other words, we keep getting the bad along with the good: no morbidity compression.

In other parts of the world, life expectancy has also improved. It increased by over a decade throughout sub-Saharan Africa since the beginning of the twenty-first century, now 62 for men and 66 for women. This is great news, but the types of morbid conditions have shifted over time. While measles, malaria, and vaccine-treatable infections have declined in frequency, HIV/AIDS, conflict and terror, and noncommunicable diseases have variably increased the disability experienced over the past few decades. Taken together, improvements to life expectancy haven't reduced the 12%–15% of years lived in poor health in most sub-Saharan African countries over the past three decades.[33] While healthy life expectancy has improved, so has the number of years living with serious health burdens.

An optimistic view stresses that disability and hardship don't characterize all of the extra years of life gained over the past four decades. But, so far, those extra years beyond seven decades also bring more years of difficulty.

2

Evolution and Aging

O Sovereign my Lord! Oldness has come; old age has descended.
Feebleness has arrived; dotage is here anew. The heart sleeps wearily
every day. The eyes are weak, the ears are deaf, the strength is disappearing
because of weariness of the heart and the mouth is silent and cannot speak.
The heart is forgetful and cannot recall yesterday. The bone suffers old
age. . . . All taste is gone. What old age does to men is evil in every respect.

—PTAH HOTEP, VIZIR TO PHARAOH ITZEZI, ~2450 BC

THE PASSAGE above, written by an Egyptian scribe 4,500 years ago,
may be the earliest written lament about old age. In his 1989 book,
History of Old Age, the French historian Georges Minois explored how
such laments were echoed throughout the millennia. These are, he wrote,
wails against the "personal and social drama" that everyone dreads and
fears. The despair is not due to concerns about death per se, but about
the withering of the physical body and the perceived scorn of a youth-
centered ageist society. But death can occur in the absence of aging. The
best life lived would arguably be spent with all our functional faculties
intact until the final cliff-fall, to "die young as late as possible" as the
British American anthropologist Ashley Montagu once remarked.

To what extent is aging malleable, flexible, even preventable—just
like other maladies that medical science eventually helps treat? The belief that aging can be cured like a disease fuels much of the biomedical

industry's focus on the proximate, molecular hallmarks of aging, from failures of protein folding to dysfunction of the mitochondrial power-houses of our cells, exhaustion of our stem cell progenitors, and other aspects underlying the *how* of aging.

But to get to the heart of the matter, we need to first grasp *why* aging occurs despite our best efforts to date. In chapter 1, I defined lifespan and summarized recent trends; here I'll deepen our definition of aging and show how aging is ubiquitous in the animal world. We'll distinguish between *actuarial* aging and *physiological* aging, and address whether aging is just another disease to be cured or something so universally human that it's effectively unbeatable. Situating humans in the vast tree of life is essential, as knowing our position can help distill our unique features (or lack thereof) and provide clues to what aspects of our aging bodies and minds may be modifiable.

————

Many strong opinions exist about how to define aging. For our purposes, aging simply refers to the accumulation of changes in the body over time that increase our vulnerability to disease and, ulti-mately, death. With aging, physiological function loses efficiency. Our ability to bounce back from shocks and a variety of assaults gradually crumbles. Take two familiar examples from our bodies: our kidneys and our hearts.

Kidneys often become less effective in filtering waste from the blood with age, resulting in excess protein in the urine. In the lab report that follows your annual physical, this change is indicated by a declining glomerular filtration rate (GFR). A low GFR usually means kidney dis-ease, and a very low GFR suggests kidney failure, which leads to death in the absence of dialysis or a transplant.

Our hearts also break down with age. Arteries often become stiffer with age as stress on arterial walls leads to fragmentation of the elastin fibers that make up the middle intima layer of the artery. Stiffness forces the heart to work harder to pump blood through more rigid arteries. This results in higher blood pressure and a greater tendency toward hypertension with age, which makes the heart work even harder. A hard-working heart can thicken its left ventricle chamber and become larger with age. As it loses efficiency, it becomes more likely to fail.

When either organ fails, we die. Some factors we can control may alter features up the chain of causality (I discuss this in more detail in chapter 9), but even under ideal conditions, aged arteries still stiffen and aged kidneys still lose efficiency. This loss of capacity to maintain homeostasis, whether in the kidney, heart, or the rest of the body, in the face of challenges has also been called *dysregulation* and is a core feature of aging.

Many people employ body-as-machine metaphors when describing how aging occurs. The idea is that if we were cars, then aging would manifest as ailing performance on the road. Regular oil changes, new timing belts, and transmission fluid inspections certainly help, but declining performance is inevitable. The car is more likely to fall apart when it hits a pothole or when there is a drastic change in temperature. Eventually, the transmission, carburetor, and other parts break down as the odometer spins forward. The physiological changes in our body are similar to what goes on under the hood of the car.

But the analogy quickly loses gas when you consider that your friend's classic candy-apple 1970 Mustang has had almost every part replaced at least once, if not twice, and at great expense. Most drivers would consider a car ready for the salvage yard when the costs of repair outweigh its Blue Book resale value. Plus, cars don't reproduce. From an evolutionary perspective, the point of life is to generate more life. Genes can strive toward immortality by being passed down through reproduction even when the body they once inhabited is no longer. An organism that lived a thousand years but never reproduced would be an evolutionary dead end.

Aging therefore includes everything that leads not just to higher mortality, but also to physiological changes that reduce the body's ability to make viable babies. Those changes to fertility may be synchronized with other physiological changes, but natural selection works in many fabulous ways. Many mammals show similar rates of decline in survival and fertility with age, but there's a lot of variation across the tree of life.[1]

Such "actuarial aging," reflecting declines in demographic rates of survival and/or fertility, is defined this way because survival and fertility are the key components constituting biological fitness—the raw materials that determine how natural selection acts on traits. All organisms are Nature's experiments—different ways of assembling bodies that compete (and cooperate) to pass genes to future generations. We're familiar with the role that selection plays in shaping the aerodynamic design of a bird's wings or in favoring genetic variants that improve malaria resistance. Selection for genes that protect against malaria is so strong that these genes are favored even when they can lead to harmful conditions, such as hemophilia.

Stepping back from thinking about one trait or another, it's thrilling that selection also shapes the life history of a species. A life history refers to the bundle of demographic traits that affect fitness from birth to death—how quickly we grow, the timing of menarche, the ages at which we give birth, how many babies we have, and the size of those babies when born, among others. The evolved life histories of species don't occur in a vacuum. They make sense only in light of the broad range of environmental conditions during which they evolved.

Consider ginkgo biloba. A popular, if not scientifically supported, claim is that ingesting its leaves improves your memory.[2] But ginkgo is remarkable for a different reason. Ginkgo trees, along with bristlecone pines, are among the longest-lived species in the world, with lives stretching beyond a thousand years. Aside from Adam and other biblical patriarchs, and the wild hopes of bold longevity fanatics,[3] no human in any environment will ever live that long. Ginkgo trees are blessed not only with longevity but also with minimal aging. Ginkgo trees that are 600 years old show similar efficiency in photosynthesis, seed germination, and disease resistance as trees that are 200 or 20 years old.[4] On the other hand, the ephemeral life of a mayfly has been known since the days of Aristotle. Once nymphs emerge from freshwater lakes to become adults, they live for about a day—long enough to reproduce and then die.

If aging refers to the mortality rate increase with age, then slower aging is only one way to live a longer life. Another way to live longer is for background mortality—experienced perhaps by individuals of all ages—to be lower. Or for there to be a delay in aging's onset. Without aging, the probability of dying remains relatively constant each year of your adult life. In such a scenario, everyone in this population still eventually dies, but there's no aging because the likelihood of dying doesn't change with your age. While we're used to seeing aging in long-lived mammals like ourselves, many short-lived critters age as well. Guppies and water fleas have lifespans numbering in days but show steep declines in fertility and survival with age.

It may seem obvious why background mortality can limit lifespan, even though its sources—such as predators, natural disasters, and infectious disease epidemics—are largely external to the internal workings of an organism. This type of external mortality is sometimes called "exogenous" or "extrinsic." It stands in contrast to "intrinsic" mortality, which reflects succumbing to internal, degenerative causes rather than Nature's arsenal of hailstorms, locusts, and butchers.

But now we're faced with a paradox. How could natural selection—Nature's blind tinkerer of organismal design with no blueprints but just an amoral reward for higher reproductive success—ever allow an increase in intrinsic mortality with age?

Dying is very much incompatible with maximizing fitness. If selection has sculpted the developmental organic machinery that results in a peak-performing adult, why would such complex design fail to maintain its luster? Isn't such deterioration counter to the idea that survival improves fitness?

———

One of the earliest "functional" explanations for aging from an evolutionary perspective was that having older adults deteriorate and die helps make room for the young. Call it the "out-with-the-old, in-with-the-new" hypothesis. It was first proposed in the nineteenth century by the German biologist August Weismann. But echoes of this idea date back to the Roman poet and philosopher Lucretius, who argued that aging and death are much-needed benefits that make room for the next generation.[5] While a "programmed death" that frees up resources for the young sounds reasonable from a good-for-the-group perspective (and to young colleagues eyeing the large offices of senior professors), it sadly assumes that which it tries to explain. You wouldn't need to eject the old if the old weren't already ravaged by aging. More importantly, the altruistic self-sacrifice of older adults is not something that evolution easily permits. Where it does seem to occur, like in Pacific salmon, may-flies, and certain eels, reproduction explodes in one big bout and then death follows shortly. Those species usually have only a small chance of reproducing more than once, thanks to a short lifespan or long migration. Evolution has wisely favored the shunting of all somatic resources to reproduction as a winning life history strategy, called semelparity.[6] Natural selection doesn't readily favor biological suicide as a means of furthering individual reproductive success. To his credit, Weismann publicly acknowledged that his ideas were wrong.[7]

Why Do We Age?

The book of Genesis suggests aging comes from God's punishment of Adam and Eve's rebellion and fall. Other myths also invoke aging and mortality as punishment for hubris or a violation of some sort. But as I have mentioned, the presumed inevitability of aging isn't inevitable. Our bodies could have a longer expiration date of the warranty period that includes high-level functioning and performance. Despite having the same genome, a queen honeybee outlives the worker bees by more than tenfold: 1–2 years for the queen versus 15–38 days in the summer and 150–200 days in the winter for the worker.[8] Aging is not a given or a constraint, per se, but is something that happens for a reason.

The modern approach to understanding the why of aging came almost a century after Darwin's critical insights into how natural selection could lead to complex design and biodiversity. Curiously enough, Darwin himself had very little to say about why we age. In a lecture delivered in London in 1951, the British biologist Peter Medawar (deemed the "wittiest of all scientific writers" by evolution popularizer Richard Dawkins) made a profound advance on a "great unsolved problem of biology." He noted that what matters for aging among sexually reproducing organisms is what happens early in life. In the wild, few might make it to old age because of predators, natural disasters, and other kinds of exogenous mortality that I've mentioned.

Recognizing that expected remaining lifespan only dwindles with age leads to an important insight: the force of natural selection becomes weaker at older ages. Its peak occurs at the onset of reproduction and diminishes thereafter in unison with the fall in reproductive potential. In the parlance of evolutionary biology, the declining likelihood of propelling genes into future generations with age is referred to as a diminishing "residual reproductive value." A lethal genetic mutation that results in the death of a child will be selected against, because that child will never pass the mutation to future generations. If the lethal mutation manifests at age 80, that's a different matter. That 80-year-old already had children, and likely passed the gene to offspring. According to this view, the actions and whims of the elders are invisible to selection! This key insight was inspiration for several giants of evolutionary biology, including J.B.S. Haldane, R. A. Fisher, and the brilliant William D. Hamilton.[9] It underpins all evolutionary theories about aging.

An allele that is harmful only late in life is invisible to selection, hiding in its shadow (figure 2.1). Mutations expressing only in selection's shadow can therefore accumulate over generations, leading to the gradual demise of our soma (Greek word for "body" used in biology to reflect the whole body of an organism except the germ cells involved in reproduction). This is the basis of the "mutation accumulation" theory of aging. It is not the accumulation of mutations over one's own lifetime that can increase cancer risk. Those types of processes can happen too. It's the snowball accumulation of harmful genes that build up over generations.

So, as J.B.S. Haldane first proposed, diseases like Huntington's disease might be maintained in the population because they appear relatively late, after age 30. Huntington's is a progressive brain disorder caused by a single extended CAG repeat in the *huntigtin* gene on chromosome 4. (The disease struck one of my heroes, the famous folksinger Woody Guthrie, who inspired generations of protest musicians and songwriters. His mother died of undiagnosed Huntington's disease[10]

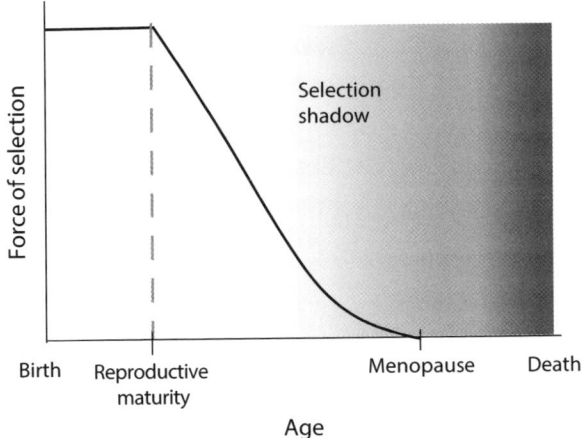

FIGURE 2.1. **The force of selection declines with age.** Natural selection is ineffective for winnowing out any mutations whose harmful effects only manifest in the "selection shadow."

and, despite reportedly having said at age 17, "There's no way I'm gonna get that disease," Woody was diagnosed with Huntington's at age 40, hospitalized by age 42, and died from it at age 55. The terrible disease destroyed his body, but only after he had at least nine children with at least three women.[11])

While the single dominant allele causing Huntington's disease may be an extreme example, it's important to recognize that new germline mutations occur naturally and somewhat randomly because of the mechanics of DNA replication and repair. Most mutations are likely to be neutral, some may be potentially harmful. Only the very rarest will give you enviable superpowers. Germline mutations are the ones made when the DNA coming from Mom and Dad experience copy errors during fetal development. These are the mutations that can transmit across generations, unlike the more abundant "somatic" mutations that appear over your life course by, say, experiencing too many sunburns or smoking too many cigarettes. The number 42 is not only the secret to life, the universe, and everything—as posited by *The Hitchhiker's Guide to the Galaxy* by Douglas Adams—it is also the average number of unique (de novo) mutations carried by each child at birth.[12] Biological aging, or senescence, is the end result of enough of these slightly harmful mutations accumulating over many generations. These alleles can accumulate because of genetic drift—the less-attention-grabbing evolutionary process by which alleles spread by chance alone, without any push or pull that eliminates them or makes them spread like wildfire.

Don't Antagonize Me!

Those late-acting harmful alleles could also spread if, instead of being neutral, they had beneficial effects earlier in life when the force of selection is strong. In the cruel calculus of natural selection, the young are valued more than the old. This principle motivates the idea of "antagonistic pleiotropy," wherein selection favors alleles that produce harm at late ages because of the way they improve survival or fertility earlier in life. Pleiotropy refers to the multiple effects of single alleles, while antagonism refers to opposing effects at different stages of the life course.

Introduced by the influential evolutionary biologist George Williams, this idea has inspired a lot of thinking about the evolution of aging. Williams uses an example of a hypothetical mutation that aids bone calcification during child development, but in adulthood leads to arterial calcification, a risk factor for hypertension and heart disease. While no such gene has yet to be detected, a recent meta-analysis suggests that calcium intake could lead to a higher risk of heart attack and stroke. Calcium helps support bone growth early in life, and supplements are often prescribed to help prevent postmenopausal osteoporosis.[13] Bone calcification and arterial calcification may indeed be correlated, but excess calcium intake is not the reason our arteries stiffen with age.[14]

Williams introduced this plausible, but ultimately incorrect, example in 1957 because no good cases of antagonistic pleiotropy had yet been identified. Good examples of antagonistic pleiotropy have since been discovered, though mostly confined to the so-called model organisms used in the laboratory to study aging: yeast, mice, roundworms, and fruit flies. Inducing selection effects requires working with organisms that have short lifespans and can be bred or manipulated easily. By knocking out genes or blocking their actions, experiments can reveal that some genes may extend lifespan, but at the cost of fertility, growth, or development earlier in life. Experts believe that antagonistic pleiotropy is nearly ubiquitous in nature.[15] And so what about us humans?

In the last few years alone, several labs have made exciting progress in identifying candidate human genes with the telltale signs of antagonistic pleiotropy.[16] For example, we now know that more than fifty genetic variants known to increase the risk of coronary artery disease (CAD) are associated with a number of reproductive traits in both men and women, at least in the United States. Having those alleles led to higher reproductive fitness, but at a cost of greater CAD risk at later ages. As a complex disease, CAD involves many genes, and so revealing the molecular signature of antagonistic pleiotropy for a modern scourge like CAD, especially under low fertility conditions due to the widespread use of birth control, is remarkable.

Another fascinating example is a gene that helps aid survival in extreme cold and high-altitude environments but may predispose folks to cancer at older ages. Inuit, Tibetans, Aymara, and other Indigenous inhabitants of these extreme environments have mutations in tumor suppressor genes that reduce cell death (called apoptosis), thereby helping cells tolerate extreme conditions. But, with reduced apoptosis, there's a greater chance that renegade cells can grow without opposition. In other words, they become cancerous. Many of these genes under selection in extreme environments have been linked to a number of both common and rare human cancers.[17] These effects aren't limited just to humans. Similar linkages have been shown in other mammals of the Tibetan plateau.[18] These types of mutations might explain the high cancer rates reported among circumpolar Inuit, and the strong associations of cancer incidence in countries with very low average temperatures.

These and other recent examples come from genetic association studies, rather than from experiments. Diehard "if it ain't experimental, it's crap" skeptics may withhold gold stars until there's more conclusive evidence showing cause and effect. But we're getting there. The ability to triangulate information from multiple sources and methods has advanced considerably in recent years, along with the growth of huge genomics datasets coupled with medical histories. I expect we will be hearing much more about antagonistic pleiotropies in the near future.

Dispose of My Hard-Earned Soma

A big part of Williams's lasting legacy was his introduction of the role of trade-offs in shaping the evolution of aging. Trade-offs are everywhere you look: the $9.95 per month you (used to) spend on getting a bona fide Netflix DVD delivered to your door is money you can't put toward finally getting curtains on your bedroom window. The idea that trade-offs structure all complex design pervades evolutionary thinking and engineering alike. Budgets are limited—whether we're talking about calories, energy, time, or money—and so spending or investing optimally requires careful attention to how you allocate these critical resources to different life functions. In order to be successful, all organisms need to grow, reproduce, and survive. If you spend longer growing, your bigger size may help you ward off predators and competitors and may even enable you to reproduce at a higher rate. But each year spent growing delays you from the action of adulthood. Humans, like most mammals, are determinate growers. That is, we grow until we don't, then, like a flipped switch, we shift energy into reproduction instead. (Some species don't like to switch it up. Indeterminate growers, like snakes and bristlecone pines, grow and reproduce at the same time.) The danger with any delay is that there's always a chance you could die before successfully reproducing.

This leads us to one of the most fundamental questions governing life history evolution: *Under what conditions is it worth investing in an immortal body that might live forever?* As Williams noted, "It is indeed remarkable that after a seemingly miraculous feat of morphogenesis a complex metazoan should be unable to perform the much simpler task of merely maintaining what is already formed."[19] Ensuring that the physical body can persist over time, and in decent shape, doesn't come easy. Fighting entropy requires constant efforts to maintain order and functionality. No matter how often I clean, my kids will trash a room in under five minutes—guaranteed.

Efficient and accurate DNA repair, immune surveillance of micro- and macro-assaults on our body, handling toxic by-products of cellular processes, wound repair from stepping on LEGOs with our bare feet— these all require fuel, and complex physiological hardware and software. Simply breathing and metabolizing energy generates physiological by-products that need cleaning up (just like combustion from flooring your car's gas pedal makes you go faster, but puts you one step closer to overpaying at Casey's Auto Repair).

What determines whether and how much it's worth to keep your cellular spark plugs shiny and sparky? The simple answer: the risks of fatal demise no matter what you do. These are the "exogenous" (to the body) or "extrinsic" mortality hazards I introduced earlier.

So even if immortality were biologically possible, hazards in the environment—from cataclysmic storms, building-toppling Godzillas, or just stumbling drunk off a cliff (but at least winning a Darwin award, a tragicomical honor given to those who help evolution by selecting themselves out of the gene pool)—will cut life short sooner or later. With enough hazards, few might still be alive by the time they reach their 60s. This puts a premium on reproduction earlier in life, to ensure representation in future generations. In a world of finite resources (in other words, the world), optimal life histories never put all the eggs in the maintenance basket to ensure a sterile eternity. This underlies the "disposable soma" theory of aging coined by the biologist Thomas Kirkwood in 1977. According to this idea, the high-level investments needed to maintain the body indefinitely will never be favored by natural selection because reproduction is almost always a more profitable way to spend limited resources and maximize fitness.[20]

The disposable soma idea is powerful because it predicts aging whenever the germline is separate from the soma. This separation is irreversible and means that germ cells do not contribute to the soma, and the soma can't make the sperm and eggs that form your sex cells. The soma is simply the vessel that serves to preserve the germline. Mutations that form in the soma can't hurt the germline. So a devastating mutation in a liver cell that triples your cholesterol levels, acquired in your 30s, won't be passed on to your kids. By four months in the womb, a human female fetus has already separated out every cell that will develop into a mature egg.

All vertebrates separate soma and germ cells, and so aging is widely expected among those with backbones. In invertebrates like hydra—not the nine-headed serpentine water monster of Greek lore but the tiny tubelike multi-tentacled freshwater creature, discovered by the microscope's inventor van Leeuwenhoek in the 1700s—there is no such separation. Even without Matrix technology, these fascinating creatures can re-form if you cut them up in pieces. Hydra are famous in my circles because they "escape senescence." Regenerative stem cells give rise to both somatic cells (buds) and germ cells capable of indefinite renewal under favorable conditions. They usually reproduce asexually by a process called budding, though some species can reproduce through old-fashioned sex.[21] One estimate shows, in a lab at least, that 5% of an adult population could still be alive after 1,400 years. Despite the great evolutionary biologist William D. Hamilton's claim that senescence should be found "even in the farthest reaches of almost any bizarre universe,"[22] we need look no further than those hydra, the tiny jellyfish *Turritopsis dohrnii*, some lobsters, planarian flatworms, and a variety of plant and fungus species. These species display what my friend, the biogerontological ombudsman and kick-ass Appalachian fiddler Caleb "Tuck" Finch, first called "negligible senescence."[23] They play by different rules.

Once we appreciate the disposability of our vertebrate soma, the search for the physiological mechanisms that underlie antagonistic pleiotropy stands on more solid ground. Damage that accumulates needs to be specified, as do the repair mechanisms and their costs. Some degree of errors in DNA replication is necessary to help generate the mutations that selection later acts on. In fact, there should be an optimal level of DNA repair that combats molecular entropy, but doesn't do so perfectly. Senescence, according to Kirkwood, is a result of "an energy-saving strategy of reduced error regulation in somatic cells."[24]

Beyond these kinds of errors, other types of damage occur—to proteins and cells across multiple levels of biological organization.[25] The powerhouse of cells (mitochondria) experience dysfunction, the caps on the ends of your chromosomes (telomeres) shorten with each cellular division, nutrient sensing gets compromised at the cellular level, the dynamic regulation of protein networks goes awry, stem cells get exhausted. Aging is the physiological manifestation of the complex interactions of these and other system components, each afflicted in some way by imperfect repair. These cellular and molecular processes have been labelled by biologists as the "hallmarks of aging," as they represent the "common denominators of aging" in mammals.[26] Most people in the longevity-extension industry see demarcating these and figuring out how they work in isolation, what affects them, and how they're intertwined as the starting point for everything. (I vowed that this book is not about cellular biology, but we'll return to thinking about how these detailed mechanisms link back to whole-organism design in chapter 8.)

Demon Attack

The hypothetical Darwinian demon is famous for its godlike superpowers: it reproduces minutes after birth, gives rise to millions of children, and, well, lives forever. Of course, such an organism doesn't exist, but its hypothetical existence illustrates a basic engineering principle: you can't maximize everything. Life, again, is about trade-offs. Evolution tinkers toward maximizing reproductive success—and if that means a long and prosperous life, great! But often it will not.

So, the logic goes: in a cruel and hazardous world, your best bet for reproductive success is to stop growing early and convert that soma into babies, as fast as you can, without compromising your kids' survival. The brutality of a high-mortality world means that you're unlikely to be around in the distant future, and so best to give everything to having and boosting your children. Don't spend too long growing, lest you might never reproduce. And don't bother paying for the extended warranty.

In a safer, more predictable world, things look different. Now it may be worth growing longer, especially if that helps you produce bigger babies who may be more likely to survive, and at a faster rate. And the greater assurance of still being around by the time your children start giving you grandchildren means it may very well be worthwhile to spend some of your life force on keeping all systems go.

One of the earliest and most exquisite illustrations of this idea involves furry Virginia opossums studied by the ex-Hollywood lion and tiger trainer turned evolutionary biologist Steven Austad.[27] He studied the same species of the animal in two very different locations in the US South. On the mainland, where Virginia opossums experience high mortality from predation by bobcats and wild dogs, the opossums age quickly, showing telltale markers just like we do: cataracts, muscle wasting, and arthritis.[28] Genetically similar opossums on Sapelo Island, just off the coast of Georgia, however, haven't had predators for at least four thousand years. Compared to the mainland opossums, the island ones experiencing no predation show all the signs of a slower life history: they reproduce later, have smaller litters, and die old. They also show slower physiological aging. Collagen fibers making up the tendons that support muscles and bones are well preserved in late life, a general indication that their bodies age more slowly.

In general, physical adaptations that also help foster a safer environment, say, from predators, including hard shell protection in tortoises, or wings enabling flight in bats and birds, tend to correspond with longer lifespans. Among birds, species that experience a faster rate of mortality with age are those with higher levels of exogenous mortality.[29]

Comparative studies like these across a wide range of species are consistent: changes in exogenous mortality seem to alter the force of selection with age in ways that affect senescence, lifespan, and pace of life. It would be even more powerful to show through experiment that a change in the force of selection alters the life history in the way predicted by the theory. Enter from stage left everyone's favorite fruit fly, *Drosophila melanogaster*. Careful selection experiments with these lab-friendly flies confirm key predictions: reducing extrinsic mortality increases lifespan after generations of exposure. It also delays maturity and decreases the birth rate earlier in life.[30]

The joint logic of antagonistic pleiotropy and disposable soma form the backbone for current evolutionary thinking about why we age. As we sketch out below and throughout the book, moving from general theories of aging to the specifics for any species, much less a single population, depends on many details. And a good understanding of a species' natural history.

Extrinsically Intrinsic

The binary partitioning of mortality into extrinsic and intrinsic makes sense, both for the theory and in practice. In principle, extrinsic mortality should affect how selection shapes intrinsic mortality. But is mortality ever truly extrinsic?

Consider the case of the northwest shores of Lake Malawi, where there's a good chance you might get struck by lightning. The chance of dying from a lightning strike is more than fourteen times higher there than in high-income countries, and five times higher than the highest rates ever reported anywhere.[31] Seems like lightning is about as random and extrinsic a source of mortality as you can find. But, one could just move to a more distant part of the shore and reduce that risk. Or one could stay inside and close the windows during a storm. In other words, your judgement, an intrinsic quality, matters.

Let's try another example. Predation by ravenous carnivores is the quintessential illustration of extrinsic mortality. But predators notoriously prey on the slow, weak, or infirm.[32] Gruesome to say, but a pregnant Bambi probably runs more slowly and is worth more calories than her younger sister. The easier to catch or harder to escape, the better for lazy predators.

What about the other commonly invoked form of extrinsic mortality, infection? Our bodies have numerous multifaceted defenses to handle the varied assailants on and in our body. We have skin as a semiporous barrier. We invest in a wide array of cellular soldiers, the white blood cells being the most familiar. But many other important immune defenses from fevers to targeted phagocytes, T cells, and antibodies help put up the good fight. How much we put into building our immune machinery will affect the likelihood a critter evades our defenses, or at least puts up a battle that leaves us tattered and worn but still breathing.[33]

The notion of a mythical class of mortality that strikes everyone equally may be misguided.[34] All causes of mortality have both extrinsic and intrinsic components. Attempting to disentangle whether deaths are due to intrinsic or extrinsic causes is like asking whether disease is caused by nature or nurture. Or if it's random versus age-related. It's always some degree of both. Our bodies become compromised whenever mortality rates increase with age.

Instead of worrying about whether a cause of death is extrinsic or not, we need to appreciate how physiological traits relate to fitness more broadly, and in different environments. The biologist David Reznick illustrates this beautifully with an experiment focusing on tropical fish.[35] Guppies living in lower parts of a river in Trinidad experience a lot of predation by larger fish. When abducted from the river and taken to the lab, these high-predation guppies both outlive and outreproduce their wussy low-predation brethren. How is this possible if harsh environments select for faster senescence?

Evading predators requires being a fast swimmer, and so a high-predation environment selects for enough repair in the relevant systems for the older fish to maintain swimming speed. Effectively dodging hungry wolffish and pike cichlids, in turn, reduces mortality. This is all to say that while the general logic of "high mortality = faster aging" is valuable, local details matter! Whether we see faster or slower aging depends on how physiology leads to death. It may also depend on how we conceptualize and measure aging. How any particular cause of death— be it starvation, hypothermia, or land sharks—affects mortality is what statistical inference folks call "endogenous." In other words, how we age depends on our built and maintained capacities. A clearer understanding of the natural history of aging will allow us to map the selection pressures shaping vigor and survival across the lifespan.

Aging in the Wild?

You might be thinking that aging makes sense in protected environments where we're released from the slings and arrows of misfortune. Hence, we see it throughout the period in our recent past because life expectancy rapidly expanded around the world. Or in domesticated animals, among our overfed pets, and in most mammals studied under favorable laboratory conditions eating chow without predators or famine. It's a common refrain that aging simply doesn't exist in the wild. Under natural conditions, we shouldn't expect to see aging in humans or in other species. Even the visionary biologist Medawar's early argument that motivated so much modern evolutionary thinking stressed the belief that senescence occurs at ages "which the great majority of the population do not reach." Medawar went on, emphasizing, "The fact is that under the exactions of natural life they do not do so [reveal innate deterioration]. They simply do not live that long."[36]

This idea has its corollary in most contemporary studies of humans. Many researchers of aging believe *you can only study aging in protected environments*, where presumably most deaths are due to aging-related (intrinsic) conditions. Given that very little mortality occurs in postindustrialized countries before age 40 (e.g., < 2% in Sweden, < 4% in the United States), this has largely meant that most studies of aging and lifespan in the biomedical and epidemiological world have been in high-income countries.

The evolutionary biologist George Williams recognized that aging cannot be absent in the wild. He pointed to the decline in athletic performance in top-notch athletes by the time they're in their 30s as evidence that aging occurs throughout adulthood. Indeed, the evolutionary theories spun above would expect aging to "begin" as early as the age of reproductive maturity. That's the point at which the force of selection begins to decline. In other words, aging is a gradual progressive process occurring throughout adulthood. And not just the outward signs of aging that begin at more extreme ages, like gray hair, frailty, and senility. Those may very well not be observed much in the wild.

As suggested by the evolutionary theories I've highlighted, animals should be expected to live long enough to reproduce successfully, and probably not much longer than that. Length of life is like Nature's warranty period with a quick expiration date in the wild. Are we currently living beyond the warranty period, or is our warranty period itself extended?

————

Aging is not at all rare in the wild. But it took years of careful long-term study, with obtaining reliable ages, to recognize that aging occurs in wild populations throughout the animal kingdom.[37] Prior to this, inferences about aging were made using population cross sections—comprising individuals of different ages at one slice of time. While having a cross section is a good start, it's hard to interpret what is inherently a longitudinal process that individuals experience—aging—by comparing *across* individuals of different ages. Ideally, analyses using cross sections and longitudinal follow-up should reveal similar patterns. But if there are cohort differences in diet or other exposures, as expected during periods of rapid environmental or cultural change, then inferences about aging will differ if based on cross-sectional versus longitudinal study. Also, older adults may be robust survivors and not typical of the younger people they're being compared against in cross-sectional analyses.

While survival is often higher in zoos or in captivity, longitudinal studies convincingly demonstrate that aging in the wild is probably the rule rather than the exception. Each year, new findings confirm that senescence is widespread. Even my daughter's favorite flying insects—dragonflies and damselflies—are more likely to die at older ages.[38]

What is fascinating is that aging rates vary more in the wild than typically observed in laboratory model organisms. Laboratory conditions shouldn't just be thought of as lifting the veil to reveal the true underlying pattern of aging. The environment shapes everything, including how natural selection will affect life history allocations to maintenance.

Too Good to Be True?

If genes exist that modify the rate of aging even in the wild, then surely targeting and manipulating those genes should help us move closer to Methuselah-hood. But evolution is usually smarter than us. Should it ever be so easy to tweak one small thing and reap untold benefits without paying some Faustian penalty? Wouldn't evolution have figured it out already? Remember that Nature's prize committee doesn't care about longevity for longevity's sake. Bodies serve the interests of their genetic overlords, and reproduction is their sole obsession.

One of the classic examples suggesting the life-extending potential of gene tinkering comes from another wonderful model organism, the nematode, *Caenorhabditis elegans.*[39] A mutation of the *age-1* gene affecting the insulin-like signaling pathway can extend lifespan by up to 80%. That's a massive increase in longevity. That pathway affects not just aging, but also development and stress resistance. Other mutations in the nematode have also been shown to extend lifespan and vigor (e.g., *daf-2*).

But Nature is smarter than we are. These mutant worms live their long prosperous life only under strict controls in the lab. When subject to conditions that mimic their natural environment—with food limitations and more variable temperatures—fertility is reduced early in life, and they are outcompeted by the more short-lived nonmutants. This has been confirmed by sending the mutants back to their native soil. Consistent with antagonistic pleiotropy, almost all mutations known so far to extend lifespan in model organisms—by 27% to 112% in *C. elegans*, 20% to 90% in fruit flies, and 19% to 70% in mice—usually reduce fertility or carry other harmful consequences, like cognitive impairment or greater susceptibility to infectious disease.[40] There's simply no such thing as a free (rejuvenating) lunch. Indeed, we should be skeptical about any promises of pure life-extending benefits that come at no cost whatsoever.

In humans, an extreme example illustrates how, like in the nematode, the currency of evolution takes precedence over maximizing lifespan alone. Based on the results of experiments with livestock, authorities within prisons and other institutions once believed that castration could make male inmates more "tractable." Castration was therefore occasionally administered in the early twentieth century to "tame the unruly," especially among those with mental disabilities. A study from the 1960s compared three hundred white inmates in Kansas who were surgically castrated against those who were spared the knife.[41] Men were castrated between the ages of 8 and 59. Though these samples are hardly randomized, inmates across the two groups didn't differ when it came to having medical conditions that could predispose them to early death, or the amount of time spent in prison. However, when we compare the survival rates of eunuchs with their matched intact counterparts, we see a huge potential trade-off between reproduction and longevity. Castrated inmates lived thirteen years longer than those who were not castrated (69 vs. 56 y). Among those living beyond age 40, the longevity boost was greater the earlier in life the castration occurred: from almost twelve years if castrated between ages 8–14 to just four extra years if castrated after age 30.

Though rare, male castration pops up throughout history, as a means to preserve the tenor voices of boys into adulthood or to reduce the threat of cuckoldry by guards in royal courts filled with concubines. While no difference in longevity was found between a small sample of sixteenth- to nineteenth-century European singers castrated before puberty ("castratis") and a non-castrati control sample, a more recent study of Korean eunuchs from the Chosun dynasty shows a survival advantage of fourteen to nineteen extra years, and a higher chance of becoming a centenarian.[42] In other words, extra lifespan can be gained, but at a cost that few would ever want to pay.[43]

If Aging Is Not Programmed, Then Why Must I Learn Python?

While mutation accumulation, antagonistic pleiotropy, and disposable soma represent the "classical" evolutionary theories of aging, there are tons of other theories about the *proximate mechanisms* of aging. Even by 1990, there was a glut of more than three hundred theories, putting Baskin Robbins' generous selection of thirty-one ice cream flavors to shame (though naysayers may protest that their flavor library includes more than 1,300 flavors).[44]

The important distinction here is that evolutionary theories try to explain *why* something occurs, whereas most theories of aging try to explain the *how* of aging. These focus on the physiological mechanisms, the details about how we age, like the hallmarks of aging I mentioned before. The multibillion-dollar antiaging industry is not as concerned with *why* we age, if slowing down aging instead just requires knowing *how*. Fair enough. But as the Nobel Prize–winning ethologist Niko Tinbergen pointed out almost seventy years ago, ultimate questions about why and proximate questions about how are not in opposition, but are complementary paths to understanding.

Ultimate and proximate can intersect. If mutation accumulation was the main string-puller behind aging, we would expect each species, and maybe even subgroups within species, to show different random mutations leading to aging. In other words, we wouldn't expect universal (genetic) mechanisms across the tree of life—which, if true, greatly reduces the relevance of most nonhuman lab work to improving human welfare. If antagonistic pleiotropy is instead a more relevant mechanism behind why we age, we should expect many mechanisms affecting key trade-offs to be conserved across species.

Regardless of the evolutionary mechanism that explains why we age, there is an important lesson to learn: aging may be a consequence of the declining force of selection with age, but aging itself is not adaptive—even if it results from an adaptive process. That is, aging is not a trait actively favored by selection. It is a by-product of the developmental program and other organic machinery, a result of what the biodemographer Bruce Carnes and colleagues call "imperfect structural designs built with imperfect materials and maintained by imperfect processes." Senescence is a "byproduct of evolutionary neglect, not evolutionary intent."[45] Therefore, we shouldn't ask "What is aging for?" Instead, we should investigate the fitness costs of aging, and the fitness benefits of other processes that contribute to aging as a by-product. We should focus on how early life exposures affect aging, and why people vary so much in the onset and manifestation of aging.

Is Aging a Disease?

Even without the evolutionary notions of trade-offs with growth or reproduction, much work on aging is compatible with the disposable soma ideas that damage is costly to repair and that a high enough assault rate in a hostile environment can render indefinite survival to be virtually impossible. Aging manifests as the loss of mystical "vitality," though recent decades have helped open the black box of what vitality might be. Various concepts over the years try to capture the "things falling apart"-ness of aging. They go by ominous, terrifying labels like allostatic load, physiological dysregulation, and regulatory decoherence. They all emphasize that, with aging, it becomes increasingly more difficult to bounce back after a jolt. Aging, according to the microbiologist Leonard Hayflick, is a natural and universal effect of "thermodynamic instability" in all multicellular animals. Consistent with these views, the more awful things that happen to you over a life course, the faster your demise.

From a clinical standpoint, heart disease, cancers, strokes, dementia, and kidney failure strike us in our late golden years. We may first be diagnosed with these chronic diseases in middle age, but we're far more likely to succumb to them in late adulthood. Many other ailments that make life worse for wear also increase with age—like arthritis, osteoporosis, cataracts, and hearing loss. Age is the common factor underlying these afflictions.

The central role of age inspired the Roman playwright Terentius to proclaim over 2,200 years ago that "old age itself is a sickness."[46] The opposite view, as argued by many, including the infamous physician of Greek antiquity, Galen of Pergamum (whose ideas about all things medical dominated for about fifteen centuries), was that aging is instead a natural process. His treatise *Hygiene* (also known as "On the Preservation of Health," 175 AD) didn't claim that there was nothing to be done— in fact, Galen argued that diet and other preventions could slow aging.[47] He's reported to have lived to age 70 about two millennia ago, so maybe he was on to something.

Where people differ is in how to interpret the manifestation of particular age-related clusters of symptoms we conveniently call diseases. Are these deviations from normal, "healthy" aging or unfortunate, often haphazard, consequences of universal aging?

The modern view that aging itself is like a disease was first championed in the 1950s by Robert Perlman, who referred to the aging syndrome as a "disease complex" and believed a normal or healthy aging was an "imaginative figment."[48] The decision whether to label aging as a disease or not may sound arbitrary. After all, pregnancy and masturbation were considered diseases at different times throughout history. Fever was considered a disease too, but switched to being more of a symptom once it was realized that different causes could elicit fever.[49]

On a practical note, how does it make sense to call something that everyone will experience a disease? Calling something universally experienced a pathology arguably adds more shame and stigma to an already belabored subset of the population. Although everyone who lives long enough will experience aging, not everyone will get heart disease, diabetes, or dementia. How do you diagnose something that affects everyone differently?

Despite these caveats, the move to view aging like a disease is gaining steam. And for more pragmatic reasons. First, viewing aging as natural and unavoidable may sound too fatalistic for some, as if there's nothing to be done—though even Galen and others argued that naturalness doesn't mean inevitability or helplessness. Second, we're used to fighting diseases. Much of the health-care industry is bent on diagnosing and treating separate diseases post facto. The enormously important field of geriatrics struggles for attention in its holistic treatment of the whole person and their well-being, rather than of one disease at a time. If you want to focus efforts on treating or curing something, better to call it a disease (and launch a War on Aging!). In the US, the Food and Drug Administration won't approve any products that could slow down aging, even if they existed. Under the current system, only drugs that treat designated diseases can be brought to market with stamps of approval.

In line with these views, a "geroscience" movement has positioned aging as the central risk factor for most, if not all, age-related chronic diseases. In other words, aging is the target. Tackling heart disease or Alzheimer's disease will surely help many people, but such a disease-by-disease approach is like placing Band-Aids on a massive wound. Eliminate one disease and others pop up like a renegade whack-a-mole. If aging is linked to all these conditions, then better to focus efforts on the fundamental source(s). In this view, aging is not a disease per se, but is still treated like one. And by slowing down aging, we can achieve the broader goal of increasing healthspan. Geroscience therefore addresses common genetic, cellular, and molecular processes underlying the basic biology of aging.

I favor a balanced view, wherein both diseases and aging are targets for intervention. In the US, most federal funding for health targets specific diseases or disease processes, not the basic biology of aging. This is likely to change, with growing funding from private foundations and industry and greater public funding from China and Singapore.[50] Disease targeting has been successful in the past for heart disease and some cancers, but Alzheimer's disease has proven challenging. Diversifying our biomedical approaches to extending our lifespan is essential, given the different types of diseases and frailties we're likely to experience in our later years.

Exponentially Yours, B. Gompertz

We've established that mortality rates increase with age in birds, mammals, and, of course, humans. And that such increase is a by-product of natural selection. The physiological machinery breaks down, but by how much do mortality rates increase per year? What affects the *pace of aging*?

Like Gregor Mendel identifying the rules of genetic inheritance prior to any understanding of genes or genetics, the British mathematician and actuary Benjamin Gompertz first realized in the early nineteenth century that mortality rates increased with age in a particular kind of way, and with striking regularity. So much so that Gompertz managed to get an equation named after him, and the highest compliment is that his equation is often referred to as "Law." An impressive achievement for someone denied access to universities for being Jewish. And who at the time was more interested in estimating premiums for lifetime annuities. Just like the field of historical demography can thank churches for meticulously recording baptisms and genealogies, the fields of ecology, biology, and gerontology can thank the early actuaries for developing new methods to figure out how much to charge for life insurance.

But Gompertz did have high aspirations: "If generalizations lead to probable theories, they should be regarded as pleasing associates, to be entertained at the feast of knowledge."[51] He noticed that mortality rates over the life course look a bit like a bathtub—starting high early in life, bottoming out to its low point around puberty, and then increasing thereafter. By carefully studying death records among adults ages 20 to 60 in England, France, and Sweden, Gompertz developed an elegantly simple equation to describe the "increasing thereafter" part. In its modern form, we would say that the instantaneous risk of mortality (called a hazard) increases exponentially with age, or $h(age) = A \times e^{b \times age}$.

What is wonderful about this equation is its surprising simplicity in capturing a complex phenomenon—just two parameters to estimate: A and b. These parameters have a straightforward interpretation. A is the "vulnerability" of a population, its initial mortality at puberty, set in place by the environment or by genetics. The b parameter is what determines how fast the mortality rate increases with age. The bigger the b, the faster mortality takes off like a rocket. This pattern of exponential growth goes a long way in representing what mortality rates look like after puberty—not just in humans, but in many species, from mice to quail.[52] It also does a decent job fitting human mortality, including those with low life expectancy e_0 such as India in 1900 to high e_0 such as Sweden in 2020. What's amazing about the Gompertz equation is that it doesn't really matter if what kills us varies tremendously across populations. Could be tuberculosis, or chronic obstructive pulmonary disease, or an aneurism. The drumbeat of exponential march doesn't seem to care.

One convenient metric easier to interpret than b is called the mortality rate doubling time, or MRDT. This is the number of years it would take for the mortality rate to double. Given exponential growth, this can be readily expressed as $\ln 2 / b$. The aging polymath and Darwin doppelganger Tuck Finch calculated back in 1990 that MRDT is about seven to eight years in humans, *regardless of the level of mortality*. MRDT is an attractive metric because it's independent of the level of mortality, consistent with a universal aging pattern in humans. But for the same reason, MRDT is absurdly limiting, and not terribly useful for those trying to improve human welfare. Doubling mortality rate from 0.05%

to 0.1% mortality per year looks a lot different than doubling from 2% to 4% per year. Mathematically, the doubling is still just doubling. But in the first low-mortality case, there's a difference of 0.05%. In the second high-mortality case, the difference is 2%, reflecting a much greater absolute number of deaths.

Many individual causes of death also show Gompertz-like mortality rates with age, at least after age 40. At least sixty-five causes of death in the US in 2018 followed a Gompertz pattern with age, with a median MRDT of seven years—similar to the all-cause mortality described above. MRDTs range from a low of three for Alzheimer's disease, to twenty-one years for asthma. Even COVID-19 mortality follows a Gompertz pattern with age, with an MRDT of seven to eight years, though the slope varies among countries.[53]

Variations on a Theme: Exogenous Mortality Revisited

If Gompertz mortality represents some idealized universal pattern, then deviations from it can offer valuable clues. Although Gompertz himself recognized that mortality rates may not increase unabated at latest ages, the main critique has been that it doesn't adequately predict mortality rates beyond age 85, where claims of slower, or "negligible" senescence have been made. When first identified, negligible senescence was viewed as an anomaly, perhaps a result of poor age estimates or sparse data. Or as an artifact: those remaining survivors beyond age 85 might just be uber-vigorous. At the population level, mortality rates can seem to slow down or even flatten with age, but this is due to a mixed population in which some folks are more robust than others. Only the robust are still around at latest ages.

Another serious deviation from Gompertzian mortality is ludicrously obvious in humans. The onset of exponential growth occurs at least a decade past the age of sexual maturity, when the force of selection is near its peak. In other words, aging, as reflected in exponential increase in mortality, is delayed in humans.

The English actuary William Makeham proposed a slight but important modification to the Gompertz equation: adding another term. Why bother? This was to account for the fact that the mortality rate seemed to increase more rapidly at later ages. So now, $h(age) = c + A \times e^{b \times age}$. Notice the c term is a constant—it doesn't change with age! The c term therefore comes closest to capturing the exogenous mortality risk I discussed above. Hypothetically, risks affect everyone at all ages equally. In practice, c usually reflects the lowest mortality rate experienced at any age, often close to age of maturity. Combining the two terms predicts actual mortality better. But the real advantage is that the two terms partition (mathematically at least) intrinsic mortality (the Gompertz exponential, $e^{b \times age}$) from exogenous mortality (the constant term c).

Math in Search of a Theory

The implication that a simple mathematical relationship can describe mortality with age throughout much of adulthood, regardless of the composition of causes, and across many species, suggests some underlying principles. Gompertz thought that the loss of some biological "vital force" might lead to "deterioration, or an increased inability to withstand destruction."[54] These formulas were believed to reflect how life, in the words of the medical statistician John Brownlee, was "dependent on the inherent energy of certain substances in the body, an energy which is gradually being destroyed throughout life."[55]

Less than a century later, the biologist Raymond Pearl, regarded as one of the founders of biogerontology, was the first to compare mortality profiles of different species in search of some fundamental biological law of mortality and lifespan. He compared survival curves of US males from 1910 with that of *Drosophila* flies, and marveled about their similar shapes, despite "extreme differences in habits of life, structure, physiology and environmental stresses and strains" (see figure 2.2). A survival curve refers to the number or proportion of a population cohort still alive at each age. By fitting both curves with the Gompertz equation, he equated an 86-year human life with a 97-day fly lifespan. He did observe, however, that humans show higher relative survivorship at most ages, due to "public health and sanitation." But *Drosophila* show lower mortality rates for the last 20% of life (figure 2.2).

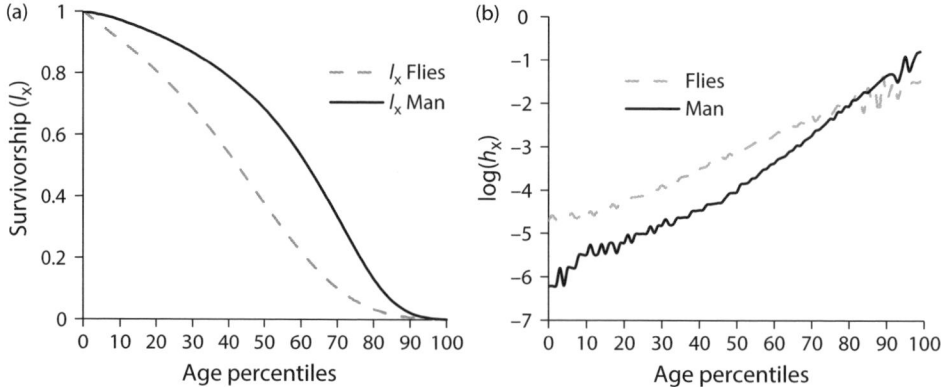

FIG. 2.2. **Mortality patterns for US men in 1910 and male *Drosophila*.** From Table ii in Pearl and Parker 1922. (a) Survivorship (l_x) for flies and males, where ages are shown relative to species-specific lifespans. (b) Mortality "hazard" at each relative age, graphed logarithmically. A mortality hazard is the instantaneous risk of dying at each age. The relatively straight lines reflect exponential growth. From this analysis, Pearl and Parker concluded that 1 year of human life = 1.1279 days of *Drosophila* life.

Despite publishing sixteen papers with the same title ("Experimental Studies on the Duration of Life"), Pearl gave up on his dream of finding a universal law, as he could never get the curves for different organisms to match up exactly. Using current terminology, he thought that extrinsic and intrinsic mortality could not be easily separated in humans, and so the pure "Gompertz" senescence is obscured by non-senescent causes of death.[56]

Ever since Gompertz, other mathematical depictions of the force of mortality have been proposed. Some may fit the data better, or fit mortality across broader age ranges. For example, the clumsily named Heligman-Pollard model can fit mortality data across the entire lifespan, including an often-observed accident bump of higher mortality in early adulthood. But it involves too many parameters to estimate (eight!), and their interpretation is not so straightforward.[57] The simplicity and general consistency of the Gompertz model, maybe with the additional Makeham parameter, makes it a fan favorite.

And, as a fan favorite, the formula has inspired many theoretical musings to help explain why the equation might work so well. One of the first noteworthy attempts to figure it all out came sixty years ago, by the bio-gerontologist Bernard Strehler and the chemist Albert Mildvan.[58] Strehler and Mildvan's theory proposed that damages from the environment accumulate and compromise an organism's ability to maintain energy production, a life force they called "vitality." If the damages imposed by environmental assaults outstrip the remaining vitality needed to overcome them, death results. Given their assumptions and some algebraic elegance, they show that death rates should increase according to the Gompertz model. Their model leads to some predictions about expected maximal survival ages, average survival ages, and even incorporates the role of chance in how stressors might destroy the "molecular bonds" of life.

One interesting prediction of their model is that its two key parameters—the initial mortality rate A and the rate of aging parameter b should be inversely correlated. That is, the greater the initial mortality rate upon entering adulthood, the lower the rate of exponential increase in mortality. This Strehler-Mildvan correlation has been widely documented across countries, and within countries over time.[59] What a wild regularity in the relationship between the two numbers that define so much about the shape and magnitude of mortality![60]

Figure 2.3 shows that lower initial mortality but higher rate of increase in adult mortality corresponds to a rightward shift of the survival curve. This shift has been called "rectangularization" because it reflects the increasing compression of mortality to the end of life (introduced in chapter 1). An implication of the inverse correlation between A and b is called the compensation law of mortality, yet another law in this lawless land. If high mortality populations show slower pace of mortality increase, and low mortality populations show a higher pace of increase, then mortality rates should start to converge at later ages—eventually you hit the same wall of death, no matter how you get there. Mortality compensation implies that populations differing in mortality rates throughout much of life should show a similar mortality rate at some later age of convergence. This late-age convergence has been interpreted to reflect a common species-typical lifespan. Taking the relationship between A and b for Swedish cohorts from the mid-eighteenth to early twentieth centuries, and substituting into the Gompertz equation, reveals an age of convergence. It's about 95 years old.

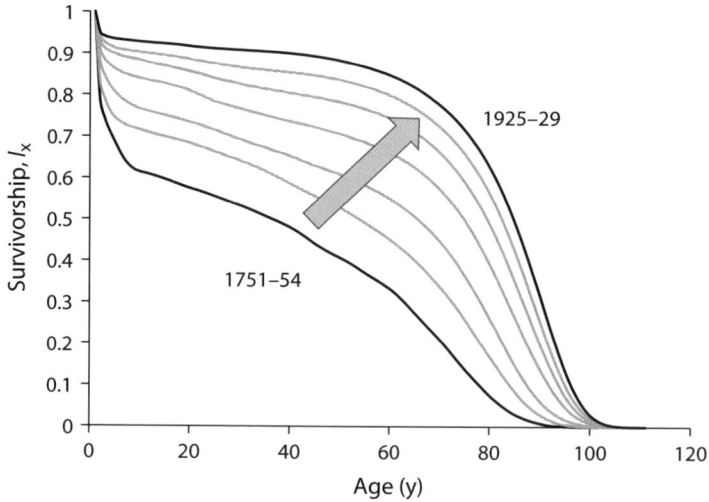

FIGURE 2.3. **Survival "rectangularization."** Each line represents a
survivorship curve for a female Swedish five-year birth cohort
(from 1751–54 to 1925–29). Survivorship represents the percentage
of the cohort still alive from birth to the ages given on the x-axis.
Over time, the survivorship curves of Swedish cohorts become
increasingly "rectangular," as non-senescent mortality declines
and deaths manifest in late adulthood. *Data source:* Human
Mortality Database.

From Whence We Came

While toying around with these somewhat obscure relationships in the
early 2000s, I felt I was unveiling some underlying order, a glimpse into
Nature's secrets about human lifespan. In spite of all the differences in
history, experience, and exposure, the notion of a species-typical lifespan is compelling. Most of the voluminous work on this topic spans
across many disciplines but gives only meager, speculative attention to
small-scale subsistence societies—despite our having lived most of our
existence under preindustrial conditions. Beliefs that the environments
of our ancestors and of contemporary subsistence populations were
harsh—a classic example of high exogenous mortality—support the
conclusion that any signal of "natural" aging in the wild will be difficult
to observe or even that trying to would be a futile, misguided exercise.

With no time machines, and well-funded missions to extend lifespan in high-income countries, most efforts to understand aging have reasonably concentrated on industrialized populations.

Now that I've laid out the relevant terms, leading theories, and the big questions about aging, lifespan, and longevity, we're ready to tap into the well of contemporary small-scale societies and our ancient past to evaluate generalizations about human lifespan under prehistoric conditions. Here's where the rubber starts to hit the road, providing evidence to propose and support an evolved seven-decade lifespan in the natural history of our species.

3

Catching Up to the Present

The days of our years are threescore years and ten; and if by reason of
strength they be fourscore years, yet is their strength labor and sorrow

—PSALM 90 V.10

WHEN ASKED by a student what she considered to be the first evidence of human civilization, the famed anthropologist Margaret Mead is reported to have said "a femur." Not any leg bone, but a fractured one—15,000 years old—recovered from an archaeological site. When the student gasped in surprise for her not mentioning clay vessels, fancy figurines, or cave art, Mead explained how a broken femur that has healed is evidence that another person cared. They helped the wounded by feeding and tending to the injured. Most animals wouldn't survive long enough with a broken thigh bone for it to heal. Injuries resulting in bone breaks like this would seriously hamper your survival. Whether this story is true or not (probably not), numerous cases from ancient bones suggest the early origins of compassion and caregiving.[1] These cases remind us that a big part of what enables a longer human lifespan is human generosity. We'll see how it all fits together soon enough.

So far, we've shown patterns and changes in life expectancy and other longevity measures consistent with long human lives, at least over the past several centuries of Europe, the United States, and the rest of the "developed" world. But now, to get closer to the heart of this book: Is human longevity limited largely to recent history? How can we best get a sense of how deep in time our longevity has existed? Better yet, to what extent does long life feature as a star player in our success as a species?

To paraphrase a lyric from an obscure R.E.M. song, there are several ways to get there from here. Among the groups who survived into the twentieth century still practicing traditional livelihoods lies an amazing fount of knowledge and wisdom. From a common ancestor with chimpanzees some 6–8 million years ago, through our Australopithecus ultra-great-grandmothers like "Lucy" with ape-size brains 3–4 million years ago, and with the dawn of early *Homo* by the middle Pleistocene 2–2.5 million years ago, on through the late Pleistocene with *Homo erectus*—all of these ancient hominins were mobile hunter-gatherers. Only until about ten thousand years ago did lifeways start becoming more diversified among the globe's roughly 5 million inhabitants.

Many of our characteristic species-typical traits came into being over the course of hominin evolution. Over the span of hundreds of millennia, we became bipedal, our brains got bigger, we became hairless (many of us at least), developed stone and wooden tools, and graffitied the ceilings and walls of cozy caves. It's much harder to see changes in life history traits that don't leave a lesion on a bone, or that don't fossilize. While paleoanthropologists lose sleep over what can be gleaned about lifespan from the fossil record (see chapter 4), my chosen path for tackling questions about lifespan and aging is among living people.

In this chapter, I'll introduce the groups whose careful study forms a large part of the empirical basis for a seven-decade lifespan for our species. We'll then trek through the evidence and explore how lives are cut short. I'll throw in some complementary data from other groups and contemplate how changing conditions could influence lifespan. We'll see how the seven-decade lifespan shines through as a clear signal, highlighting the long-standing existence of a postreproductive lifespan. But first, how reasonable is it to rely on subsistence populations in the first place?

Elephants in the Room

The advantages of working with living peoples should be obvious, but let me address any potential skepticism or protests from the outset. There are very few populations left today that don't interact with regional or global markets, that aren't tied to national health-care systems or enmeshed in local politics and religions in ways that affect the very kinds of questions we'd like to answer. While all living people today are "modern," yonder days of scholarly past glibly used the pejorative "primitive" to refer to these populations. Another term, "tribe," also suggests a primitive timelessness, a pre-civilized state that reeks of the nineteenth-century ranking of societies. I use the more neutral, but also awkward, term subsistence populations.[2]

Subsistence populations today are not living fossils, and so cannot speak directly to what past populations were like. There are no written records or alien time capsules that can give us direct stories, so all methods of inquiry about the past must be indirect. Living populations of hunter-gatherers can give us a good glimpse of what living, aging (and dying) might have been like. They are not replicas of the past, but are analogues. If contemporary hunter-gatherers still practice a traditional livelihood and feature other traits usually associated with many past foragers, such as mobility, egalitarian social structure, food sharing, and cooperation, then we can see what lifespan and aging might look like in the presence of those cultural features. Certainly we want to pay attention to the existence of "outside influences" that could impact the foods people eat, how they treat illness, whether they gave birth in a clinic, and so on. But rather than dismissing certain groups that incorporate modern amenities meant to improve life as "not primitive enough" or "unrepresentative," I much prefer the approach of comparing groups that vary in lifestyle. Such comparisons can give us direct insight into how certain amenities affect health, aging, and lifespan.

The invisible protests I can imagine come from several camps. The first comes from a good place—that any equating of living people with past population harks of racism, discrimination, or stigmatization. Or that relying on groups today involves some misrepresentation of lifeways. For over two decades, I have taught an undergraduate class on hunter-gatherers. One of the key take-home messages is that representations and interpretations of living hunter-gatherers often tell us more about ourselves than about the lives of real people. Popular and scholarly generalizations alike about hunter-gatherers have ranged from viewing foragers as peaceful, highly cooperative vegetarians to violent, carnivorous, and patriarchal "road warriors." Health and well-being have been viewed as either Rousseau-like bliss—free of the cramped sardine-living and Frankenfood-eating of hegemonic, capitalist society—or as a Hobbesian "war of all against all," where life is "solitary, poor, nasty, brutish and short." How to separate our conceptions from the reality?

Put simply: *there is no quintessential hunter-gatherer population*, either now *or* in the past. When the Ju/'hoansi of the desert savannahs of Botswana and Namibia were widely studied by scholars involved in Harvard's Kalahari Research Project in the 1960s, they became everyone's ideal pristine hunter-gatherer, later popularized in the public imagination by the 1980 comedy, *The Gods Must Be Crazy* (see box 3.1). In that movie, when a glass Coca-Cola bottle tossed out of a passing airplane falls from the sky, all hell breaks loose and Ju/'hoansi society unravels. This binary view of subsistence groups as being either pure or tainted by "acculturation," and thus less interesting to the public imagination, is a simplification that denies groups their varied histories and identities. When studied by the Kalahari Project, the Ju/'hoansi were widely viewed by the anthropological community as a timeless exemplary preagricultural society, whose lifeways were preserved owing to their isolation in the Kalahari Desert. They were believed to have been untouched by colonialism and other major movements of people across Africa. They were also portrayed in the media as peaceful, egalitarian, living an idyllic life working few hours a day, and valuing generosity, sharing, and socializing—the "original affluent society" with limited needs and desires for commercial goods.[3]

Box 3.1. Dobe Ju/'hoansi (Botswana/Namibia)

Nancy Howell's study of the **Dobe Ju/'hoansi (!Kung)** in the Kalahari
Desert of Botswana and Namibia from 1967 to 1969 was the first
demographic account of a foraging society. The people were highly
mobile, moving camp frequently throughout the year. They were
politically egalitarian and fiercely independent, known to "vote with

FIGURE 3.1. **Ju/'hoansi (!Kung) of Dobe, Botswana.** (*above*) Matriarch
of Dobe village in the 1960s. (*next page*) Women returning from a day's
gathering. *Photo credits:* Richard Lee. (Richard Lee and Nancy Howell
have generously left a wonderful legacy, including hundreds of publicly
deposited photos from their fieldwork in the 1960s, including follow-up
visits. See https://tspace.library.utoronto.ca/handle/1807/10394.)

FIGURE 3.1. (*continued*)

their feet" (that is, walk away or relocate camp) when things didn't
go their way. At the time of study, they lived in small groups of fewer than
fifty people and maintained gift-giving exchange relations (called *!hxaro*)
with relatives living in distant camps as a way of ensuring steady access
to food, given its high daily, seasonal, and annual variability. The majority
of adults had spent most of their lives hunting and gathering, although
they would occasionally visit relatives at cattle posts. Ethnic San
populations like the Ju/'hoansi were heavily involved in the ivory trade
with Europeans and Afrikaaners in the nineteenth century and also
engaged in trade with pastoral and agricultural populations. The period
of study covers some cultural change. An "early" retrospective sample
covers the period before the 1950s, when the Bantu influence in the
Dobe area was minimal. A "later" sample covers the prospective time of
study when the lifeways of the Ju/'hoansi were changing rapidly. By the
1980s, Ju/'hoansi livelihood was more a mix of herding, wage labor, and
farming, with lifestyle change more evident. As a pioneer in the field,
Howell determined ages through a combination of relative age lists,
known ages of children and young adults, and other demographic tools.[4]

Later studies by other scholars, including archaeologists, revealed that the Ju/'hoansi had anything but an isolated existence. In the nineteenth century, they were actively involved in the ivory and ostrich feather trade. Ostrich feather hats were a hot commodity in Europe and fetched a good price. As far back as a thousand years ago, the Ju/'hoansi were surrounded by herding and farming neighbors. For hundreds of years during this time, they may have even had a more mixed economy. Based on archaeological history of the region, the historical "revisionists" Ed Wilmsen and James Denbow argued in the 1980s and 1990s that the Ju/'hoansi may have even been pushed toward a more meager existence living as nomadic foragers *because* of political and economic turmoil wrought over centuries. When the anthropologist Richard Lee and his Harvard buddies showed up in the 1960s, all the mercantile capital had disappeared, elephants had been mostly eliminated, and rinderpest virus had wiped out much of the cattle and other livestock in the region. This led to the appearance of isolation and the impression of a professional foraging lifestyle. But, according to the revisionists, both are instead just evidence of the Ju/'hoansi as the downtrodden rural proletariat.[5] Much academic debate, between Lee and other "traditionalists" and Wilmsen and other revisionists, over these kinds of views struck hard at the core of anthropology, pushing the question of how best to understand hunter-gatherers.[6]

And this question isn't just about the San populations of southern Africa. Many other foraging groups are enmeshed in relations with other groups, what has been called *complex interdependence*. The Aka, Mbuti, and other "Pygmy" groups of central tropical Africa have long been trading meat for rice and other goods. The Penan of Borneo and other tropical foragers of southeast Asia were likely farmers a millennium ago, but gave up farming because of a rising Chinese demand for valuable products, such as rattan and fragrant woods only found in the humid forests. A heated debate in the 1990s centered on whether it was even possible to live a self-sufficient existence in a tropical forest, not because of the difficulty of procuring protein-rich meat, but—au contraire!—because carbohydrate-rich fruits and vegetables are often in short supply in tropical rain forests. Though tropical forests are notorious for all their vegetation, sadly, much of it is inedible.[7]

Will the Real Hunter-Gatherers Please Stand Up?

While Ed Wilmsen's case may be overstated for the specific Dobe region where Lee and others began their studies, the general points about nonlinear contact histories and economic livelihoods remain. But that's okay. If it weren't, I wouldn't have bothered to write this book. In addition to "there's no single 'prototypical' hunter-gatherer" I add two other simple but important declarations: no group is pristine, and no group is a Stone Age relic. All practicing hunter-gatherers are modern humans subject to the same evolutionary forces as everyone else. None lived in the land that time forgot, not now, nor even before European contact.

We don't need hunter-gatherers to have had endless unbroken traditions in order to learn from them. Part of human nature is to be opportunistic and adaptively shift strategies when circumstances warrant. The extreme revisionist view deprives individuals of agency and renders them as subordinated victims of powerful people and forces. It assumes that everything is transformed because of the proverbial Coke bottle. Some things do change, like religious beliefs after long-term presence of missionaries. But, as Lee argued in his defense, groups like the Ju/'hoansi still maintain the "irreducible core of their culture." They still retain some autonomy over the daily decisions of their lives.

The demands of foraging will still result in patterned behavior and psychology, regardless of prior history, a Rainbow Brite T-shirt on their back, or the Penn State hat on their head—as revealed by decades of exciting work on subsistence strategies, mobility and residence patterns, cooperation and sharing behavior, and much more—the bread and butter of the field I identify with called human behavioral ecology.[8] As the brilliant demographer of the Hadza (see box 3.2), Nick Blurton Jones, remarked about the debate, "All those foreign visitors are very interesting, but people still have to eat, day after day."[9] I'm a strong believer that if contestants on TV's *Naked and Afraid*[10] had to live not just twenty-one days in the wild as hunter-gatherers, but instead over years, if not decades, much less a couple of generations, then we could learn a lot from studying them, too.[11]

Box 3.2. Hadza (Tanzania)

The **Hadza** live in the eastern rift valley of Tanzania, in close proximity to Olduvai Gorge and the famous fossilized footprints at Laetoli. Now numbering about a thousand, their demography has been studied in the mid-1980s up through 2000 by the meticulous Nick Blurton Jones. Trading with neighboring herders and horticulturalists has been

FIGURE 3.2. **Hadza of Eyasi basin, Tanzania.** (*top*) A group of women returns to camp singing. (*bottom*) A group of men hunting zebra together, 2006. *Photo credits:* Brian Wood.

sporadic among Hadza over the past century, and the overall amount of food coming from non-wild sources varies from 5% to 10%. The Hadza have been exposed to a series of settlement schemes over the past sixty years, but none of these has been very successful. Although some Hadza have spent considerable time living in a settlement with access to maize and other agricultural foods, many have not and continue to forage and rely on wild foods. To this day, the Hadza remain one of the few groups where many are still subsisting largely from hunting and gathering. Ethno-tourism has picked up since the 1990s, but hasn't been a reliable source of income. Like others, Blurton Jones aged the Hadza using relative age lists, combined with a core cluster of folks with known ages. He also used records from past researchers and a statistical method introduced by Kim Hill (see Box 3.3). The age profile of Hadza mortality has remained remarkably similar over a period of decades, starting from before the 1960s.

Once we move away from relying too heavily on any single well-worn example, we can appreciate that there is much variability in the lifeways of even contemporary hunter-gatherers, much less those living in the prehistoric past (as should be expected, when you consider the broad role that environment, culture, and history play in shaping behavior, physiology, and life history). That variation is key for seeing what patterns might be universal—and which ones might not be. It is also key for understanding how people adapt to their environments—not in spite of culture, but often because of it. The effective social transmission of information from one brain to another is one of the most important aspects of our success as a species.[12]

As Robert Kelly, an archaeologist who has written the best go-to synthesis of hunter-gatherer lifeways, explained, "hunter-gatherer is a category we impose on human diversity. It is not itself a causal variable."[13] In other words, if we can understand how different conditions affect health, aging, and mortality, then our task if we wish to make educated guesses about these things in the past is to figure out how prevalent

Box 3.3. Ache (Paraguay)

After reading everything I could about the Hiwi and writing up my
first lengthy paper (that, in retrospect, was all middle with no beginning
or end), I was hooked. I wanted more. Luckily, I was fortunate to live
and work among the **Ache** in Paraguay for my dissertation. The Ache
were full-time, mobile tropical forest hunter-gatherers until as recently
as the 1970s. My adviser Kim Hill started working with them as a Peace

FIGURE 3.3. **Ache of Mbaracayu Reserve, Paraguay.**
(*above*) Woman carries her toddler during a forest trek.
(*facing page*) Ache sharing meat when in camp in the
1980s. *Photo credits:* Kim Hill.

FIGURE 3.3. (*continued*)

Corps volunteer soon after contact. Hill and Magdalena Hurtado wrote a beautiful, inspiring book on Ache life history, published the same year I started graduate school. They separate Ache history into three time periods—a precontact "forest" period, prior to 1970, of pure foraging with no permanent peaceful interactions with neighboring groups; a transitional "contact" period (1971–1977) in which epidemics had a profound influence on the population; and a "reservation" period during which they have supplemented the foraging they still do on periodic treks with slash-and-burn farming, living in relatively permanent settlements (1978–1993). During this latter period, the Ache have had some exposure to health care. The precontact Ache period shows marked population increase, due in part to the open niche that was a direct result of high adult mortality among Paraguayan nationals during the Chaco War with Bolivia in the 1930s. And, back to demography: Hill and Hurtado improve on Howell's methods of age estimation by taking averages of different informants' rankings of people's age, using informant guesstimates of the absolute age differences between people. They also used a statistical technique called "polynomial regression" that permits estimation of year of birth based on the age ranks.

those conditions may have been throughout our evolutionary history. As an extension of Occam's razor, where the simplest explanation is usually the best one, we rely on the demographic "uniformitarian" principle that cause-effect relationships should be as similar in the past as they are today.[14] As we move forward, we should remember that the overreliance on "essentializing" a category like "hunter-gatherer" is where we start to run into trouble. The commonality is about what people eat, and how people obtain their food. While this is just one stream of daily life, it nevertheless gives us somewhere to start, opening a window to the past, albeit one with distorted glass.

Linking contemporary hunter-gatherers to past hunter-gatherers does not have to be condescending, imperialistic, or discriminatory. In many ways the linkage between the past and present could be viewed as respectable, something to be proud of, by artists and cultural icons throughout the ages. Henry David Thoreau returned to the "wild" at Walden Pond not just to "live deliberately" in nature, but to pursue "absolute" freedom and "wildness" amid his perception of private enterprise and urbanicity chewing away at our humanity. Many in the Western world have similarly turned to foraging as a lifestyle choice, not just because of frugality or out of necessity but, as Richard Lee himself confessed, from the "feeling that the human condition was likely to be more clearly drawn here than among other kinds of societies."[15] Honestly, it was a similar motivation that made me eager to live and work with the Ache and Tsimane.

Introducing Anthropological Demography

The sociologist Nancy Howell's detailed analysis of Ju/'hoansi hunter-gatherers took the uniformitarian principle to heart. She combined rich ethnography with rigorous demographic study, creating a tour de force that revolutionized our understanding of the Ju/'hoansi life course from birth to death. Her groundbreaking work helped inspire similar projects with other subsistence populations in South America, Tanzania, and the Philippines (and it inspired my own work in Bolivia with the Tsimane). It is the meticulous study of just five groups that best informs a quantitative understanding of the world of hunter-gatherers: the Ju/'hoansi

(!Kung) of Botswana and Namibia, Hadza of Tanzania, Ache of Paraguay, Agta of Philippines, and the Hiwi of Venezuela.[16] While snippets
of information also exist for other groups, these five I single out for
prime time (and showcase in boxes throughout this chapter). When
studied in the twentieth century, they were living as full-time hunter-
gatherers with minimal interactions with more regional or national play-
ers (not that some degree of outsider contact didn't occur at other
times, or would be a deal-breaker, as I argued above). These were also
focused investigations, in which demographic topics like lifespan and
mortality were main areas of interest.

Why so few studies? As Nick Blurton Jones explains, about why it
took him over two decades to publish his comprehensive book on the
Hadza, "demographic research among very mobile hunters and gatherers simply takes a long time." It might also have to do with a keen observation by the geneticist and biodemographer Ken Weiss back in the
early 1970s: "Field anthropologists usually are less interested in demography than in almost any other aspect."[17]

Similar studies have also been done with a few other groups practicing
more varied forms of subsistence. This includes horticulturalists—a
fancy label that refers to a livelihood dependent on low-tech slash-
and-burn farming. Subsumed under the same label of horticulturalists
is the recognition of hunting, fishing, and gathering (leading to awkward
and confusing variants like hunter-horticulturalist or horticulturalist-
foragers and the like). The Tsimane and Yanomamö do all of the
above: they farm, hunt, fish, and gather (see boxes 3.3, 3.4). I also include
pastoralist herders, who often supplement their meat-and-milk-based
diet with farming or foraging. Many pastoralists are nomadic, moving
with seasonal migration of the herds, and in relation to water sources.
In dry areas, these are often semipermanent waterholes.

Horticulturalists and pastoralists include plant and animal domestication as vital elements of their livelihoods. For example, Tsimane rely
on sweet manioc and plantains for their staples, and also sometimes raise
chickens and pigs. Manioc, or cassava, is a potato-like root crop first
domesticated in the Americas, probably in the southwest Amazon, some
four thousand years ago. Today it pervades the tropics and is ranked as
the sixth most important crop plant worldwide.[21] It's hardy, easy to

Box 3.4. Yanomamö (Brazil/Venezuela)

Yanomamö and Tsimane are both forager-horticulturalist groups in Amazonian South America. Several different demographic studies of the Yanomamö have been carried out over the past thirty years. Although often construed as hunter-gatherers, Yanomamö have practiced slash-and-burn farming of plantains for many generations, which in the arcane classification system makes them "horticulturalists," just like Tsimane. They mostly live in small villages, in protected round houses (*shabono*) of fewer than fifty people. The effects of the rubber boom and slave trade before the eighteenth century on Yanomamö were minimal. The Yanomamö remained mostly isolated until missionary contact in the late 1950s. The first life table was published in a classic paper in 1975, by the geneticists James Neel and Ken Weiss.[18] Theirs, built through the use of model life tables and other methods, shows higher mortality than most other contemporary subsistence populations.

But the most complete demography comes from the two Johns, John Early and John Peters, based on prospective studies of eight villages in the Parima Highlands of Brazil, along the middle Mucajai River.[19] Births and deaths were recorded by missionaries and the Brazilian Indigenous protection agency FUNAI since 1959. A precontact period (1930–1956) predates missionary and other outside influence. The contact period (1957–1960), "linkage" (1961–1981), and Brazilian periods (1982–1996) saw increased interaction with miners and Brazilian nationals, and more infectious disease. Ages during this period were estimated using a chain of average interbirth intervals for people with at least one sibling of known age, and relative age lists in combination with estimated interbirth intervals.

FIGURE 3.4. **Yanomamö of Venezuela, Brazil.** (*top*) A ~55-year-old man makes *curare* poison. Curare is a toxic substance from plants applied to arrowheads to make them poisonous to prey. Making curare is a dangerous venture, as boiling can result in dangerous fumes. As the local expert, Simodowa's curare is distributed to all the local hunters in the village of Mishimishimabowei. (*bottom*) In a residential *shabono* collective residence, the elder shaman Dedeheiwä, known for curing sick kinsmen and wreaking supernatural mayhem on his enemies, is groomed by his granddaughters. *Photo credits:* Raymond Hames.

plant, and easy to tend. (Even a black thumb gardener like myself got this to grow in Bolivia.) More importantly, Tsimane and many other Amazonian groups ferment manioc and drink it as a beer called *shocdye'*. The catalyst for fermenting comes from the saliva of women who chew the root and spit the goopy mash into a tree-bark container or a pot mixed with water. The frothy blend is then covered with palm leaves for several days to build the brew. It's rich and heavy, and when strong, kin and neighbors come in droves to drink from the shared *erepa* bowl hollowed out from the large green melon of a calabash tree (*Crescentia cujete*), to hang out, gossip, and share stories.

Box 3.5. Tsimane (Bolivia)

The **Tsimane** inhabit tropical forest areas of the Bolivian lowlands, congregating in small villages near large rivers and small tributaries. I first started working with the Tsimane in 1999, soon after my work in Paraguay with the Ache, and made demographic surveys my first priority. There are now more than 17,000 Tsimane living in more than ninety dispersed settlements in the Beni region. The Tsimane have had sporadic contact with Jesuit missionaries since before the eighteenth century, although weren't successfully converted or settled. Evangelical and Catholic missionaries set up missions in the early 1950s, and later trained some Tsimane to become teachers in the more accessible villages. However, the influence of missionaries on daily life has been minimal. With improvement of roads and cheaper outboard motors on canoes, market integration has been increasing, as are interactions with loggers, merchants, and colonists. But everywhere, Tsimane still continue to fish, farm, hunt, and gather for much of their subsistence. The demographic sample used here is based on reproductive histories I collected in eighteen remote communities from 2002 to 2005.[20] Changes in mortality are evident over time, and so I contrasted an "early" period (1950–1989) with a "later" period (1990–2002). In 1990 the Tsimane government was formed, as was a mission-sponsored health clinic just for Tsimane.

FIGURE 3.5. **Tsimane horticulturalists of Beni, Bolivia.** (*top*) Village elders and community members hanging out. (*bottom*) An elder woman, her daughter, and granddaughter make shocdye', a fermented beverage of manioc root and corn, while her granddaughter helps and tastes the ground mash. *Photo credits:* Michael Gurven.

Tended crops and animals raised for food, eggs, and milk are more recent additions to the human menu. The transition from foraging to farming is referred to as the Neolithic or Agricultural Revolution, occurring between 4,000 and 12,000 years ago in different parts of the world. The notion of "transition" is a bit of a misnomer, better described as an addition or supplement, a mixed subsistence mode like we see with contemporary horticulturalists. The Neolithic period witnessed the domestication of animals like goats, cattle, and pigs. These helped provide reliable sources of meat, transforming hunting into animal farming. Domesticated animals also served other uses, like providing their manure and labor for farming, and wool and hides for clothing.

Greater sedentism and population growth led to limited foraging and hunting options, with dietary diversity narrowing as a result. Wherever people relied more on a limited number of carbohydrate-rich but nutrient-poor staple crops, they increased their susceptibility to famines from crop failure.[22] As a general rule, settlements became more permanent, with intensive irrigation and economic specialization. The potential for stored surplus, though, helped afford greater divisions of labor and increases in social complexity that were often tied to sedentism and economic intensification. Along with the benefits of dense populations, however, comes the potential for virulent contagious infections, often arising from contact with domesticated animals and having a large enough (human) reservoir population for disease to spread. For example, measles, which continues to be a major cause of mortality in unvaccinated children, likely derived from rinderpest virus affecting cattle. Smallpox, deadliest of deadly pre-Edward Jenner's vaccine, likely originated in east Africa some three to four millennia ago from a cowpox-like ancestral virus.[23]

This complex bundle of features related to domestication varied in composition, sequence and timing in different world areas, with the earliest timing in the Levant (now eastern Mediterranean region of western Asia) some 12,000 years ago, in Southwest Asia ~9,000–10,000 years ago, to later appearances like in the Americas between 4,000 and 8,000 years ago.[24]

But the important point is that throughout the Holocene period of the last twelve millennia, humans came up with new ways of producing food. You are what you eat, and so as diet shifted, along with many other factors (clumped together as the all-encompassing summary of a life-time's health-related exposures called the "exposome"), we can ask whether lifespan and aging changed as well. After all, the Holocene transformed how we lived our lives.

The tendency to compare our exposome and health with that of hunter-gatherers is obvious, because of our long evolutionary legacy of hunting and gathering. But, contrary to common perceptions, evolution did not stop at the dawn of the Holocene. In fact, the pace of genetic change accelerated with the increases in population growth experienced during our more recent evolutionary history. As people spread around the globe, they adapted to their local environs. As might be expected, many recent changes in genomes reflect differences in local diet and immune function.[25] So it's valid to ask how aging and lifespan may have changed during this period. Are pastoralists as mismatched to the foraging lifeway as urban New Yorkers? More so than horticulturalists? Or is it the massive urbanization and industrialization of the last century that transformed how we live, age, and die?

Reading History Sideways?

This approach of relying on the study of contemporary subsistence populations to ponder the historical trajectory of humans from past to present may sound familiar. It mirrors the debate about how we consider hunter-gatherers in the present versus the past. The same caveats apply for thinking about prehistoric farmers or herders. Inferring a temporal trajectory from a cross section of populations is what the sociologist Arland Thornton calls "reading history sideways."[26] Thornton argues that the common practice of ordering populations as if in some developmental trajectory proceeding from lower to higher complexity reeks of ethnocentrism. Starting with, say, Aboriginal Australians, moving to Tahitians and onward to the pinnacle of western Europe are the kinds of dated comparisons Thornton had in mind. Relying on the non-European present to inform the history of the European past on many dimensions of human life is certainly misguided on many levels.

First and foremost, there is no inevitable trajectory when it comes to most things, including lifespan. Conditions do not proceed simply from worse to best. The 1.5-year decline in life expectancy at birth in the US in 2020 (due largely to COVID-19) should be testament to such ups and downs, just as was the 2.9-year drop during World War II.[27]

In a limited number of cases, we can compare lifespan in the same populations over different time periods. This "longitudinal" perspective is ideal, but often hard to come by, as it can require decades of study. What we'll find, though, is that even over the short-term, no group is a closed system of timeless tradition. People always reflect on how things used to be, and how they're different now.

When it comes to survival and aging, what matters is the full totality of exposures and experience—not a generic label, or a specific time period, unless those reflect differences in the exposome. This is why we need to look at as many groups as possible, to evaluate the merit of claims that the health of populations with certain characteristics can deliver insights about the health of past populations expressing those same lifeways.

Lastly, when trajectories about modes of subsistence are obnoxiously viewed as historically inevitable, then nonindustrial societies come off as anomalies, mistakes, or backward. We're left wondering why foragers might still exist today. Instead, path dependency of history, power of local culture, and the multifaceted features of environment and ecology have a lot of sway on who does what across the landscape. We shouldn't be deluded into thinking that postindustrialized society is the "best" on all fronts, the end point of some unavoidable trajectory. We may be long-lived, but are we as healthy as we could be? And at what cost come our longevity gains?

Counting the Rings, and Clocking Age

"Do people have those too?" my daughter Evie asked me. She was 8 years old at the time, pointing at the concentric rings circling around the face of a stump about 1.5 meters in diameter of one of the "gentle giants" in Hendy's Woods. These are the immense coastal California redwoods that can live over two thousand years. I explained to her how counting the growth rings can tell us the tree's age, and the thickness between rings tells us whether each was a good or a bad year. Were it only that easy to determine the ages of people!

It's worth spending a moment here to explain why you should believe anything I say about the ages of people who don't usually pay much attention to knowing their age. In the old days, investigators would just guess people's ages based on what they looked like, or maybe cluster people into a comfortable bin, like "age 45+." Major sleuthing is required to estimate people's ages in the absence of written records, or perfect memory, especially when there's not much salience for such arbitrary things as knowing how many times Earth has spun around the sun since your birth.

The approaches that anthropological demographers have taken differ in their details but share a similar spirit. First you have to be able to identify people by name or some genealogical relationship. The late anthropologist Napoleon Chagnon (whose office I now inhabit) had to wrestle with taboos against saying the names of the dead when working with the Yanomamö of Venezuela and Brazil. But even being able to speak someone's name doesn't always solve the problem. The patriarchal Tiwi of Australia rename the children each time a widow remarries, which typically happens multiple times over one's life.[28]

One of my first goals when the Tsimane project started was to get decently reliable ages on people. You can't study aging without knowing ages. Some folks had some ballpark awareness of their age, and that was usually a good place to start. Understanding the linear tick-tock of years advancing along a number line beyond eight or so does not come easy among Tsimane without some formal schooling. Occasionally, people had an ID card ("carnet") with a birthdate that was often based on the "guesstimation" of whichever bureaucrat administered the form.

The process for me begins by detailing a person's reproductive history. In eliciting the names and timing of each child born to women or men, and everyone's siblings and half-siblings, you can quickly come up with a relative age list for everybody in a village. This involves a lot of back-and-forth among kin and spouses, and is a lot of fun. As a gauge to help guide those I was working with, I used photos of people with known ages. That way, for example, someone could point to a person to refer to the age that someone died, without needing to voice a number. After doing this across several villages located deep in the forest, I came to realize that many folks had migrated there decades ago from far upstream along the Maniqui River to escape debts from crooked merchants, who would exchange limited clothes and goods for roof panels woven by Tsimane from palm thatch, at abysmal rates. The merchants would usually give the goods first, then return later to pick up the panels, giving more goods away while demanding more and more, pushing Tsimane further into debt (a process referred to in local Spanish as *habilito*, or debt peonage). When I went to visit those upriver villages some months later, I immediately knew who people were and how they were related. The Tsimane living there were amused and a bit spooked that I already knew their names, and could update them about how their relatives were doing many miles away. Such immediate familiarity made it easier to conduct more demographic surveys.

Once armed with a relative age list, you need to peg down some exact ages throughout, so you can fill in all the gaps along the list of names. Some methods allow you to estimate the gaps pretty accurately, especially the more known ages you have.[29] Often you can do this by using an event calendar—memorable events that coincide with births, deaths, or marriages. I used the arrival of important roads in the region, inaugurations of Catholic and evangelical missions, some notable bad flood years, and the timing of a few well-known local murders. I was also lucky to discover that the Catholic mission Fatima kept good records of all baptisms, as required by the Archdiocese of Nuestra Señora de La Paz. These baptisms occurred shortly after birth, or when older adults today were fairly young. As many Tsimane inhabiting the upper Maniqui region had at one point lived or passed through the mission, those records provided independent information on ages and birth dates. The Tsimane Father Javier Canchi, righthand man to the Alsatian priest who lived in Fatima for half a century (Padre Martín), kindly let me handcopy their records: 1,284 people registered between 1952 and 2003. (On top of all this, there are simple rules that also help reconcile discrepancies: people can't be born before their mother, women can't give birth at age 60 or age 5, and six-month intervals between births ain't legit.)

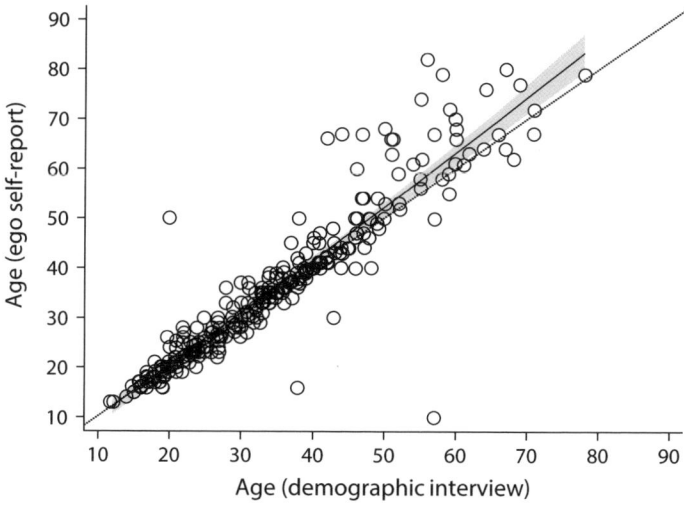

FIGURE 3.6. **Age estimation: self-reports versus demographic interviews.** Self-reported ages are plotted against ages estimated from demographic interviews among Tsimane, 2002–2005. Tsimane are more likely to overestimate their ages, as indicated by all the points above the diagonal.

Figure 3.6 compares what people told me about their age against the final age estimate resulting from my laborious sleuthing. Many age estimates agree with what people said, but the numbers shift when older adults are reporting. And disagreement is greater for folks living in more remote areas without schooling. But what's the gold standard to compare against? When I checked the baptism records, my estimates for older adults were pretty close, off by just a couple years here and there. That was reassuring, but then came another opportunity to assess the age estimates.

In the summer of 2014, I was giving a fun talk on "the evolution of human longevity" as part of the RAND Institute's "mini-medical school," and during a coffee break I was approached by an eager Steve Horvath. Steve is a geneticist and biostats whiz who at the time had just come up with what seemed like the holy grail of biomarkers of aging: a universal measure of biological age, based on methylation patterns in blood, or any tissue for that matter.

Methylation is a biochemical process that alters how genes are expressed in the body. By measuring the accumulation of methylation changes to particular regions of the genome, Horvath's algorithms constitute an "epigenetic clock" that gives a staggeringly close match to chronological age. While the match is somewhat exact, deviations of predicted ages seem to provide a window into biological aging. For example, having an epigenetic age older than chronological age (called "age acceleration") has been shown to predict several cancers, poor lung function, frailty, and cognitive decline, among other health outcomes. Most importantly, new second-generation clocks with darling names like GrimAge (after the Grim Reaper, not McDonald's purple Grimace) also predict all-cause mortality.[30]

Always in search of new tissues, populations, or species on which to test-drive his epigenetic clock, Steve and I teamed up to look at the methylation age of sixty Tsimane, including a man who at first glance I took to be a 10-year-old child. I first met Severín during one of my first trips along the upper Maniqui River to a small community of two extended families. He was shy and would look at me askance, out of the corner of his eyes, before trekking down the cliff by the river's edge and poling upriver in a dugout canoe. I hadn't seen a ten-year-old do this alone before, but Severín seemed wise beyond his years. I was even more curiously shocked when he returned some hours later with seven *vonej* fish. Tsimane are precocious masters of their domain, but already at age 10?!

During this time, I was interviewing all adults about their families, where they had been born and lived, and their reproductive histories— the same interviews that would allow me to get a better estimate of everyone's age. After interviewing Severín's siblings and parents, it was undeniable that Severín had to be over 30 years old. Looking more closely at him later on, I could see that the uncanny lines in his face were the marks of a life long-lived. Yet he was not just the height of a child— his proportions were those of a child as well. His methylation age turned out to be 29.2. In other words, he did not have an obscure disease with delayed or accelerated aging. He most likely had pituitary dwarfism, where the pituitary gland doesn't produce enough growth hormone. Severín doesn't hunt, but he fishes competently, and a few years back he became chief (called *corregidor*, or corrector) of his village. While

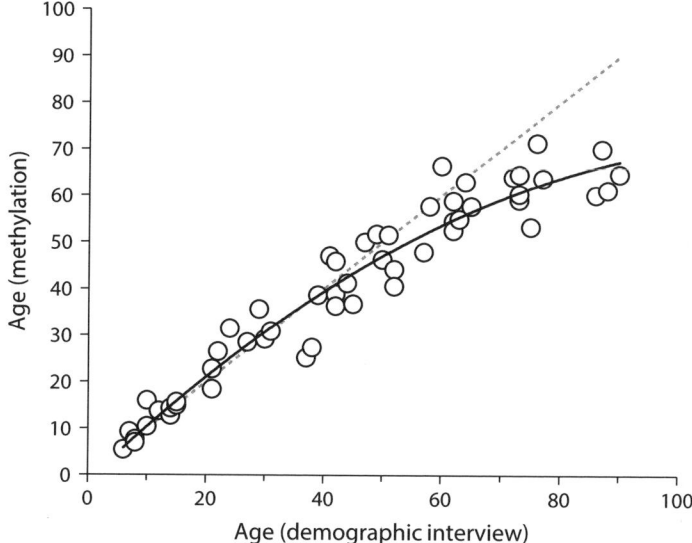

FIGURE 3.7. **Age estimation II: DNA methylation versus
demographic interviews.** The overall correlation is strong
(r = 0.96). The relationship in the figure suggests that older Tsimane
adults are biologically "younger" than their chronological years.
Dotted line represents when estimated ages from both methods
are identical.

old-fashioned detective (demographic) work helped confirm his age,
the use of epigenetic clocks helped distinguish actual biological aging
from the mere appearance of youth.

Figure 3.7 shows the relationship between the DNA-based methylation ages estimated using Horvath's algorithm and the ages estimated
from my demographic interviews. The correlation between the two is
an astounding 0.96, where 1 means a perfect correlation. While the
match is pretty tight, older adults seem to be biologically younger than
their chronological age. In a comparative analysis, we found that these
older Tsimane were also biologically younger than both US whites and
Hispanics from Los Angeles. It remains to be seen whether this younger
epigenetic age reflects better health from the Tsimane lifestyle or, instead, reflects the signature of the more robust, healthier subset of Tsimane who survive to middle age.[31] In any case, these types of clocks are
the closest we've come to counting human tree rings.

Box 3.6. Hiwi (Venezuela)

The **Hiwi** (formerly known by the derogatory term, Cuiva) are neotropical savanna foragers of Venezuela near the Colombian border, studied in the late 1980s by the same anthropologists who worked with the Ache (Kim Hill and Magdalena Hurtado). The Hiwi were contacted in 1959 when cattle ranchers began encroaching into their territory. Although living in semipermanent settlements, Hiwi continue to engage in violent conflict with other Hiwi groups. At the time of study, more than a hundred folks were living together in one community. Almost the entire diet was wild foods, including the large aquatic rodent capybara, feral cattle, and a variety of roots and mangoes. Shamans had important roles, with hallucinogens featured in rituals. Nearby Guahibo-speaking peoples practiced agriculture, while the Hiwi inhabited an area poorly suited for farming. As among the Hadza, repeated attempts by missionaries or government schemes

FIGURE 3.8. **Hiwi of Venezuela.** (*left*) A woman nurses her baby while cracking palm nuts. (*right*) A man fishes with bow and arrow from his dugout canoe. *Photo source:* Kim Hill.

to introduce farming had long failed, until recently. It's very possible that there are now no longer any Hiwi living a nomadic foraging life. I've never visited the Hiwi, but my first virtual (armchair) foray as a first-year graduate student was spent obsessing over their food-sharing patterns, using rich data collected by my advisers. Sharing is widespread among Hiwi, like in other hunter-gatherer groups, but favors kin and close associates who reciprocate. In other words, sharing is not indiscriminate. We'll see in chapter 4 how sharing and cooperation are critical for making us who we are today, and in facilitating the evolution of longevity.

Hunter-Gatherers in the Twentieth and Twenty-First Centuries

The rich ethnographic record includes information on hundreds of cultures, but only fifty or so have been well studied, and among these are the five with carefully collected information that allow us to explore lifespan and aging among hunter-gatherers. You've already learned about Ju/'hoansi, Hadza, and Ache hunter-gatherers (see boxes 3.1, 3.2, 3.3), and about Yanomamö and Tsimane horticulturalists (see boxes 3.4, 3.5). Next, I highlight relevant background information about the Hiwi of Venezuela and the Agta of the Philippines—two hunter-gatherer groups whose lifespans when studied were shorter than that of other groups (see boxes 3.6, 3.7).

Adding to the relatively unacculturated foragers and horticulturalists we've showcased so far, I briefly describe just a few more groups to expand the diversity of cultures I cover and because the reliable data on their survival patterns across the life course are precious.

The Northern territory of Australia has the largest proportion of Indigenous population on the continent, with over 40 aboriginal language groups in the region, including Warlpiri, Gunwinggu, Yolngu, and Pitjantjatjara (figure 3.10). In the late 1950s, the sociologist Frank Lancaster Jones compiled mortality data among roughly 17,000 "full-blood"

Box 3.7. Agta (Philippines)

The Casiguran **Agta** of the Philippines were studied by the missionary-turned-anthropologist Tom Headland from 1962 to 1986. The Agta live on a peninsula close to a mountainous river area and the ocean. There were 9,000 Agta in the eastern Luzon territory, and demographic study was focused on both the Casiguran and the San Ildefonso groups of about two hundred people. Although the Luzon area is itself very isolated, Agta have maintained trading relationships with lowland horticulturalists for at least several centuries. One of the remarkable features of Agta culture is the widely acclaimed observation of women hunters. The women did not just hunt with dogs and a machete, as has been observed in other groups, but hunted solo with bow and arrow. Images of women stalking prey and shooting arrows may seem like it contradicts an age-old universal dictum, that men = hunting and women = gathering, but foragers don't care about our quaint generalizations. Men are not hardwired to hunt, just like women aren't hardwired to gather. What people do to make a living is flexible, depending on how far to venture out on trek, available technology, other options available, and the steepness of the learning curve. If pretty steep, men and women usually specialize, but if not so steep, we see some degree of men and women switching back and forth between activities. And the divisions of labor make it so we don't have to do everything.

Among the Agta, the circumstances favoring women's hunting are revealing. First, Agta women's hunting was not very common in the past, and does not occur much now. But when it does, women hunt with the aid of dogs, at a close distance to camp, and when they don't have infants breastfeeding on-demand. And, here's the real clincher: there aren't abundant roots or other starchy carbs to feast on in the local forest, so women hunters trade game they kill for rice—as if women are gathering by hunting.[32]

Agta age estimation was achieved through reference to known ages of living people and calendars of dated events. As in the Ache and Yanomamö studies, the Agta demography is divided into a "forager" period (1950–1965), a transitional period of population decline (1966–1980), and a "peasant" phase (1981–1993). These latter phases

FIGURE 3.9. **Agta of Luzon, Philippines.** (*top*) Sharing meat with children. *Photo credit:* P. Bion Griffin. (*bottom*) Three generations living together. *Photo credit:* Daniel Major-Smith.

are marked by guerilla warfare and subjugation by loggers, miners, and colonists. The twentieth century introduced schooling among the Agta, and brief skirmishes during the periods of US and Japanese occupation.

FIGURE 3.10. **Northern Territory Aborigines, Australia.** (*top*) Warlpiri elder artists of Lajamanu use painting to connect to country across generations. *Photo credit:* Warnayaka Art. (*bottom*) Gumatj clan of the Yolngu of northeastern Arnhem Land during the annual Garma (or "two-way learning") festival in 2023 dancing the *bunggul*. Garma celebrates cultural traditions, builds connections between clans, and fosters new opportunities. *Photo credit:* Peter Eve/Yothu Yindi Foundation.

Northern Territory Australian Aborigines.[33] At that time, few Aborigines in the region were still full-time foragers. These data can give a sense of what mortality looks like among acculturated former foragers. A significant amount of age-clumping occurred at five-year intervals, and so a smoothing procedure was done on the age distribution of the population. It is likely that infant deaths and more remote-living individuals are undercounted, although Lancaster Jones made some adjustments to account for this. I view these data with caution but include them because no other reliable data exists for the land down under, apart from a Tiwi sample, culled from the same author.

The **Gainj** are swidden horticulturalists of sweet potato, yams, and taro in the central highland forests of northern Papua New Guinea (figure 3.11). At the time of study by Patricia Johnson, James Wood, and colleagues in 1978–1979 and 1982–1983, roughly 1,318 Gainj were living in twenty communities.[34] Peaceful contact with outsiders was fairly recent, in 1953 with formal pacification in 1963, and genetic and linguistic evidence confirms their relative isolation. Prior to contact, population growth had been zero for at least four generations. An A2 Hong Kong influenza epidemic reduced the population by 6.5% in 1969–1970, and probably accounts for there being so few older people. Data were obtained from government censuses from 1970 to 1977.

The **Herero** of northwestern Botswana are Bantu-speaking pastoralists studied by Renee Pennington and Henry Harpending in the late 1980s (figure 3.12).[35] They were traditionally cattle and goat herders in the Kalahari Desert of the Ngamiland District, numbering 10,000–15,000 during the time of study. They had migrated to this area in the early twentieth century, due to displacements from the Herero-German Wars. They lived in extended family homesteads without running water or electricity, had rarely married outside the group, and, at the time of study, were very successful cattle herders. They also raised more drought-resistant goats and other livestock. Fertility was low, increasing from just under three births per woman in the first half of the twentieth century up to seven in the 1980s; the lower earlier fertility was likely due to pelvic inflammatory disease stemming from sexually transmitted infections.[36]

FIGURE 3.11. **Gainj of Papua New Guinea.**
(*above*) Elder man. (*next page*) Three generations.
Photo credits: Kenneth L. Campbell.

Death (with and) without Weeping

The anthropological demographer Renee Pennington, who worked with the Herero, once quipped, "Demography is ultimately about sex, but never so much fun in its details." Details about death are never fun, but what we're really interested in is life.

So let's dive right in.

Remember that the survival curve (l_x) tells us the probability an individual survives from birth to each age x across the life course. It starts at

FIGURE 3.11. (*continued*)

1 with birth, then declines thereafter, eventually reaching 0. Another way of thinking about the survival curve is if you have a population with 1,000 people born at the same time, how many will still be alive at each age? Eventually, at some late enough age, no one escapes eternal slumber.

From birth to adulthood, despite so few populations, notice that the trajectories of the survival curves vary quite a bit (figure 3.13). None show the very rectangular-like shape typical of modern nation-states since the mid-twentieth century that I introduced in chapter 2. A rectangular survival curve means that everyone lives until very late in life. Instead, we first see a steep drop early in life, reflecting the high mortality experienced by infants and children. On average, 12%–40% don't survive their first year of life, while 24%–49% don't make it to their fifth birthday.

If we generously consider age 15 as the transition to adulthood, then, on average, only 57% and 64% of children ever born reach adulthood among hunter-gatherers and horticulturalists. Let that sink in for a minute. For every 100 children born, between a third and under a half won't make it to the point where they'd be past puberty. This survival is lowest among the Agta (45%), and highest among the Ache (66%).

FIGURE 3.12. **Herero.** (*top*) A Herero man poses with his two wives and their children in front of their house. (*bottom*) Elder Herero man watering cattle. *Photo credits:* Renee Pennington.

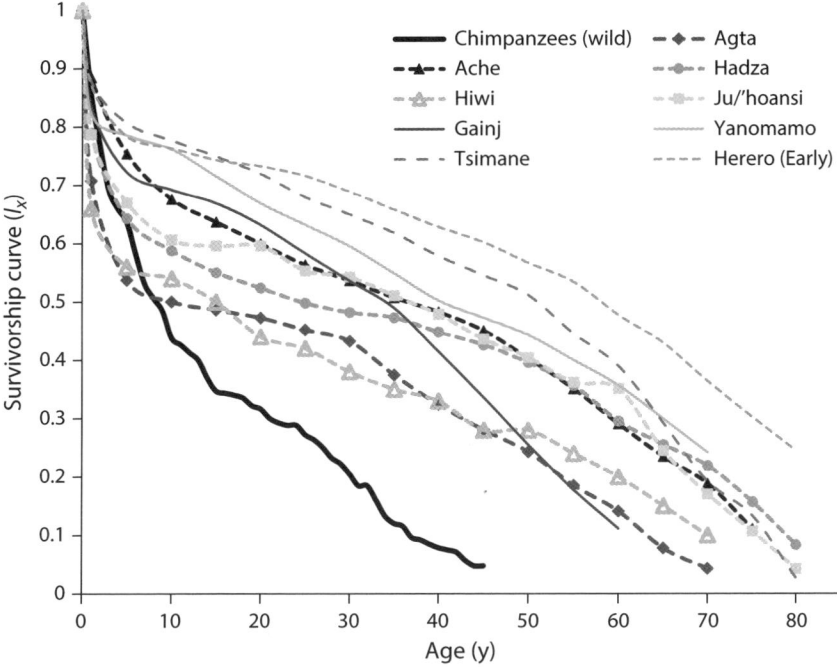

FIGURE 3.13. **Survivorship and mortality.** Survivorship (l_x) curves for contemporary subsistence populations described in the boxes. Groups differ most
in how large the drop in survival is during the first five years of life. Survivorship
for wild chimpanzees is shown for comparison.

It may seem crass to make comparisons, but doing so gives us perspective: many complain about the inefficacy of US health care and
society, in part, because for being a highly developed superpower that
put a person on the moon over half a century ago, US infant and child
survivorship is lower than in much of Europe. But the difference is
nothing compared to what foragers experience. Roughly 8 per 1,000
births in the US versus ~3 in Europe don't make it until their fifteenth
birthday. Sure, that's a 167% higher risk of dying in the United States
compared to much of Europe. But hunter-gatherers lose a whopping
430 out of every 1,000 born, and horticulturalists 360—over 14,000%
higher than Europe. For additional perspective, the three countries
with the lowest life expectancy in the world today lose between 95 and
138 per 1,000.[37]

While we've largely limited our focus to the subsistence populations for which we reliably know mortality patterns across the entire life course, data on infant and child mortality alone are a lot easier to come by. Infant mortality in a larger sample of twenty hunter-gatherer populations with minimal to low outside influence shows a similar pattern of very high rates, ranging from 14% to 40%, with an average of 27% dying in their first year of life. From this larger sample, we can check the likelihood of dying before age 15 as well. That ranges from 22% to 56%, or 49% on average.[38] Sedentary hunter-gatherers, with some level of acculturation, show lower infant mortality (about 15%), but horticulturalists and pastoralists show rates of infant death (about 21% on average) very similar to more nomadic hunter-gatherers. On the other hand, child mortality seems to be a little lower in horticulturalists (39% on average) and pastoralists (34%).

Overall, though, the story is clear: life is cut very short among subsistence populations. Death is an affair not of old age but, sadly, mostly of young children. For most of us in the so-called developed world, births represent new life, not the looming threat of premature death. Among Tsimane and many other groups, newborns aren't given names until at least a year or two. While parents sometimes say they're still deciding, or want to wait to see who their babies might remind them of per naming conventions, it's hard not to interpret such behavior as a psychological defense against probable tragedy. In *Death without Weeping*, the anthropologist Nancy Shepard Hughes explains how the high incidence of early-life deaths in northeastern Brazil's poor shantytowns seemed to normalize infant deaths in such a way that there was little outward grieving; an outsider might even assume indifference.[39] She highlights how, even in Europe and North America, the recognition of infant death as a medical "problem" that could be solved, rather than as part of the natural order of things, was a recent social invention of the late nineteenth century.

Even though the Tsimane have lower child mortality than hunter-gatherers, one in four children never see adulthood. From my records, 58% of Tsimane mothers have lost at least one child. Of the women who have reached the end of their childbearing years (~40 y of age), 86%

have lost a child.[40] Let that also sink in. Almost every Tsimane mother and father will have directly experienced one of the worst experiences imaginable during their lives, the loss of a child.

Back in 1998, a dear Ache friend's newborn died suddenly late at night, despite our attempt to take the baby to a decent hospital across the Paraguayan border into Brazil. The next day, I went to visit my friend to commiserate his loss, but he wasn't at home with his wife where I had expected him to be. I found him playing volleyball, giggling with his friends. This jived with the Ache philosophy of *kuame*, or let it be/forget about it, a mantra to help one persist amid tragedy that you can't do much about.

Tragedy of Lost Lives

Figure 3.14 shows how early life mortality is so much higher in hunter-gatherers than in the industrialized world, but somewhat similar to that of mid-eighteenth century Europe. Why is the death rate so high in early life? Causes of death are difficult for investigators to assess without physicians, direct witnesses, detailed medical histories, or autopsies. When causes of deaths are elicited during interviews with family members, it can be difficult to get a clear story of what happened, much less one that would meet the usual criteria for causes of death we now write on death certificates. While accidents, animal attacks, and homicides may be easier to identify, infections and other diseases common in nonindustrial societies are diagnosed with a large margin of error. What people say, or what has been called the "emic" perspective, is often hard to categorize. For example, at least two dozen Tsimane women told me that their babies succumbed to coughing fits because their husbands were not around, were observed flirting, or were believed to be having sex with other women. Or they were just terrible husbands. Many more people claim sorcery by jealous neighbors, by menacing outsiders (non-Tsimane Bolivians are referred to as *napo'*), or malicious spirits who have been upset in one way or another. Reasons also include violating certain social norms, like food taboos and menstrual taboos. For example, according to widely held cultural beliefs, a menstruating woman

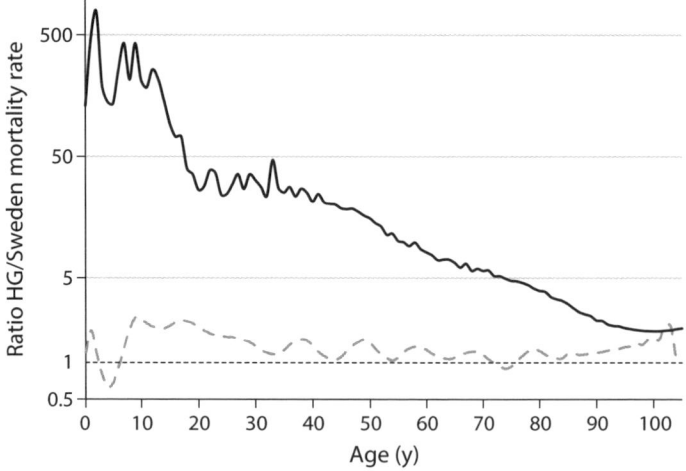

FIGURE 3.14. **How high is hunter-gatherer mortality?** The
ratio of mortality rates among hunter-gatherers to those of
Sweden in 2019 (black) and in 1751 (dashed gray). Compared to
contemporary nation-states in the industrialized world, hunter-
gatherers suffer higher risk of dying at all ages, especially early in
life. The difference diminishes by late adulthood, with a fourfold
difference by age 80. Hunter-gatherer mortality is much closer to
agrarian Sweden from 270 years ago.

should never make the shocdye' beverage that children sometimes
drink, lest misfortune fall.

I compiled causes of death as reported for as many of the subsistence
groups as possible. This resulted in 3,328 deaths, summarized in table
A3.2 (see Appendix).[41] Given limitations of the source data, I use broad
categories to group causes: illness, degenerative disease, accidents, or
violence.

Overall, two-thirds of preadult deaths are from illness. Often these
involve synergies of infection and malnutrition, sometimes leading to
diarrheas and fevers that can be lethal to small children. The combination
of malnutrition and poor health can be deadly, as was likely the case for
the Agta. Cases of gut obstruction, pneumonias, influenzas, and hard-to-
identify illnesses that bring high fevers were commonly observed across
groups. Malaria and tuberculosis were the major killers of Ju/'hoansi

children. Among Tsimane, over half of infant deaths were due to infection, and of those, half were respiratory-related. Respiratory illness was
also important among the Ache, whereas gastrointestinal illness was more
common than respiratory illness among the Hiwi, Hadza, and Agta. This
type of morbidity can be common after introducing supplementary
foods while breastfeeding, or after infants are fully weaned. When
pathogens run rampant, the introduction of new foods, combined with
contaminated water and poor hygiene practices, can make for a deadly
combination. Also common are abscesses that later lead to sepsis, which
left untreated can be deadly. These are especially common in the tropics,
familiar to most people who have spent any time in a humid rain forest.

Of all the Tsimane infants who succumb from one dreaded horror or
another, over a third don't survive their first month.[42] As also among
the Hiwi, some of these are likely due to congenital anomalies and complications related to pregnancy or childbirth.

Sadly, violence accounts for about 17% of child deaths. Many of these
are cases of infanticide and neglect. In my interviews with Tsimane
women about all of their births, I heard about twenty-six confirmed in-
fanticides: ten the mother chalked up to accusations of infidelity by her
husband, whereby the baby was believed to be fathered by another man.
In most, but not all, cases like this, the murderer was the husband. Five
were deemed "unwanted," mostly by young mothers still in their teens.
Two were killed because they were "born too soon" after the previous
birth. Two other cases were of babies born with physical deformities.
Ache also reported killing infants they perceived as too weak or having
defects (*pura*) at birth. Among Tsimane, I heard stories as well about
how the weaker of two twins would also be killed. For this reason, there
are many precautions believed to prevent a woman from having twins
(for example, never eat joined plantains, which are themselves referred
to using the same word for twins, *epoj*[43]). Infanticide also occurs sometimes in preference for one sex over the other, more commonly against
girls, as among the Hiwi and Yanomamö. Contrary to a popular old idea,
infanticide against female infants was never an organized means of population control in light of limited resources. It's almost always an individual
decision, such as to speed up timing to the next pregnancy.[44]

A violent end was not uncommon for precontact Ache children. When important men would die and be buried, it was customary to sacrifice a living child into the open grave, so that the deceased's angry spirit wouldn't seek revenge and take another life. These *chape* children were usually under 5, often girls. They were also sometimes injured or orphaned, and sacrifice was chosen as an alternative to a life of suffering and begging. A dying man might make a request about who should join him, but more often, band members would decide.[45] Having close relatives nearby often meant the difference between life and death—some literally pulled children out of the grave. One survivor gave a haunting recollection of his horrific experience to Kim Hill: "My mother told the Ache to throw my brother and I in the grave to be sacrificed with our father. There was a big fight. Some dragged us to the grave site and others pulled us back out again. There were those who defended me. They were not all relatives, some of them were just my defenders. They liked me. . . . They tugged and pulled for a long time. Finally my defenders won. They said to me, wow you must have been strong. Why didn't you get torn apart when we were pulling so hard on your arms and legs?" When this same lucky person's stepfather later died, his baby sister was sacrificed: "My brother and I held on tight to her but they pulled her away from us. We tried to defend her, but some big men pulled her away. My brother and I cried a lot. Our sister screamed. They didn't kill her, they just buried her alive."[46]

Another 8% of preadults died from accidents. This includes snake bites among tropical groups like the Ache and Tsimane. Leeches, ticks, biting ants, lice, and mosquitoes are annoying, painful, and can be debilitating, but are usually not lethal for adults. But I've seen large nasty stinging bullet ants and fuzzy caterpillars that could floor an adult for hours, and I imagine they could be lethal to a small infant. It's rare for this to happen, fortunately, as small infants are usually in very close proximity to others at all times, especially with more dangerous surroundings. Common to most groups are fatal falls from trees, cliff edges, and river drownings, severe burns from falling into fire, and choking on grubs. Though rare, there was one case among both Agta and Tsimane of an older child accidentally dropping their baby sibling.

And a sad case of a 7-year-old Ju/'hoansi boy who perished from "hunger, thirst and cold" after getting lost in the bush, unable to find his mother.[47]

While the percentage of deaths due to different causes is revealing, we need to keep in mind how common death is in the first place. In the US in 2019, 19% of deaths among those under age 15 were "unintentional injuries," including traffic accidents, drownings, fires, poisonings, and other gruesome ways to shed this mortal coil. This is a high percentage of deaths. In absolute numbers, this can still be large, as the US is a nation of 340 million, with more than 2 million total deaths that same year. But with only 0.8% dying before age 15, that's about 21,000. All horribly tragic, mostly preventable deaths. But the likelihood that an American succumbs to one of these accidents before adulthood is very low: fewer than 2 per 1,000. For subsistence populations, the risks are much higher for most causes, because death rates are a lot higher. Since half of those born don't make it to age 15, and 8% of preadult deaths are from accidents, the risk of dying from accidents is about 40 per 1,000, or twenty times greater than in the United States. Doing the same exercise for violent deaths, hunter-gatherer children are about 350 times more likely to die from the harmful actions of others.

Surviving Adulthood

By age 15, mortality rates in hunter-gatherers and horticulturalists bottom out, ranging from 1% to 1.5% per year for two decades, then climbing to about 2% by around age 40, where it then increases exponentially. The chance of dying doubles by age 60, and again by age 70. While early life mortality patterns vary a good deal between populations, there's a clear nonindustrial pattern over much of adulthood. Relatively low mortality persists over a good chunk of adulthood before mortality accelerations come into play. That flat period of low mortality is an important departure from the more typical Gompertz-like mammalian increase soon after sexual maturity. As we first saw in chapter 2, only after about age 30 does the mortality rate in subsistence populations double every seven to ten years.[48]

As shown in figure 3.14, adult mortality in hunter-gatherers is much higher than in the industrialized world. At age 20, hunter-gatherers are over twenty-five times more likely to die than Swedes in 2019, about twenty times more likely by age 40, eight times by age 60, and four times by age 80. The gap narrows with age. The chance of dying for hunter-gatherers is fairly similar to that for eighteenth-century Swedes.

Making it beyond adolescence, what's the chance of surviving throughout much of adulthood to the time of menopause? On average, hunter-gatherers have a 54% chance of making it to age 50, ranging from 35% to 65% across groups. Horticulturalists have a similar chance of 57%. Tsimane show 69%, and Herero pastoralists 75%. These numbers are very low when compared against say, the US in 2021: Americans have a 93% chance of surviving from age 15 to 50. But these are very high compared to our nearest primate cousins. Chimpanzees in the wild have a paltry 3% chance of making it to age 50 from 15. Even under improved conditions of captivity, removing predators, ensuring available food, and treating sickness, that probability increases only to about 18%.

The Life You Expect Is Not Really the Life You'll Lead

We're now in a good position to revisit life expectancy, the most commonly used lifespan indicator, introduced in chapter 2. Life expectancy at birth (or e_0) is low in hunter-gatherers, ranging from 21 to 37 years. And we are now armed against flawed thinking that everyone lives to their 30s and then drops dead. We've seen how so many born don't make it to adulthood, and so we recognize that these low life expectancies at birth mostly reflect years lost from such devastatingly high mortality in the first decade of life. We need to look at life expectancy beyond scary childhood to reflect the age people can expect to live to if they survive running the early life gauntlet. Figure 3.15 shows what this looks like, adult life expectancy for all of the populations that by now you're hopefully starting to remember.

As you might expect, the Agta, Hiwi, and Gainj are on the lower end of life expectancy, while Yanomamö, Tsimane, and Hadza are on the higher end. Differences are greatest at birth, and get smaller with age.

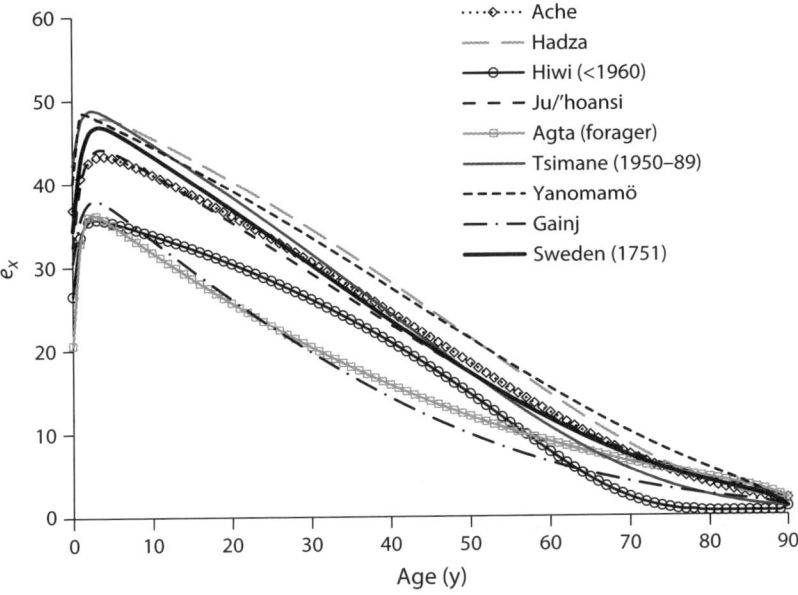

FIGURE 3.15. **Life expectancy (e_x) across the life course.** Curves refer to specific hunter-gatherers or horticulturalists in the legend. Sweden in 1751 (black bold) shows e_x age profile that overlaps with subsistence populations. Note that e_0 is lower than e_5 because of high mortality early in life. Upon reaching age 40, people could expect to live an additional fifteen to twenty-eight years.

Those able to survive to age 15 can expect to live an additional twenty-nine to forty-three years, or well into their 50s. Once making it to age 30—the average life expectancy at birth for hunter-gatherers—people don't suddenly vaporize into the mist. Instead, they've got twenty to thirty-five more years of life left to live, taking them well beyond their 50s into their 60s. If a woman survives to the age when she's likely to reach menopause (about 50), she can expect to live another twelve to twenty-one years, on average. That takes her close to 70 years. By age 65, folks have another eight to eleven years, on average, taking them well into their eighth decade. The age profiles of adult life expectancy for these remote-living hunter-gatherers and horticulturalists are remarkably similar, despite inhabiting very different environments and succumbing to different ailments.

Note in figure 3.15 the overlay of Sweden's life expectancy profile from back in 1751, which we first saw in chapter 1. On average, Swedes back then fared a little better than hunter-gatherers and horticulturalists, but by age 20, there are no differences in remaining life over the rest of the life course. In other words, small-scale subsistence populations living under some of the harshest conditions of the present with minimal to no access to health care exhibit a very similar life course as eighteenth-century preindustrial Europeans.

What Do Adults Die From?

Infection

The causes that cut life short in adulthood are similar to the causes affecting children and adolescents. Infection is also the major killer of adults, followed by violence and accidents. Respiratory illnesses, such as bronchitis, tuberculosis, pneumonias, and influenza, account for a fifth or more of illness-related deaths. Among the Ju/'hoansi, for example, pneumonia and tuberculosis were major killers of adults. Epidemic diseases like measles, dengue, and cholera were absent in newly contacted Amazonian groups, consistent with the notion that small, mobile populations couldn't support these contagious vectors (though they can be deadly after contact). Other diseases are endemic and so have been around for a while, such as malaria, yellow fever, yaws, and the flesh-eating leishmaniasis. Infectious disease, especially respiratory illness, may be even more common in older adults, as is the case with the Tsimane. Gastrointestinal illnesses account for 5%–18% of deaths in traditional human societies.

Childbirth

Among women, dying in or shortly after childbirth is much more common in subsistence populations practicing natural fertility than in the industrialized world today. Estimates exist for five populations: Agta (3,520 per 100,000 live births), Hiwi (~1,333 per 100,000), Hadza (1,022 per 100,000), Tsimane (702 per 100,000), and Ache (670 per 100,000).

The numbers of actual cases here are few—a total of 76 deaths out of 7,391 live births. (We don't know how common maternal mortality was among the Ju/'hoansi. Nancy Howell says it was "not a common event," but then reports a few cases. The Ju/'hoansi custom of "noninterference" by those who may want to help perhaps lowers the risk of postnatal infection.) That yields a conservative average of 1,028 per 100,000. This means that for every birth that a mother has, there is about a 1% chance she dies. If she were to have six births over her lifetime, her chances of dying are about six times as high, equivalent to the chance of flipping a coin and landing heads just four times in a row.

This level of maternal mortality is extremely high. A 1% chance of dying because of childbirth is higher than that observed in all but three of the 186 countries in the world where this is measured, surpassed by a narrow margin by Nigeria, Chad, and South Sudan. To put it another way, a hunter-gatherer or horticulturalist woman is forty-nine times more likely to die in childbirth than a US woman (in 2020), 129 times more likely than a French woman, and 206 times more likely than a Swedish woman.[49]

The relatively high probability of dying in childbirth in subsistence populations stems largely from bacterial infections of the reproductive tract in the absence of hygienic obstetrical practices. The Hungarian physician Ignaz Semmelweis realized only in the mid-nineteenth century, prior to Louis Pasteur promoting the germ theory of disease years later, that washing your hands with a chlorinated lime solution, especially after doing autopsies, could reduce mortality from "childbed fever" in maternity wards.[50]

Violence

Violent death accounts for 15% of adult deaths among the groups in table A3.2. Hiwi (~1,018 per 100,000) and Ache (~500/100,000) suffered from very high levels of homicide. Though lower among Agta and Yanomamö, homicide rates are still pretty high (129 and 166 per 100,000, respectively). Homicide generally affected adult males disproportionately, except among the Hiwi, where females were just as likely to be victims as males. Homicide is much lower among the Ju/'hoansi, Hadza, and Tsimane (29, 37, and 46 per 100,000, respectively).

Violence often comes in waves, and can vary over time. It shifts as neighbors move around, and with state intervention. For example, Ache display a very high level of homicide, although primarily from skirmishes with rural Paraguayans, not with other Ache or surrounding Guarani, the more dominant ethnic group in the region. A similar situation occurred with the Hiwi. Over half of the violent deaths, which accounted for almost a third of all deaths, were caused by non-Hiwi *criollo* neighbors. These colonizing conflicts explain why death rates from homicide are about ten times higher among Ache and Hiwi than among African hunter-gatherers. The Ache and Hiwi rates include infanticide and suicide, and so their high rates are inflated. Tsimane homicide rate would almost double, to 81 per 100,000, if I likewise include infanticide and suicide.

The likelihood of dying from the violent intentions of another human has decreased with greater state-level intervention, colonialism, and missionary influence in many small-scale groups around the world.[51] At least postcontact, not including the deaths from conquest, deliberate genocide, or displacement. For some perspective, the homicide death rate was 6.3 per 100,000 in the United States in 2022; the highest rate in the world is in El Salvador (52 per 100,000).[52]

Bumps in the Night

The composition of accidental deaths varies among groups, including falls, river drownings, accidental poisonings, snake bites, burns, and getting lost. These are avoidable incidents that get discussed a lot in small communities, and come close to the exogenous or extrinsic mortality discussed in chapter 2. Early and Headland tell about a drunk Agta man who went fishing with a net loaded with lead weights. He went to urinate off the side of his dugout canoe, when the boat tipped over. The heft of the lead weights prevented him from resurfacing. Among Tsimane, drownings, tree falls, animal attacks, especially poisonous pit vipers, and other accidents together contribute more than double the risk of dying than from homicide. Together, accidental and violent deaths account for an average 19% of all deaths (ranging from 4% to 43% across groups).

Dangerous predators are culturally salient in many societies, judging by myths, stories, songs, and games, and by the fear people express when sights and sounds of predators loom in the distance. Despite all this, or maybe because of this, death by predation is actually rare among foragers. Among the Hadza, leopards and lions are heard frequently, and many have had encounters, especially when hunting near water holes at night. Elephants are greatly feared but there are no cases of deaths from stampeding elephants in living memory. Even scavenging prey from other predators has made for some injuries and close calls, but fortunately no tragedy.[53] I've heard stories of numerous encounters with jaguars by Tsimane and Ache hunters, usually resulting in hiding, fleeing, or, on occasion, attempting to kill it.

Many groups apply their cumulative knowledge to reducing predation risks. Grouping patterns, warning displays like fires, weapons, and other cultural means of avoiding and deterring predators contribute to the low impact of predation on human survival.[54]

Old Age, or Natural Causes?

Lastly, there are the so-called degenerative diseases of aging. These are hard to diagnose, and evidence is scant. The absence of evidence certainly isn't evidence of absence, but, as we'll explore in chapter 9, chronic diseases like atherosclerosis, and the heart attacks and strokes that come with it, were and probably remain rare. If so, that hugely contrasts with rates in industrialized nations today and, increasingly, over the rest of the world, where heart disease, cancer, stroke, and diabetes are leading killers. Overall, degenerative disease seems to account for about 9% of adult deaths up through the 50s, and about a quarter of deaths for those older than 60. Many cases of degenerative deaths, however, are attributions of "old age" in the absence of any obvious symptom or pathology. It's anyone's guess what that might mean. Even in the US, it wasn't too long ago that "old age" or "natural causes" was scribbled on death certificates when there was no obvious cause. (As recently as 2020, famed actress Betty White was reported to have died in her sleep of "natural causes" two weeks shy of her one hundredth birthday, though investigations later revealed that a stroke was the true cause of death.) As might be expected, the more acculturated Northern Territory Aborigines show the highest prevalence of aging-related deaths. Possible heart disease and neoplasms characteristic of cancer each accounted for nine of the forty-nine deaths due to degenerative illness in adults over age 60.

A Seven-Decade Human Lifespan?

In chapter 2 I talked about how the modal age of adult lifespan gives us a good sense of the "typical" potential lifespan one can aspire to. We saw how this mode increased in Europe over the past couple hundred years, especially in the last century. The modal age of death reflects an important stage in physiological decline regardless of the specific causes of death that entail our demise. So let's look at modal ages of adult death and the variation around these modes, to provide insight into the stability in adult lifespans among subsistence populations.

The effective end of the human lifespan under traditional conditions is about seven decades. Figure 3.16 shows the proportion of deaths occurring at each age, among those who already survived to age 15, for hunter-gatherers and horticulturalists.[55] These show an average modal adult life span of about 72 years of age, with a range of 68–78 years (see Appendix). For hunter-gatherers alone, the mode is 69 years, but the shape of the curves looks similar across hunter-gatherers and horticulturalists. The mountain peaks are not steep and narrow like the ones we saw in chapter 1 for recent decades in longevous Sweden. Instead of the steep, narrow Death Mountain observed in all high-income countries these days, the climb is very gradual on the way up. This means that there is much more variability, and *inequality*, in the duration of adult lifespans in these high-mortality subsistence populations than seen in modern industrial populations. The gently sloping hill is due to all the different ways of dying that may be considered different from the "normal" course of aging. At least that's one

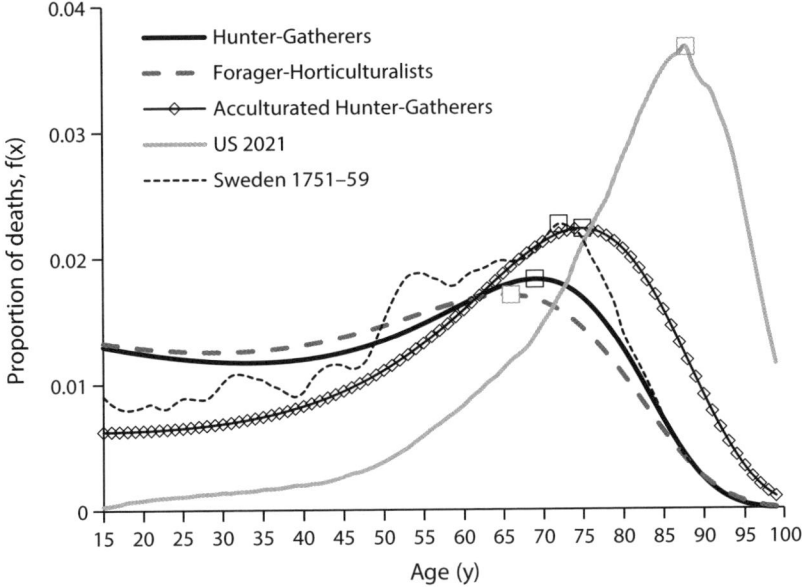

FIGURE 3.16. **Modal age of adult lifespans.** The proportion of adult deaths, f(x), organized into three categories: hunter-gatherers (Ju/'hoansi, Ache, Hadza, Hiwi), horticulturalists (Gainj, Tsimane, Yanomamö, Xilixana), acculturated hunter-gatherers (Northern Territory Aborigines, Ache [reservation]). Eighteenth-century Sweden and United States in 2021 shown for comparison.

hypothesis—that these other deaths cloud what might be caused by aging alone. An alternative hypothesis is that subsistence populations may just age and succumb to the weary toil of life more rapidly than today's urbane city folk.

Apart from the mountain's shape, the mode itself is a striking contrast to that of high-income countries. The US mode in 2021 peaks at age 87, and is a pretty steep crag of a mountain. This US mountain is right-shifted almost two decades. In other words, US adult lives are indeed longer—62% of US adult lifespans are longer than 75 years. Far fewer hunter-gatherers and horticulturalists have such long lifespans—just 16% and 14% of adult lifespans, respectively. The modes for subsistence populations are much less peaked, with those peaks accounting for less than 3% of adult deaths. If you were standing on the top of Death

Mountain and looking at the spread of adult lifespans covered by five years above and below the peak you're standing on, you'd find that 35% of US adult deaths are covered in the range 82–92. Among subsistence populations, it's about half: 20% of hunter-gatherer lifespans fall within the range 64–74 y, and 19% of horticulturalist lifespans fall within the 60–70 y age range.

Acculturated hunter-gatherers show a slightly later peak (75 years) than more isolated hunter-gatherers and horticulturalists. The peak is higher elevation—24% of adult lifespans fall within the 70–80 year range, and 31% of lifespans are longer than 75 years. This steeper climb to the peak and greater density of deaths at the later ages suggests that the release from some causes of death, such as violence and warfare, along with better access to antibiotics and other types of medical care, can improve both the average duration of adult life and the proportion of folks that achieve it.

Overall, these patterns of adult lifespan duration in small-scale societies occupying diverse environments, eating widely different diets, and experiencing different hazards suggest that, by about seven decades, most people experience sufficient decline that if they do not die from one cause, they're soon to die from another. Again, this pattern applies equally to mid-eighteenth-century Europeans, who were largely agrarian but certainly not living as hunter-gatherers.

What does this overarching similarity say about our evolved human longevity? Does the difference between then and now reflect merely the removal of different hazards and the introduction of seatbelts, penicillin, and more readily available food? Or do those advances just make the climb to the peak steeper, but not shift the mode the extra decade or two we see today? In other words, has that elusive process of aging indeed slowed down? We'll return to these related questions again in chapter 8.

Postreproductive Lifespan: Uniquely Human?

Let's be clear: in all human groups we've seen so far, survival is evident well beyond the usual age of menopause, around 50 years. This period of living has been identified as a novel life stage of humans, a period of "postreproductive" or "postmenopausal" life. If natural selection primarily rewards reproduction, then humans are indeed in a weird position to count postmenopausal lifespan as a species-typical trait. Or, if push comes to shove, humans sometimes share the stage with a few privileged mega-mammals. More skeptically, others have argued that postreproductive lifespan is a general mammalian trait, nothing much to see here, please move along.[56] So, is a seven-decade lifespan with human postreproductive life substantial enough to be considered unique, and thereby demand a special explanation? (Mostly yes!) Or is it what you might expect for a long-lived mammal of our body size? (Mostly no!)

To dive into the specifics, we can borrow a clever metric coined by the evolutionary biologist Daniel Levitis. Here comes another acronym: PrR, or "postreproductive representation." It's the proportion of adult life spent postreproductive.[57] If one can expect to live well beyond menopause, but few adults ever make it to menopause, then PrR would be small.

Figure 3.17 shows PrR for many human subsistence populations, along with other primate and mammalian contenders for comparison. PrR varies between 30% and 55% in human subsistence populations: roughly a third to over half of the adult lifespan is lived postreproductively. This is below the 60%–75% we witness in high-income countries but is substantially higher than the other mammalian species known to have postreproductive life. For example, experienced Asian elephant grandmothers hanging around their adult daughters are known to improve the survival of their grandchildren. But only 12% of their adult lives are spent "post-fertile." Chimpanzees at Ngogo have the highest life expectancy of any chimpanzee group and may even experience true menopause.[58] But even their PrR is just 14%. While post-fertile survival is not rare among lucky mammals, a post-fertile life stage is.[59] Such a life stage *is* unique to humans (well, and a couple other species). Orcas, which are technically dolphins, and short-finned pilot whales come the closest to human levels—but more about them in chapter 5.

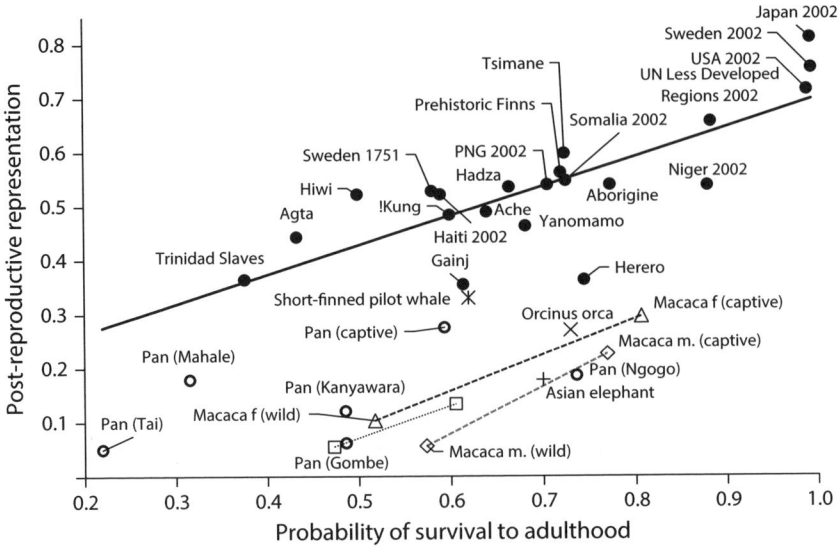

FIGURE 3.17. **Postreproductive lifespan across humans and other species.**
Most nonhuman species show a relatively short postreproductive lifespan
compared to humans. Humans in filled-in black circles, chimpanzees in open-
circle black. Other species with observed postreproductive lifespans, including
killer whales (*Orcinus orca*), short-finned pilot whales (*Globicephala macrorhyn-
chus*) and Asian elephants (*Elephas maximus*), are shown using symbols (×, *, +).

Really, Nothing Known about Other Populations?

We focused a lot of attention on five to ten populations because those
are the ones that have been best studied using systematic methods to
give us the most confidence about ages and lifespans under the closest
to traditional conditions. These populations are our best glimpse of the
past in the ethnographic present. But is there really no information about
other subsistence populations that might either support or tear down
what I've shown so far? As I highlighted earlier, information on infant
and child mortality from a larger set of populations shows a similar range
as among our select sample. But what about beyond childhood?

In the early 1970s, the geneticist-demographer Ken Weiss published
a landmark monograph. It's one of those amazing books that everyone
cites but no one reads. It heroically aimed to create model life tables for

"anthropological populations" to provide mortality profiles among small-scale societies past and present. It was an ambitious quest to best-guesstimate survival across the life course given the very limited information available at the time. For example, you may have only a single census of who was living in a small community in a given year, and you didn't really know people's ages well. Or maybe you only know what mortality looks like for a limited part of the life course, like ages 5–9. But nothing comes for free. In order to say a lot with so little you have to make a lot of assumptions. You often have to assume the population is neither growing nor shrinking, and that there's no net migration. And that the rates of mortality and fertility have been unchanged for some time, like, at least a century.

Given the belief that cultural change is slow, Weiss assumed that traditional populations would already be at some equilibrium and that these assumptions should therefore be okay. Even though populations might be in flux, at least he could estimate what a population mortality profile *should* look like. He correctly assumed that many old censuses often missed young kids and that ages of older adults were wild guesses. Part of his solution was to ignore information about people over age 55 altogether.

So with the limited information available, he used some demographic techniques that incorporated the assumptions above, and threw in a little Gompertz mortality modeling for extra dazzle, to infer the age profiles of mortality and fertility that could generate the numbers of people observed in each age class of the population. This process he described casually as "delicate . . . involving much feedback and adjustment, as well as some subjective judgement."[60]

He applied his techniques mostly to assemblages of skeletal material, but also considered living populations, including the Yanomamö. He used information that the Yanomamö were modestly growing, and that the average Yanomamö woman would give birth to about eight children if she survived throughout her reproductive years. Both of those observations are within the domain of possibility from my sample.

But his life table shows a low life expectancy e_0 of just under 20 years. One big difference with what I've shown so far is that this Yanomamö life table shows huge infant mortality (43% for girls, 27% for boys). Infanticide was not uncommon in the Yanomamö, due in part to a stated preference for sons. But, overall, the life table doesn't look much different from what we observed with the Agta during their "foraging" phase. It does look less favorable than the Mucajai Yanomamö studied by Early and Peters that I reported on earlier. If correct, the discrepancy shows that the Yanomamö are "not a continually high-mortality population," and that mortality varies by region, including more warfare in the southwest region, and during times of epidemics.[61]

Weiss's goal was to give the study of "primitive" populations a quantitative basis. As he himself noted, the methods do not apply to populations who have been "disrupted in the recent past." Rather than come up with a life table for actual populations as they exist in a slice in time (called period life tables in the jargon), his approach idealizes population archetypes, giving us long-run best-guesses of what vital rates might look like over long stretches of time. But, as we've seen so many times, the "average" may not represent *any* population.

Even if one had a decent estimate of life expectancy, many different age trajectories of mortality are compatible with that same life expectancy, such as higher child mortality but lower adult mortality, or lower child but higher adult mortality. For example, a population with a very low life expectancy at birth of 20 is possible under different combinations of age-structured mortality rates showing a wide range in adult survivorship: survival from ages 15 to 40 can vary from 23% to 53%, adult life expectancy (from age 15 onward) can vary from 22 to 35 years, and percentage of the population over age 50 could vary from 12% to 20%! All with e_0 of 20.[62]

Using these kinds of life tables primed with spotty, partial information is certainly better than nothing, and for many purposes may be a decent start. In fact, life tables were devised to be used precisely this way, for areas with poor or no vital registration system. Ken Weiss concluded his book advising that "we can only live with the fear of uncertainty" because no other data existed at the time. But other reliable, more fine-grained data *do* now exist. As ambitious as Weiss's efforts were, they do not contradict what I've shown about the seven-decade lifespan.

Other Low-Mortality Foragers Out There?

Some keen observers may wonder why I've neglected to include other populations that have stirred the public's imagination over the past century. Low life expectancies at birth have been reported for several groups still pejoratively referred to as "pygmies."[63] So fascinated was the ethnocentric West with exotic African Pygmies that an Mbuti and eight Batwa were presented as living dioramas in cages during the 1904 World's Fair in St. Louis, then again in the Bronx Zoo.[64] A low e_0 has been reported for other Pygmy groups, like the Batak of the Philippines (~21 years), and even lower for other Pygmy groups, like ~16 years among the Efe, Mbuti, and Aeta. These groups also all show lower life expectancies at age 15 than what we've seen, ranging from 20 to 33.[65] Are these neglected populations somehow better examples of what hunter-gatherer lifespans could be like?

If the answer were yes, I wouldn't have written this book. Groups like the Batak were experiencing population decline because of massive disruptions like "intermarriage with lowlanders, land loss and decline in cultural identity."[66] Indirect methods assuming unchanging fertility and mortality rates, and no growth or shrinking of the population, were inappropriately applied to estimate life expectancy.[67] Direct measure of Batak infant mortality (28%) suggests a lower mortality rate than that reported by those claiming very low Batak life expectancy.[68] Another group, the Aeta of the Philippines, were forcibly removed from their territory to live in urban settlements, and were no longer able to hunt and gather. Their demographic data can hardly be representative either.

For the pygmy groups in central Africa, there's another reason why those life expectancies may not be reliable. These all applied Ken Weiss's model life tables to census information. Remember, those make major assumptions that are likely invalid in these cases. Quite simply, no source data exist among pygmy populations using methods designed specifically to assess mortality rates by age, except for the Agta.[69] The data that do exist, as among the Efe, suggest *lower* infant and child mortality rates, which look closer to Tsimane levels, than those suggested by their low life expectancies: 14% infant mortality and 78% survival to age 15.[70] Maybe those mortality rates are too low and unrepresentative, but hard to tell without more reliable information.

Let's explore a few more examples. The Bismam Asmat were hunters, fishers, and sago palm gatherers inhabiting the swampy forests of Irian Jaya, studied in the 1970s, but first contacted in the 1950s. Tribal warfare was not uncommon, and settlements were small and semipermanent. Infant mortality was high, about 30% overall across multiple regions, which led anthropologist Peter Van Arsdale, applying Weiss's life tables, to come up with a life expectancy of 25 years. Fertility was high (~7–8 births per woman), leading to positive population growth (>1% per year), despite the high mortality. Van Arsdale remarks that it was "unusual" for Asmat to live to age 60, but he also mentions the 90-year-old former warrior Saati, whose exact birth date was confirmed by missionaries.

The Semai Senoi of the tropical Malaysian peninsula were studied in the late 1960s by Alan Fix.[71] Applying Weiss's life tables, he reports a life expectancy at birth of 30 years, with adult survival from age 15 to 50 being about 45%. A 15-year-old can expect to live another thirty-three years, a 40-year-old another twenty-three years. Overall, these patterns suggest the Semai fared worse than the Ju/'hoansi, but better than the Agta and the Gainj, and somewhat similar to the Hiwi.

Unfortunately, there's very little we know about longevity in the Arctic. Anecdotal claims in the nineteenth century asserted that "primitive Eskimos" were short-lived, with few living past age 50. Medical missionaries, however, seemed to think the opposite, that "the Eskimo of the far North was healthy [and] . . . lived to a very great age." To resolve these contrary views, the Arctic explorer and ethnographer Vilhjalmur Stefansson obtained records from the Russian church concerning Alaskan Aleuts over the period from 1822 to 1836. Of 497 recorded deaths, 197 (40%) were of adults age 45+. About 11% were age 70 and over. Stefansson concluded that neither view of "non-Europeanized" Aleutians was correct, but instead, the situation "accord[ed] with the Biblical 'the days of our years are threescore and ten.'"[72]

Among pastoralists, we only considered the well-studied Herero. But a limited study of nomadic Turkana pastoralists of northwest Kenya in the late 1970s reported a life expectancy at birth of 34 years, with 24% of infants dying in their first year of life. Among Fulani pastoralists of central Mali, life expectancy was estimated at 37 years among a semino-madic subgroup, and 25 years among an agropastoral subset living in an infection-prone area of the inner Niger Delta.[73] Such high mortality overlaps with what we've seen among hunter-gatherers. And like the case with greater acculturation and sedentism among hunter-gatherers, Turkana who have partially or permanently settled show life expectancies in the mid- to upper 40s, with infant mortality rates of 13%–15%. Settled Turkana practiced farming, and had the benefits of a health clinic set up in the mid-1970s. While the infant mortality levels were based on real data from reproductive histories, as you may have guessed by now, the rest relied on using model life tables.[74]

To summarize, we can say that the spotty data that exists from other populations is largely consistent with the pattern we've seen in much better-studied groups (or they're based on indirect methods with wildly violated assumptions, and so they're hard to swallow at face value).

Even under the best of circumstances, mortality estimates in subsistence populations are based on such small samples, it's enough to make industrial-strength demographers wince and cringe. Statisticians speak about "confidence intervals" as a way to assess how sure they can be about whether an orange is an orange. If an infant mortality rate of 20% is based on just 100 births, the 95% confidence interval is a wide road, between 12% and 28%. If based on 1,000 births, that road narrows to 18% and 22%. You'd need about 5,000 births to be sure enough about that 20% mortality rate give or take one percent. Unfortunately, that's a much larger sample than we ever see in this business.[75]

Butterflies and Coke Bottles

I've argued so far that contemporary subsistence populations living relatively traditional lifestyles provide our best window into assessing ancestral mortality and fertility levels. But moving beyond the many examples we've already discussed, it's possible that twentieth-century postcontact groups are simply not representative of the past. At best, these groups we've discussed may cover only a small subset of what demography may have looked like in our deeper past. For example, we saw how the Ache were growing during a pioneering period in Paraguayan history after the Chaco War in the 1930s left more than 40,000 dead. The Hadza similarly benefited, whereby colonial pacification of invading Maasai and other herding neighbors reduced conflict and competition for land. And we saw that warfare is also reduced in many postcontact settings, as implied by the very word "pacification." Escape from unpredictable raids certainly improves your well-being. As a Yanomamö man living in a pacified mission village asserted, a life with endemic warfare means living with an almost constant state of fear, even during periods when raids are infrequent.

As mentioned, no area is left untouched. When the missionary-turned-anthropologist Tom Headland started fieldwork with the Agta in the early 1960s, he thought he was at "the end of the world," but was amazed to hear an Agta women singing in clear English: "Oh, come, come to the church in the wildwood." And to discover that Agta had "patron-client" relationships with neighbors with whom they traded meat to get tobacco and rice. He also realized that Agta "hunter-gatherers" knew how to farm. They hadn't just learned to farm either: there were reports of Agta having small fields at least as far back as the 1740s.[76]

On the other hand, the common tendency to think that recent conditions must be *better* than the past may itself be flawed. Is it so far-fetched to think that conditions may have been *more* favorable in the distant past?

First, measles, cholera, and other nasty infectious diseases known to decimate their hosts usually require large reservoir populations in order to take hold and spread. These were unlikely to have been common among small bands of nomadic hunter-gatherers, except after contact with large Eurasian populations.[77]

Second, folks occupying the most fertile lands and richest resources—like the lush salmon streams of the US Pacific Northwest, or coastal areas almost everywhere—were either pushed out or wiped out by dominant populations wielding "guns, germs and steel." So those groups still remaining in the twentieth and twenty-first centuries may be living in marginal areas poorly suited for farming, less favorable than those occupied when the planet was of hunter-gatherers among hunter-gatherers. Is it the case that foragers occupy poorer habitats than the neighboring farming groups that presumably displaced them? The answer, according to at least two studies, is a firm "no." Comparing the territories occupied by foragers with those occupied by non-foragers reveals no differences in their productivity.[78] In other words, there's no evidence that contemporary foragers are living any more marginally than contemporary farmers.

All that said, acculturation over the past half century is likely to have affected people's lives differently than during earlier transitions to sedentary, agricultural, or peasant life in the more remote past. Even when foragers become the new underclass of national society, and livelihoods suffer as habitats are crisscrossed with roads and airstrips, recent postcontact recovery periods are often accompanied by amenities: immunizations and public health and sanitation infrastructure that can substantially improve survivorship, just like it did for the industrialized world by the end of the nineteenth century. It is likely that physical health gets worse when these benefits are lacking or unavailable, as may have been the case for the Batak and others.

What If?

As a last resort, we can build hypothetical scenarios to tackle difficult questions. If the Yanomamö or the Tsimane eliminated infanticide completely, or if deaths related to childbirth could miraculously be a distant memory, then perhaps these changes could move Death Mountain to resemble the current US and Swedish ones.

Fortunately, there is a simple but charmed approach that considers how the age profile of mortality might look if we got rid of infections, accidents and violence, and deaths from birth complications—basically everything but degenerative diseases. By making a few simplifying assumptions, like that people die from just one cause, and that causes of death are independent of each other, we can with Godlike deus ex machina powers adjust our life table to see the demographic impact of removing those "exogenous" sources of mortality.[79]

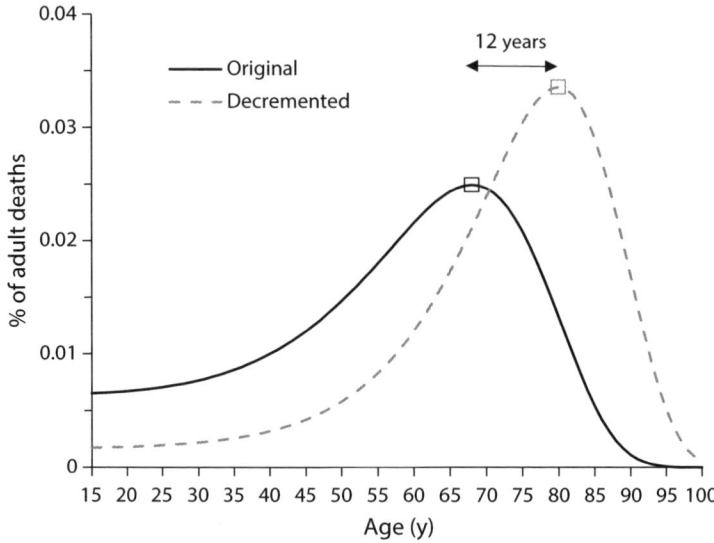

FIGURE 3.18. **What if all exogenous sources of mortality could be eliminated?** A comparison of the modal ages of adult death for Tsimane before and after the removal of all exogenous causes of death. Before the hypothetical removal (see text for details about this "decremented" life table), Tsimane from 1950 to 1989 showed a life expectancy at birth of 42 years, and modal adult lifespan of 68 years. Life expectancy at birth after removing common hazards would be 64 years, and the modal adult lifespan 80 years.

By removing the common hazards of Tsimane life, life expectancy increases at all ages, up from 42 to 64 years at birth—like the United States in the 1940s, or Bolivia in 2004. Upon reaching age 15, Tsimane could gain over a decade more life expectancy (e_{15}), and almost another decade upon reaching age 40 (e_{40}). By age 70, however, only one more year is gained from removing common hazards. Comparing before and after this hypothetical removal of all types of environmental harshness, the Tsimane would gain over a decade more of adult life. As shown in figure 3.18, the modal age of adult death for our hazard-free Tsimane shifts from 68 to about 80 years. In this new hazard-free, hopefully still-forested world, 82% of adult deaths would occur after age 60.

That Was Then, This Is Now

Another useful way to ask a "what if?" question is to examine the effects of recent changes on traditional small-scale populations. Rather than dismiss these cases for not being "primitive" enough, it is the effect of changes that we want to evaluate. Here, I compare mortality profiles of the same groups at different times. Luckily, these kinds of comparisons can be made for seven groups. Mortality profiles exist before and after some critical period, be it contact, like for the Yanomamö and Ache, or acculturation and transition to peasant status, like for the Agta and Ju/'hoansi. Although the periods covered here are on the order of decades, not centuries or millennia, relatively large differences are evident. In a majority of the populations, survival improved with acculturation at most ages. Among the Ache, the period of contact brought catastrophic diseases to the population, and about 40% died in under a decade. In the postcontact period, during which Ache were living in more permanent settlements and combining slash-and-burn farming and animal husbandry with foraging treks, infant mortality increased a little, but mortality at other ages decreased by a third to a quarter. Most improvements seem to be concentrated at younger ages, with less improvement at older ages. Lower mortality among settled Ache comes from reduced homicide and accidents in the forest, influenced in part by missionaries and state intervention. Medical attention has also helped lower mortality among Ache.

Survival improved at all ages among the Ju/'hoansi. Although settled !Kung frequently complain about meat scarcity, and greater stinginess by the haves toward the have-nots, they've also benefited from improved access to milk, protein-rich weaning foods, and a more predictable diet through association with cattle posts and government rations.[80] Among the Hiwi, contact increased infant mortality, but lowered mortality by half or more at all other ages. Among Tsimane, survival didn't improve much early in life, but did so substantially at all other ages. Among the Herero, survival is higher at all ages except for early childhood.

These patterns are striking because the large survival gains observed historically throughout much of the industrialized world usually occurs early in life at first, due to reductions in infectious diseases, followed later by more gains in adulthood. In many small-scale societies, novel or introduced infections may harm infants and small children, especially when some acculturation is not accompanied by sufficient medical care.

A key lifestyle change leading to greater connections with the outside world is a reduction in nomadism. People benefit from modern amenities by being in close proximity to navigable rivers and roads. Close proximity often comes with changes in residence, like living in semipermanent villages. Just like we've seen with the Ju/'hoansi, Ache, and Herero, greater sedentism leads to lower child mortality in other groups, too, when access to health care and other amenities offsets any increased hazards from denser living.[81]

Recognizing that recent experiences with modernization are not always progressive in a positive direction, two cases suggest worse conditions over time—Agta and Yanomamö. Child and adult mortality are much higher among "peasant" Agta and Xilixana Yanomamö. Infant mortality may be buffered by protective effects of breastfeeding, and so postweaning mortality seems to worsen more in acculturated settings among both Agta and Yanomamö. What conditions lead to poorer survival?

Peasant Agta are landless farmers living in more populated and degraded environments with few foraging options. They no longer maintain close trading relationships with nearby farmers like they used to in the past. It's possible that the synergistic effects of malnutrition and new vectors of infectious disease, like measles, increased child mortality during the peasant phase. In addition, malaria, tuberculosis, and other infectious diseases were largely absent in the forager phase for both Agta and Yanomamö, but reached epidemic proportions in later years. Even here, though, it's not all bad news. Lower early adult mortality among Yanomamö may be due to reduced warfare and homicide in recent years.

To wrap up, let's see how these changes affect our favorite metric, the modal age of death. For the five groups with improved survivorship, the modal lifespan increased by about seven years (range 6 to 8). For the Yanomamö and Agta who experienced worse conditions, no differences in modal lifespan were discernable.

The Take-Home Messages on Longevity in the Nonindustrialized World

The research I featured in this chapter provides us with the foundation we'll need to explore what longevity was like long before industrialization. As I've argued through the examples in this chapter, hunter-gatherers and other small-scale societies offer a unique window into what survival over the life course, and human lifespans, may have looked like in our ancestral past. With positive and harmful aspects of globalization affecting once-remote groups everywhere, we're unlikely to ever obtain new demographic data on hunter-gatherers.

Humans everywhere are endowed with the capacity for long lives. The differences between the present compared to the past is that many more of us now reach our fuller potential. I boil the lessons learned down to six main takeaways:

1. Among hunter-gatherers, life expectancy at birth (e_0) is low, largely because of high mortality in early life. Upon reaching adulthood, adult life expectancy is higher.
2. Modal age of adult lifespans is about seven decades.
3. Postreproductive lifespans exist in all human populations.
4. Differences in survival patterns across human populations diminish with age.
5. People die largely from infections, violence, accidents, and childbirth.
6. Improved conditions in rural settings shift modal lifespans by about a decade, to look like those of high-income countries.

Part II

4

The Long Road to Longevity

SO FAR, we've learned that among contemporary hunter-gatherers, farmers, and herders, life expectancy is low and mortality levels are high, relative to what's been seen throughout the industrialized world since the nineteenth century. At the same time, people in these societies can still expect to live about seven decades, if they survive childhood. This was also the case going back to mid-eighteenth-century Europe. But what was the life expectancy of people before this time? Do the patterns I highlighted in chapter 3 extend back to the Middle Ages or biblical times? Do they apply to humans who lived before the Agricultural (or Neolithic) Revolution some 5,000–10,000 years ago, when people began domesticating plants and animals and living in more permanent settlements? How about even farther back, to our hominin ancestors? If the potential lifespan is at least seven decades for all humans, when did that first come about? And how?

I've found that the musings of the English scholar Thomas Robert Malthus offer a provocative summary about ancient longevity. Many people remember him for his dire warnings about the wars, famines, and pestilence that follow when population growth outpaces food production. But he also studied demography. In 1798, Malthus noted that, "with regard to the duration of human life, there does not appear to have existed from the earliest ages of the world to the present moment the smallest permanent symptom or indication of increasing prolongation."[1] This view of longevity stasis has been echoed many times since. In *Farewell to Alms: A Brief Economic History of the World*, the economist Gregory Clark sums this up succinctly: "There was probably little change in life expectancy in the preindustrial world all the way from the original foragers to 1800."[2] This idea is reflected in the popular and cynical Hobbesian refrain that ancestral life in the "state of Nature," was "nasty, brutish and short." This has also been the version long pushed by many experts.[3]

Indeed, when Ken Weiss concluded his thorough analysis of ancient demography, based primarily on the spotty skeletal record, he wailed into the wind, "Why are there no old folks?!"[4] The narrative bespeaks a past with few adults over age 40.

In this chapter, I carefully examine this claim and present the evidence that shows that longevity is an ancient trait of our species, and that old people have existed for many millennia.

If it's the case that humans have had the potential for longevity long before the amenities of recent centuries, then before we search for clues about how and why such longevity came to be, we must address a fundamental concern: How far back can we claim seven decades?

From Ancient to Modern

The timing of when our familiar seven-decade lifespan first appeared is critical to address because of what that can tell us about the origin of our species (a topic we'll tackle in chapter 5). In chapter 2, I gave a sense of longevity patterns over the last couple of centuries. In chapter 3, I showed what we know about contemporary small-scale foragers and farmers, with a squinty eye toward inferring the past from the present. Now let's work our way backward in time to see if we can uncover more direct evidence of past longevity.

Demographic evidence can be hard to come by prior to the widespread growth of public recordkeeping in the eighteenth century. Before that, we are limited to only snapshots of lifespans. These snapshots are not representative of whole populations, nor do they give us a good sense of patterns globally. Unfortunately, most of these cases are confined to Europe, with some of the earliest records coming from England. As with the skeletal record, making inferences about lifespan from historical records is a challenge. Despite limitations, they do tell us something about how good things could get, since the records that remain are largely those of the elite.

The study of these records involves herculean efforts by historians and demographers to re-create population history by poring meticulously over obscure records created for other reasons, usually to facilitate the collection of taxes or the saving of souls. Entire books have been written on how this research is done, and for our purposes, we're going to trust these experts and believe their estimates. The primary sources I refer to here were compiled from records kept on English monks at Westminster Abbey and other monasteries, and other material focused on British royals and children of kings. Another source involves extensive genealogy records for important lineages throughout Europe and China. And, more recently, the Mormon Church has digitized their massive and thorough family histories. Those exist in the first place because, in the Mormon religion, soul-saving by baptism can occur even after death. Lastly, thanks to our obsession with famous folks, scientists, and scholars (and their birthdays), we uncover yet another opportunity to evaluate adult longevity throughout history.

Table A4.1 in the Appendix summarizes what I've gleaned from these sources, showing the life expectancy at birth (e_0), at the end of childhood (e_{15}), after adolescence (e_{25}), and in adulthood $(e_{40}$ to $e_{60})$ at various times in human history. As a friendly reminder, contemporary hunter-gatherers have a life expectancy at birth (e_0) of about 31 years, ranging from 21 to 42. From age 15, they live an extra thirty-eight years, ranging from 29 to 43. From age 25 they live another thirty-three years (range 23–37). From age 45, they live twenty-one more years (range 12–25), and upon reaching the glorious age of 60, they will live another thirteen years (range 6–15).

Let's start from the time of the Renaissance, then back to the Middle Ages.

One of the first "kosher" life tables was put together in 1693 by none other than comet-discoverer Sir Edmond Halley. When not chasing comets, Halley compiled a life table for Breslau (Wrocław in Polish), a city in historical Silesia, now southwestern Poland. With his "Breslau table,"

Halley wanted not just to estimate mortality, but also to assess how many men would be available "to bear arms" (his answer was a precise "9/34ths of the population"), and to best calculate the price of life insurance, which he argued "ought to be regulated" in relation to the odds of survival.[5] His life table covered the period from 1687 to 1691 and shows life expectancy at birth for men was about 28 years, with an extra thirty-seven years at age 15, twenty-two more years by age 40, and twelve more years upon reaching age 60.[6] From his analysis of deaths in Breslau during that time, you can glimpse a profile of modal adult lifespans in the 50–75 range.[7]

A little farther back, the best representational study (again, of men) is of medieval English men from the mid-sixteenth through the eighteenth centuries, based largely on more than four hundred parish registers. Like weaving together many haphazard scraps into a beautiful blanket, this painstaking approach to constructing a life table from these registers is referred to as "family reconstitution." Among other cool achievements, the historical demographer (Sir) Edward Wrigley and the historian Roger Schofield calculated life expectancies throughout this period. They found e_0 to hover around 36 years, slightly below what we saw in mid-eighteenth-century Sweden, but slightly better than it was in France.[8] Infant mortality was 13%–17%, with about 67%–71% surviving from birth to age 15. With some respectable wizardry, they later generated estimates of adult life expectancy.[9] From age 25, these literal late Renaissance men were estimated to have lived another thirty-two years (range 28–35 years). By the early nineteenth century, e_0 increased just as I described in chapter 1.

Next we move to royalty, covering roughly the same time period. The British "peerage" system refers to the exalted nobles with lavish hereditary and lifetime titles. Among nobles, e_0 for those born in the mid-sixteenth century is at the high end of what we observed in subsistence populations. It hovers in that high end for a few centuries, before improving in the eighteenth century. Prior to the seventeenth century, life expectancy at 40 (e_{40}) was about two decades and, at age 65, was just under a decade.[10]

Now let's go back even farther—by a few more centuries.

In the 1940s, while much of Europe was in the throes of world war, the US historian Josiah Cox Russell was busy building life tables for English landholders using postmortem "inquisitions" and other "magnificent collections of past records."[11] The death of feudal vassals often meant money for royal officials, creating a keen desire to record relevant information on their lives and property. Russell's life tables show life expectancies in the 30s from the thirteenth century through the fifteenth century, with a slight dip due to plague between 1350 and 1425. Adult lifespans are on the low end of the range observed in contemporary subsistence populations. Upon reaching age 15, British men could expect to live another twenty-three to thirty-three years between the thirteenth and fifteenth centuries. By age 40, they'd live sixteen to twenty-one extra years. And, by age 60, another nine to fourteen years. Roughly one in seven made it to age 60, except during the plague years, where it was one in fourteen. Russell notes, however, that even during the plague period, lifespan didn't change much. Even centenarians existed throughout these episodes of high mortality.[12]

During much of the same period, we find a rare opportunity to gain insights courtesy of the English monastery. Benedictine monks from Canterbury, Durham, and Westminster Abbey are a captivating case study.[13] These institutions were not isolated religious outcrops, but centers of learning that tapped into broader medieval society. The lifestyles of monks and the general characteristics of these urban monasteries changed very little over time. Once a novice professed into an order in his early twenties, his diet, housing, hygiene, and sanitation improved, and he was mostly spared the difficult physical labor required of his nonreligious peers. On the flip side, monks were more susceptible to infectious diseases, due to living communally in dense, compact quarters and their frequent contact with the surrounding public.

Though these monasteries are dispersed throughout England, there's a remarkable similarity in their historical mortality profiles. Upon reaching age 25, monks survived an additional thirty years—as far back as the late fourteenth century, but with a nosedive to only twenty extra years in the fifteenth century, due to the bubonic plague pandemic, before returning back to three decades by the start of the sixteenth century.

Durham was the best place to live a long life in the service of the church, whereas novices at Westminster Abbey fared worst, a difference of about five years of extra adult life throughout this period. Westminster Abbey was in a town, close to London, referred to by locals at the time as a "death trap." Nonetheless, monks did experience higher survival rates than their non-monk neighbors during the same time period.[14]

These kinds of life course reconstructions during periods of epidemic disease are invaluable, not easily substituted by the indirect methods I introduced in chapter 3. For example, epidemics can affect mortality at different ages in ways that model life tables do a poor job of fitting. When two-thirds of the tuberculosis victims at Canterbury were under age 30, the closest fitting curve from a model life table (called West Male Level 6) does a poor job because it underestimates early adult mortality.

That's a Hell of an Act. What Do You Call It? The Aristocrats.

A grand opportunity to test our ideas about lifespan covers an impressive thousand-year period, moving beyond England, though still limited to European aristocrats and noble men—more than 115,000 of them, from across Europe, whose genealogies and information have been tracked by the Mormon Church.[15]

The average age of death for European elites oscillates between 45 and 50 from the period AD 800 to 1400, rises to 50–55 years until 1700, then continues to rise thereafter.[16] While many nobles died of plague, especially in the fourteenth and fifteenth centuries, almost a third of noble men died in battle prior to 1550. Subtracting out the violent deaths raises the average lifespan by a few years for most of this period. Economic historian Neil Cummins speculates that the steep decline of violence after this period led to the "transformation of warrior nobles into gentlemen courtiers," a key part of the "civilizing process" leading up to the Industrial Revolution. The rise in age at death definitively pre-dates the Industrial Revolution, when longevity benefits began to spread beyond nobles to the rest of the population.

While *average* ages at death span across the 45–55 range, examining the full distribution of lifespans reveals the same nuance we've seen before: modal ages of death are in the 60–80 year range for nobles born between the sixteenth and eighteenth centuries, but more in the 50–60 year range prior to that. But, in all centuries, there are notable numbers of deaths in the 80–100 age range.[17]

Moving beyond Europe

The richest historical demographic study outside of Europe comes from remarkable Chinese genealogies of the Wang clan, covering approximately 30,000 people over a thousand-year period.[18] The Wang clan largely reflects a more privileged subset of the population. It is likely that early age deaths are underreported, and so we focus on adulthood. Mortality profiles by adult age are surprisingly consistent across birth cohorts spanning seventeen centuries! Life expectancy at age 30 (e_{30}) is an additional twenty-nine to thirty-four years, by age 40 another twenty-three to twenty-six years, and by age 60 another eleven to thirteen years. If representative, these numbers reflect better conditions than in Europe, and much steadier over a longer stretch of history. The mortality rates in China over this long swath of time looks more like England starting from the eighteenth century.

Based on the mortality information in the adult age range, I apply model life tables to make an educated guess about what mortality might look like across the full life course. Doing so yields a life expectancy at birth (e_0) of about 36 years, and a population with a modal age of death of 70 years, where a third of deaths are among those age 60 and older.

When in Rome

Let's go even farther back—to the early first millennium.

Information recorded in the *Corpus Inscriptionum Latinarum* provides ages of death for thousands of Roman citizens based on inscriptions on gravestones in Rome, and the Roman provinces of Hispania and Lusitania (today parts of Portugal and Spain), around AD 300–400 (table A4.1).[19] Ancient Rome was a rough place, with life expectancy e_0 of about 22 years, e_{15} of seventeen years, and e_{40} of twelve years. Yet, life expectancy at age 60, e_{60}, was thirteen years, just as we saw in China and on the high end in medieval Britain. Prospects among those living in Hispania and Lusitania were much better: e_0 of 38, e_{15} of twenty-six, e_{40} of twenty. A third group of more than 10,000 Roman colonists living in northern Africa showed even more favorable survival across the life course: e_0 of 47, e_{15} of thirty-seven, e_{40} of twenty-eight, and e_{60} of eighteen. These more longevous colonists were believed to be a "selected class, possessed of vigour and enterprise" who were "engaged in the healthy occupation of agriculture."

A final example comes from none other than the inventor of the statistical "correlation" and other convenient biometric tools, socialist refuser of knighthood and amateur eugenicist, the British mathematician Karl Pearson. Pearson examined ages of death of 141 Egyptian mummies from inscriptions on their cases during a period of Roman occupation over two millennia ago.[20] With good reason, he presumed these to reflect the "better classes" in an advanced ancient civilization, and not a "barbaric tribe." He calculated e_{15} to be about twenty-two additional years, and e_{40} to be about seventeen extra years.[21]

Fame, Fortune and . . . Longevity?

Another fun historical exploration across the lifespan frontier focuses on a different group of the well-to-do. These include about 300,000 famous people, ranging from King Hammurabi of Babylonia, born ~2300 BC, to Albert Einstein, born in the late nineteenth century. To be included in this database, you'd need to have lived long enough to be famous, so what we can glean tells us only about lifespans under the best of circumstances—among the richest, most influential, educated, and powerful.[22] This extends beyond the nobles and clergy we saw above, to include rich merchants, artists, statesmen, authors, and such.

Longevity among this celebrated group was about 60 years up through the mid-seventeenth-century cohort, increasing consistently thereafter to almost 70 among those born in the late nineteenth century.

As in the study of European nobles, the increase in lifespan began over a century before the Industrial Revolution. This increase came from a reduction of middle-age adult mortality more than any extension of maximal lifespan, which economic historians David de la Croix and Omar Licandro estimated to be about eight decades.

And now, let's take one last historical demographic joy ride.

Imagine more than 30,000 scientists from the territories of the Holy Roman Empire, now known as Germany, Austria, Belgium, and other countries of central Europe, but born between the fourteenth and nineteenth centuries. The economic historian Robert Stelter and colleagues compiled lifespan information of these scientists from detailed catalogs of professors and membership lists from scientific societies. They found a similar privileged longevity of living to about age 60, among those who survived to age 30.[23]

There are a few interesting exceptions. First, unlike the study of famous people, here we do see a dip in adult life expectancy during calamitous times. Military conflicts like the Thirty Years' War and epidemics like the Black Plague again knocked scientists out of their labs and cut lives short. Second, medical professionals experienced slightly lower survival and slower gains than nonmedical folks. Prior to the germ theory of disease, existing knowledge and treatments weren't enough to offset the dangers from exposure to sick patients. Who knew that bathing in mercury doesn't treat syphilis, or that the powdered feces of kings doesn't make for good medicine?

And Now, for Some Prehistory . . .

Once we venture farther in the past, to the time before written records, identifying unobservable traits that don't directly preserve (like lifespan) requires serious detective work. There's a big difference between studies of living versus dead populations. Living folks can be identified and tracked to see who lives or dies after a period of time. But burials containing skeletal remains only tell us who died. What's the difference, you ask?

If we followed 5,000 people until no one was left alive, we could come up with the usual life tables we made for the Ache, Tsimane, and other groups. If we uncovered a burial of 5,000 people, we could also develop a life table of sorts. But to determine the comparable distribution of lifespans from such a life table we would need to make some awkward assumptions. We'd have to assume that the living and dead populations had a similar age structure, and that folks died during a period of stable mortality and fertility rates, even if people belonged to vastly different cohorts. We'd also have to assume there was no net migration. In other words, the age-at-death distributions will look the same only if we assume some things that are often not true.[24]

In fact, these ideal conditions are rarely upheld. Add to that the mind-blowing fact that changes in *fertility* impact the final age-at-death distribution much more so than mortality! And of course, we can't forget the usual methodological issues surrounding skeletal samples. The first is that bones of infants and older adults tend to disintegrate at higher rates, and therefore they are less likely to be preserved, because of lower bone mineral density. And second, the ability to determine the true age of a skeleton is a holy grail that no one has yet discovered. These thorns have led to some gnarly debates over the years, in a field called paleodemography. With inspiring upbeat titles like "Farewell to Paleodemography," "Paleodemography: Not Quite Dead Yet," and "Paleodemography: Expectancy and False Hope," the residue left from debates about method and inference makes folks quick to dismiss any finding they think disagreeable.[25] As Nancy Howell once remarked, the study of ancient demography is "intensely interesting, but devilishly difficult."[26]

Given all these limitations, it's tempting to wave a hand and dismiss all the ink spilled over prehistoric lifespans. But that would be a harsh injustice against the many smart minds who have worked on the problem. So what does the wisdom of the bones (and their wise interpreters) reveal?

Prehistoric Mortality Tells a Different Story

There's an archetypal pattern observed in many prehistoric skeletal samples from the past ten thousand years, first described in the mid-twentieth century by the British American anthropologist John

Lawrence Angel, in his reconstructions spanning the Mediterranean. Prehistoric burials tell a different story from the one we've seen in living hunter-gatherers, farmers, and early Europeans. First, infant and child mortality seem surprisingly low—lower than what we've seen in subsistence populations, and yet mortality in middle adulthood, say ages 20–40, is quite high, higher than we see elsewhere.

And here's where it gets really strange. Very few specimens seem to be of adults over age 45, much less age 60. So, with this classic pattern, life expectancies are very low, about 12–35 years at birth (averaging around 20), e_{15} ranges about fifteen to thirty years, and e_{45} about three to seven years left of life.[27]

Here's a classic example. One of the most celebrated and well-studied archaeological sites is the Libben site in the Great Black Swamp of northern Ohio.[28] Every archaeologist worth their salt can rattle off ten interesting facts about the place. With more than 1,300 skeletons, it's one of the largest single-occupation cemetery collections from the Eastern Woodlands of the US, covering the precontact period AD 800–1100. Though the people there were farmers, they trapped muskrat and white-tailed deer, and fished with weirs. Given the rarity of blunt trauma or projectiles embedded in bone, violence was believed to have been modest.

So here are the stats: life expectancy at birth was estimated at 20 years, with an additional nineteen years expected once you'd reached the age of 15, and six more years upon reaching age 45. Infant mortality was believed to be relatively low (at least by Hobbesian standards), at 18%, while survivorship to age 15 is on par with high-mortality foragers, about 53%. But it's a rapid downhill slide to doomsday in later adulthood. Few survive beyond five decades.

The anthropologist and anatomist Owen Lovejoy and his colleagues were the first to study mortality patterns in Libben, and were well aware of the discrepancy between this Libben life course and the types of patterns described in chapter 3. To explain the difference, they proposed that novel infectious pathogens from European contact could have increased deaths among the vulnerable young, which would have then made the survivors more immunologically robust. Voila, higher child mortality and lower adult mortality in ethnographic populations, and the opposite prior to contact-related epidemics. While plausible, this explanation can't explain why the skeletal record looks so similar over many millennia in the Old World and the New World, where people experienced presumably different timing and histories of exposure with diverse pathogens.

Indeed, this Libben-like pattern has been seen in most skeletal series analyzed by brave paleodemographers, spanning thousands of years of modern *Homo sapiens* lifeways around the globe. Either average survival was much lower than contemporary subsistence groups and the historical populations of poets, scribes, and nobles, or something is amiss.

Bones of Contention

Were the story to stop there, we'd have to conclude that longevity increased fairly late in human history over the past two millennia— barring the occasional Methuselah or Peng Zu (born around 1900 BC during the Yin dynasty and believed to have lived more than 830 years). At the risk of offending some bone experts, I argue that multiple lines of evidence support a very different view: life expectancies rise and fall, but the *potential* for human longevity is universal to our species, and some longevity may even be present prior to our species origin some 250,000–300,000 years ago.

To gain some insight about why, let me introduce you to a remarkable study by my late colleague Phil Walker. As an early founder of bioarchaeology, Phil helped marry analyses of bones with living populations. At the Purisima Mission Cemetery near Lompoc, California, Franciscan priests kept detailed records of all native Chumash burials in the early to mid-nineteenth century. This presents a unique opportunity— to compare who died with who was actually recovered from the burials. The estimated ages of death from the cemetery record looks a lot like those for Libben: few young people, a glut of middle-adulthood skeletons, and few older adults (figure 4.1). But here's the amazing part: from the written burial records we see the exact opposite: a large

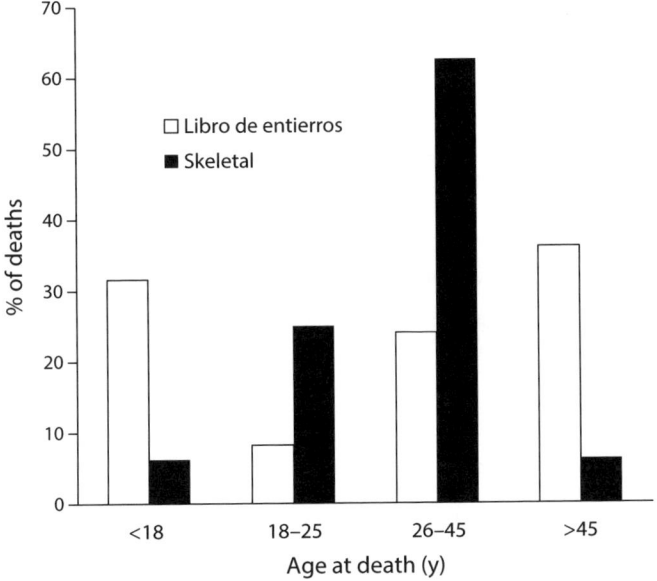

FIGURE 4.1. **La Purisima Mission, 1813–1849.** Comparison of age-at-death distributions from the written burial records ("Libro de entierros"), and from cemetery skeleton collection. Whereas the cemetery collection shows relatively few older adults, the written burial records show that older adults make up a large share of the total deaths. Adapted from Walker et al. 1988.

number of deaths early in life, and many in later adulthood; 37% of the recorded deaths were among those aged 45+, yet the skeletal record showed only 6%![29]

This simple example, afforded by the rare opportunity to compare what's in the ground with eyewitness records, illustrates two key problems with making inferences about lifespan in prehistoric skeletal samples. First, infants and elderly are grossly underrepresented—not because infant mortality was low or because no one reached elderly ages. Above and beyond effects of soil pH and moisture, which can deteriorate young and brittle elderly bones, the bones from young adults consistently preserve better. The bone mass and density of infants and children are low, hit their peak in early adulthood, and decline thereafter, especially by around age 40. This explains why bones of the very young and the old are more likely to disintegrate. Those lives remain invisible centuries or millennia later when shovels lift the dirt. Despite the burials at La Purisima being relatively recent, archaeologists reported that a third of the materials were so fragmented that they were left in the ground. Young adults were just more likely to have recoverable long bones. You can imagine how things might fare over much longer stretches of time. This story about differential preservation is simple, but it's a good one: multiplying the distributions of actual deaths from the burial records in figure 4.1 by the likelihood of preservation (assuming it to be proportional to skeletal mass with age) results in the telltale inverted-U age-at-death distribution, just like the one based on skeletons uncovered from the ground.

Bones of older women, which show declines in bone mass and density after menopause, may also preserve less well than those of older men. Other biases also result in the undercounting of women in skeletal samples. When relying on cranial features, for example, more "masculine" appearing crania are often misidentified as male. Phil Walker astutely conjectured that an earlier idea once believed to be universal—that males had higher survivorship than females in prehistoric populations—was wrong. The shortage of females could instead be attributed to these types of biases.

Though unlikely in the particular case of La Purisima, it's possible that, in other cases, bodies were not all buried in the same site. Burial customs in some cultures might place infants or elders in different locations, unbeknownst to the archaeologists, who given the usual constraints often excavate a sample. But getting a representative sample may be difficult, if not impossible. Indeed, the thirty-two skeletons disinterred at La Purisima made up just 2% of the 1,491 people recorded to have been buried there.

To conclude: if you were to assume that the age-of-death distribution from ancient cemeteries reflected the true age-of-death distribution of prehistoric populations, you'd think adult life expectancy was too low—that there were no old people.

And you'd be wrong.

Ages of the Unknown Soldiers

The issues raised above are mainly about preservation and recovery biases: the factors leading up to what generates the materials we have to work with in light of what's in the ground. But let's turn to something more fundamental: the accuracy of ages of individuals recovered from the ground. We saw how tricky this could be with illiterate folks without written records or with little use for precise ages. It's even worse with voiceless skeletons.

Luckily, given the gravity of the problem, a lot of brain power has fixated on how to improve age estimation of skeletons. Most methods rely on visual inspection of bone joints, sutures, ends of ribs, or teeth. Differences in key features of these skeletal indicators reflect developmental changes with age. Though these don't always match reliably with chronological age because factors other than age can affect the pace of biological processes. Developmental indicators—like the timing of tooth eruption or root completion, and changes in limb bone lengths—help determine decent ages of children and adolescents. Determining the age of adult skeletons, however, is more complicated, because indicators can reflect both age and aging. Senescent changes vary among populations, individuals, and across anatomical sites, leading to huge margins of error when trying to clock adults using traditional markers, like the amount of closure of cranial sutures, or morphological changes in the pubic symphysis (that's the joint between your left and right pelvic bones). As aging itself is variable, this error only increases with age.[30]

How big a problem can poor age estimates be?

Libben Revisited and More

Returning to the Libben example, a reanalysis of ages from the original study tried using a better age marker, the part of the iliac (pelvis) known as the auricular surface. Doing so raised adult ages by about five years on average. Whereas < 6% of the Libben skeletons were believed to have been ages 50 and up, the new age estimates alone shifted this percentage to 20%, a level comparable to that found in contemporary subsistence populations.[31]

Nowadays, multiple markers are used (when possible), and more sophisticated statistical techniques have been devised to make best use of these markers, and in a more probabilistic way that directly incorporates prior beliefs. This "transition analysis" as it's called involves an increasingly popular mode of models, referred to as Bayesian methods.[32] One major advantage of these methods is that they help resolve the problem of age mimicry from reference samples. Newer analyses may also help improve age estimation by incorporating how differences in physiological aging affect rates of degeneration in joints and bones. But, as things stand now, no skeletal age indicator is consistently highly correlated with age, especially at later adult ages, though there may be some promising new methods on the horizon. As of yet, however, leading experts conclude that "skeletons do not have a wrist watch."[33]

While transition analysis may not be as mainstream as one would hope, it has been applied to a number of older skeletal samples that had the classic Libben-like pattern. And the result is consistent: after adjusting age estimates, more old people suddenly materialize from the statistical ether. For example, two samples in Mexico, intensive agriculturalists at Xochilmilco around the time of conquest (AD 1521) and the urban center of Cholula during the Postclassic Period (AD 900–1500), both show the usual mountain of early adult deaths and shortage of older adults when using traditional methods. Applying the magic wand of transition analysis shifts the ages of death towards more older adults. More to the point, the modal age of adult death shifts to around age 70.[34]

Recently developed methods for improving age estimation and modeling mortality have been applied to three other well-studied prehistoric sites: Indian Knoll, Loisy-en-Brie, and Averbuch. Indian Knoll is an Archaic site in the shell mound region of Kentucky, occupied 2000–3000 BC. Loisy-en-Brie is a Neolithic rock-cut chamber in northern France dated to around 1670 BC (figure 4.2). Averbuch is a late Mississippian period site in the Nashville basin of Tennessee, occupied from the thirteenth to fifteenth centuries AD.

These newer methods have considerably revised our thinking about lifespan in two of these populations.[35] Whereas they used to resemble

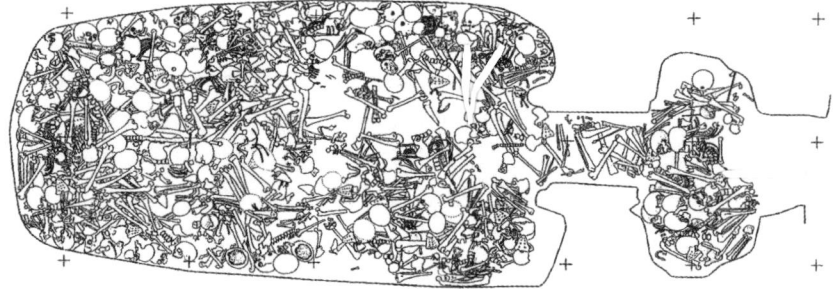

FIGURE 4.2. **The Neolithic rock-cut chamber in northern France (Loisy-en-Brie).** Dated to around 3690 BP, with some 170 well-preserved skeletons. Updated analyses reveal adult life expectancies similar to ethnographic cases of chapter 3. *Source:* Bocquet-Appel and Bacro 1997.

a more Libben-like pattern, Loisy-en-Brie and Averbuch now show mortality trajectories across the life course similar to our ethnographic cases: e_{20} is about thirty-two years, e_{40} is twenty to twenty-five years, e_{60} is twelve to eighteen years. Only Indian Knoll still showed much higher mortality levels despite using newer methods. Lyle Konigsberg, one of the masterminds behind reanalysis of these older skeletal samples, argues instead that at Indian Knoll, "many older individuals were not returned for burial in this seasonally migratory population." So, in this case, it's underrepresentation of older adults rather than underestimation of age that is still a problem.

A final example extends these same modeling approaches to Japan, but relies on a completely different age indicator. The volume of dental pulp in canine teeth reduces with age due to the continuous growth of dentin tissue. It can be assessed with elaborate technology like microfocus X-ray computed tomography (micro-CT). What's nice about this method is that the amount of error among investigators analyzing the same teeth is very low. Low "inter-observer error" is invaluable for a tool to always hit a nail on its head. Dental pulp also preserves better over long periods of time than most bones, and so relying on this indicator should buffer us against the preservation bias that makes us miss out on older adults whose bones may have disintegrated.

Here's where it gets exciting. Among Jomon hunter-gatherers from around 280–3380 BC, before the introduction of wet-rice agriculture, it was long believed based on older methods that life expectancy at age 15 was about sixteen years, with < 5% surviving to age 50. Once again, the newer methods suggest a survival pattern much more akin to ethnographic hunter-gatherers.[36] Using the newer methods, life expectancy at age 15 among ancient Jomon is now believed to be at least thirty-two more years and possibly as high as fifty-two more years, with at least 40% surviving to age 50.

Age-at-Death Tells Us More about Births Than Deaths

As if concerns about age estimates and other biases weren't enough to derail our efforts to understand the past, there's another problem with making inferences about lifespans from skeletal collections that is even more insidious. Remember that, from a burial, all you have is a collection of ages from individuals who died, the age-at-death distribution. A shattering revelation is that fertility patterns affect the age-at-death distribution more than does mortality!

This idea is so counterintuitive that many didn't take it seriously at first, but common sense often leads us astray. Mortality's effects are usually concentrated in early and late life and, to a lesser degree, in midlife, but fertility introduces new births into a population. So changes in fertility end up affecting the overall age structure of a population more than do changes in mortality.

Imagine a population in which the mortality rate is held constant, but we vary the number of children a woman bears over her lifetime. When fertility is high, there will be many young children, who typically die at higher rates. When fertility is low, few kids percolate through the population, which leans heavier toward adults. In this case, there will be a larger concentration of deaths at older ages. Comparing the age-at-death distributions between these two conditions, the difference is profound, and in the way you'd expect.[37] With lower fertility, more deaths show at later ages—higher fertility, more deaths at younger ages.

Now imagine instead we hold fertility constant but vary mortality from low (say $e_0 = 50$) to high ($e_0 = 20$). Over such a huge range of mortality differences, the resulting age-at-death distribution changes

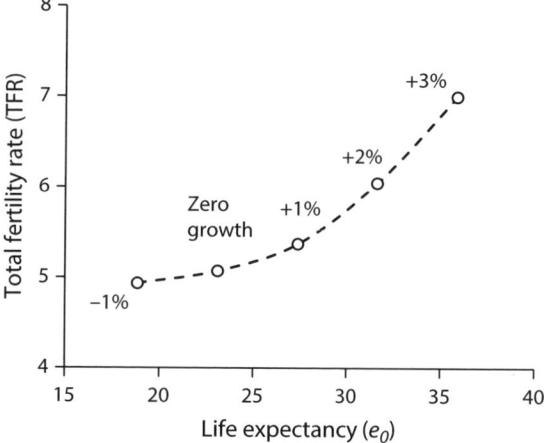

FIGURE 4.3. **Will the "real" life expectancy at Libben please stand up?** A reevaluation of prehistoric Libben site life expectancy, when considering departures from stationarity, where a population is neither growing nor shrinking. Higher population growth due to higher fertility results in a longer life expectancy. *Source:* Meindl et al. 2008.

surprisingly very little over the age range most visible to paleodemographers, that is from ages 5 to 60.

In other words, *life expectancy at birth is often not the same as the mean age of death in a skeletal sample.* The simplification to equate the two as identical has been a convenient rule of thumb in paleodemography. But whenever a population is growing or shrinking, that simplification falls apart.[38] As is now appreciated, one needs to know something about fertility or population growth to cobble together anything reliable about average lifespans.

Going back to the prehistoric Libben site, if fertility were high, say 6 births per woman, instead of a lower 5.5 births per woman, this slight change would shift estimates of survival derived from the age-at-death distribution (figure 4.3). Instead of a life expectancy of 23, it would be 31; e_{15} would move from twenty-five to twenty-nine, and survival to age 15 would move from 55% to 69%. In other words, allowing a more reasonable assessment of fertility in a classic site like Libben pushes its mortality pattern much closer to those observed in contemporary subsistence populations.[39]

To conclude, any strong belief that lifespans were terribly short throughout prehistory falls apart once you take into account the mix of sampling biases, better age estimation, and more realistic assumptions about fertility.

Going Farther Back in Time: Our Hominin Past

If these problems pervade the study of burials from recent millennia, what hope is there for understanding the deeper history of lifespan in our species lineage? Many believe it's impossible to ever know how long ancient hominins lived. Fossil specimens are hard to come by, and certainly don't reflect a "population" in any real sense. They're thinly spread over space and time.

What can save us, perhaps, is the scientific principle Ju/'hoansi demographer Nancy Howell borrowed from geology to apply to understanding past human demography: *uniformitarianism*. Recall from chapter 3 that this big word conveys a simple concept: that the processes generating observable variation in mortality, fertility, and other aspects of life history have not changed much over time. This doesn't mean mortality rates or lifespans were identical over time and space. Rather, like begets like when conditions are the same.

To apply Howell's principle to answer the vexing question of past hominin lifespan, we need to remember first and foremost that we're mammals. That fact alone helps us in making some predictions about our species. Comparing across mammalian species, many traits scale with body size. Not just physical traits like lung size or resting heart rate, but also demographic traits like gestation length, age at reproductive maturity, and many more. Larger species tend to experience lower mortality, slower development, and give birth to fewer babies—and in a patterned, predictable manner. Many of these traits are often tightly linked to each other.

The field of biology called "allometry" focuses on both identifying and attempting to explain these mathematical scaling relationships

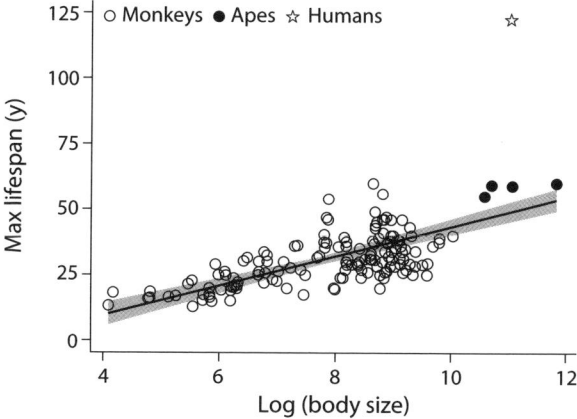

FIGURE 4.4. **Lifespan and body size.** Maximum lifespan is longer for larger-sized primates (n = 149 species). Though not shown here, primates are longer-lived than similar sized mammals. Among primates, humans have a longer lifespan than expected for their body size. *Data source:* AnAge. Data restricted to species where data quality was deemed "acceptable" or "high."

among biological traits. For example, maximum lifespan tends to increase with body size raised to the ¼ power. Since ¼ is less than 1, doubling body size from 20 to 40 kg won't double lifespan. It would increase it instead by just 19%.

If we're trying to explain human lifespan, we need to first see whether there's anything really to "explain." In figure 4.4, note that humans have a longer maximum lifespan than you'd expect just based on the typical body size of a modern human. We saw in chapter 1 that maximal lifespan is not an ideal measure, but because it's relatively easy to observe in many species (no life tables necessary!), it is the preferred metric of longevity in cross-species analyses.

Given that we're primates, it shouldn't be surprising to realize that the primate order is longer lived than other mammals. If lifespan scales with body size, is our long life explained by humans just being large primates? The answer is, Hell no! A quick analysis of a rich public database providing body size, lifespan, and other information for 976 mammalian species confirms this to be true.[40] Adjusting for body size and accounting for the fact that longer life may be detected in species that have been studied longer, apes live longer than monkeys, and humans live longer than apes. If we were typical well-studied mammals, we humans should live a maximum of only 26 years. If we were just typical monkeys, we'd live a maximum of 49 years. And, as typical apes, a maximum of 62 years. Both of those are well under the maximum observed 122.5 years, and the ~90 years observed in subsistence populations. So that's three to six more decades than expected if we really were just hairless apes!

Climbing Our Ancestral Tree

Our species is longer-lived than it should be, but is it possible that maximum lifespan was lower among our Paleolithic ancestors?

The tree of our hominin origins is a complicated one, as it is revised and updated after every new discovery. No single tree is universally agreed upon by experts. Here is what we can say to briefly orient us to the deep time depth of our ancestral history: Several species of hominins likely lived around the same time in eastern Africa several million years ago. Many died out long ago. But a few smaller apelike Australopiths (literally, "southern apes") flourished during the period known as the late Pliocene and early to mid-Pleistocene, some 2–4 million years ago. This includes the famous Lucy, named for her prescient love of the Beatles. It is likely that Australopiths gave rise to our genus, *Homo*. Perhaps you've heard of *Homo habilis* (handy man), so named because of the early association with the making of stone tools. And *Homo erectus* (upright man), a bit on the tall side, proliferated over much of Africa and Eurasia between about 2 million and 110,000 years ago.

Neanderthals, our closest extinct human relative known to have roamed throughout Europe and western Asia, and whose DNA makes up to 2% in descendants today,[41] were not the brutes they're sometimes portrayed as being. They buried their dead, and even made art, which arguably is the usual indicator of "behavioral modernity."[42]

Intended burials make it easier to recover a larger number of specimens. From a landmark analysis of Neanderthal mortality patterns, paleoanthropologist Erik Trinkaus argued that roughly half died before age 20, with only 10% living past age 40. Just like with many prehistoric skeletal samples, the peak numbers of dead were young adults, with very few older adults.[43]

One approach to skirting the problems with poor age estimates, poor preservation of infants and children, and the geriatric loss of skeletal material in older age is to clump individuals into crude categories, like young adults (~20 to 40 y) versus older adults ($\geq$ 40 y). The percentage of specimens from the Middle and Upper Paleolithic (300,000 ya to 30,000 ya) that are older adults is low—lower than observed in contemporary subsistence populations, and even lower than many of the prehistoric populations I described in the previous section. Trinkaus argues that high mobility would've made it hard for older adults to keep up, and so they were left behind to die. Their remains would've been a feast for hungry carnivores, not just hungry paleoanthropologists millennia later.[44]

Other attempts to construct ratios of older to younger adults based on dental wear patterns in larger samples also seem to show minimal representation of older adults among Australopiths, early *Homo*, and Neanderthals. Only among Early Upper Paleolithic Europeans (~40,000–50,000 ya) does the ratio become more favorable for older adults, consistent with an increase in lifespan.[45] During the upper Paleolithic, populations were expanding and spreading. It's tempting to attribute population growth and the behavioral innovations underlying modernity to all the benefits that come from more adults having longer adulthoods, a time where grandparents really start to matter. As we'll explore in chapter 5, longevity in humans, along with multigenerational cooperation, symbolic and tool-making behavior, and cumulative culture are distinguishing features of our species.

Not Just Brawn, but Brains

Human longevity is an outlier compared to the longevity of other mammals, and even primates. But there is another feature that is also an outlier—and it is more tightly linked to longevity than body size. I'm talking about brain size. It has long been recognized that longer-lived

FIGURE 4.5. **Brains and longevity.** Fukurokuju, the
God of Wisdom and Longevity, is having his large
head shaved to ensure a long, prosperous life. This folk
painting by Shiokawa Bunrin is part of a large hanging
scroll made in 1871. *Photo source:* Michael Gurven:
Santa Barbara Museum of Art.

species tend to have larger brains. And the larger-brained of two species
of the same body size will have a longer life. (This link between brain
size and lifespan even comes up in Japanese mythology, where the God
of Wisdom and Longevity, Fukurokuju, was distinguished with a very
large head [figure 4.5]. He's often depicted holding a book containing
the lifespans of all humans.)

Over a half century ago, the mathematical biologist and allometrics
guru George Sacher realized that brain size alone is a better predictor
of longevity than body size, and that both together can explain much of

the variation in lifespan across species, and in maturation rate, too. For this reason, he referred to the brain as the "ORGAN OF LONGEVITY." (He did the capitalizing, not me.)

It's fascinating that an organ that makes up about 2% of adult body weight is so critical. That smallish organ soaks up about a quarter of your daily energy supply. Our human brain is almost triple the size of that estimated for Australopiths, and for chimpanzees, our nearest primate cousins. It's also about 30% larger than that of *Homo erectus*. Armed again with the power of uniformitarianism, Sacher's insight was to take equations relating a species' maximum lifespan to its brain and body size and apply them to what was known at the time about extinct hominins. He reckoned that maximum lifespan had doubled from the time of Australopiths to *Homo sapiens*, from 47 to 90 years of age.[46]

Building on the latest estimates of ancient hominin brain and body sizes, I argue that we can make educated guesses about the maximum lifespans of the long-extinct ancestors of our ancestors.[47] Figure 4.6 presents lowball and highball estimates from *Ardipithecus ramidus* living around 4.4 million years ago, several species of Australopiths, through early *Homo* species 2.6 to 0.7 million years ago, covering all the fan favorites: *H. ergaster, H. habilis, H. rudolfensis,* and *H. erectus.* Alongside these more distant ancestors, I include Neanderthals and *Homo sapiens* throughout the Pleistocene. Much as Sacher showed decades ago, figure 4.6 shows that the maximum lifespan likely increased over the course of hominin evolution. The range of mid-40s to low 50s among our earliest ancestors over 3 million years ago resemble observed ranges among wild chimpanzees, as to be expected given their similar brain sizes (yet another reason that chimpanzees are often used as a heuristic model of Australopith life history).

Maximum lifespan among early *Homo* species rises about a decade between 1 and 2 million years ago. Some have argued that a more humanlike lifespan has its origins among *Homo erectus* over 1 million years ago, when our ancestors began exploiting roots and other underground storage organs in the open savannas[48] (see chapter 5). By the Middle Pleistocene roughly 200,000 to 800,000 ya, predicted maximal lifespan

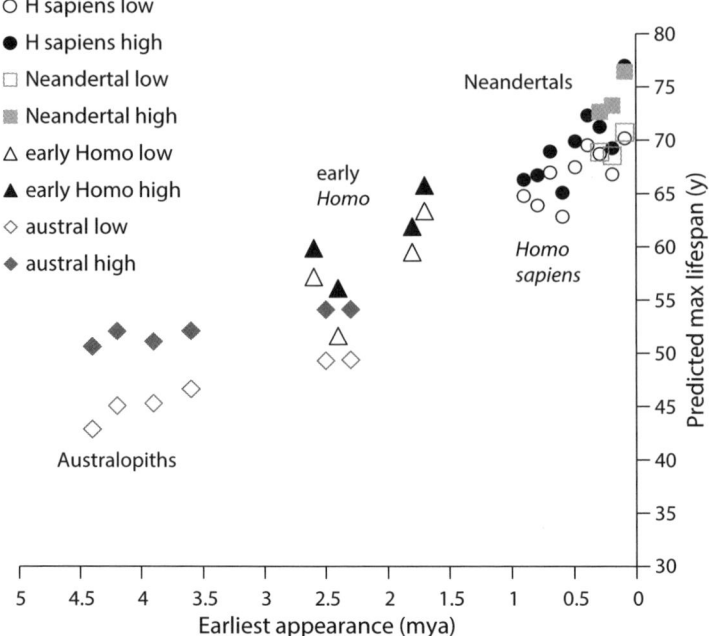

FIGURE 4.6. **Predicted maximum potential lifespan among hominins.** Estimates of maximum potential lifespan (in years), based on average estimated body size and brain size for (a) *Homo sapiens*, (b) Neandertals, (c) early *Homo* species, (d) Australopiths. Low and high estimates based on using different equations. See endnote 47 in chapter 4 for details about samples and analysis.

is squarely in the 60s to 70s. And it is certainly in the 70s by the late Pleistocene about 130,000 ya.

So, from these analyses, we would predict a humanlike adult survival curve at the latest by the time of the earliest evidence of our species, believed to be about 300,000 ya. This older date is based on the discovery in 2017 of a mandible and flint artifacts found in the earliest site at Jebel Irhoud in western Morocco.[49] A more optimistic, or hopeful, claim pushes humanlike longevity even further back to the large archaic *Homo* expanding out of Africa. This would include Neanderthals, who these days many experts classify under the same genus as us.

Since we're having fun speculating, let's explore what these predicted maximum lifespans for ancient hominins might mean in terms of more familiar measures: life expectancy at birth and at age of adulthood (e_0, e_{15}). One approach, which admittedly stretches uniformitarianism a bit, is to look at the relationship between maximum lifespan and life expectancies over the same 270-year period of Swedish history that we've now come to know and love. I add the subsistence populations from chapter 3 for good measure, and also seven well-studied primate species.[50] The latter is critical because no living human population has a maximum lifespan lower than seven decades. Our best shot here is to include nonhuman primates, because we know both maximum lifespan and life expectancies for these close relatives of ours. Here I define maximum lifespan conservatively as the age at which only < 1% of the population is still alive. I say "conservative" because the top 1% is always younger than the oldest individual whenever the population is larger than 100 individuals.

Figure 4.7 shows e_0 and e_{15} plotted as a function of maximal lifespan. In humans, maximum lifespan ranges from about mid-70s to over 100. This three-decade increase in maximal lifespan coincided with a six-decade increase in e_0, from about 20 to above 80, and a four-decade increase in e_{15}, from about twenty-five to about sixty-five additional years of life. This mostly confirms what we've seen already: that maximum lifespan varies a lot less than life expectancy. *And all the improvement beyond a maximal lifespan of ~95 y occurred only after the mid-twentieth century!* An adult lifespan that would surpass the age at menopause (that is, e_{15} > 35 years) is observed when maximal lifespan reaches about eight decades. From the guesstimates I showed in figure 4.6, we can speculate that lengthy postmenopausal lifespans might only have been common in anatomically modern *Homo sapiens*.

In addition, the range of life expectancies observed in subsistence populations (e_0 of 28 to 38) corresponds to a maximum lifespan of about eight to nine decades. This is just what Sacher predicted maximal lifespan to be for our species, and meshes with my field observations among Ache and Tsimane, and what others report for the Hadza and Ju/'hoansi. The 15–20 year e_0 range long believed to represent prehistoric humans over much of the past 10,000 years corresponds to maximum lifespans that are just too low—in the 45–60 range. That's the low range predicted

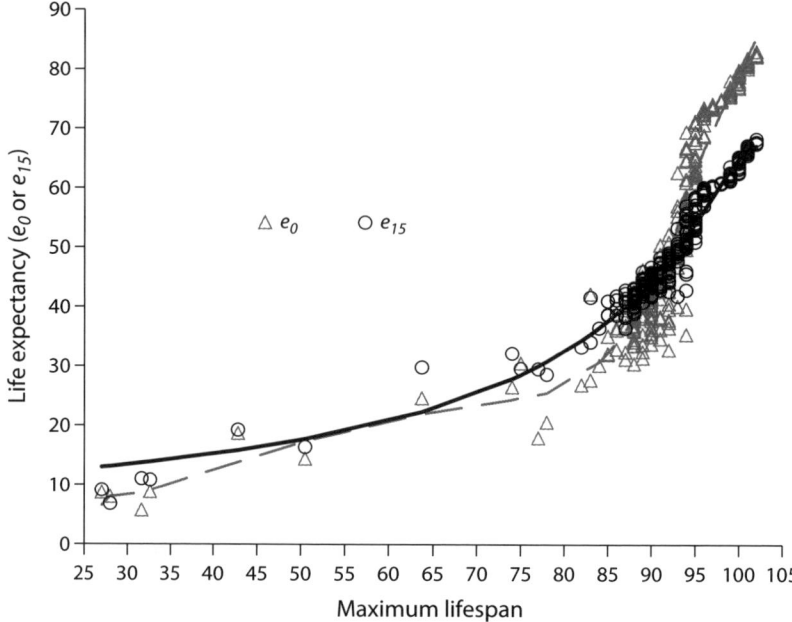

FIGURE 4.7. **How does maximum lifespan relate to life expectancy at birth (e_0), and at age 15 (e_{15}).** Maximum lifespan for humans defined here as the age x when proportion surviving in the population (l_x), is <1%. Data include Sweden 1751–2019, ethnographic hunter-gatherers and horticulturalists from chapter 3, and seven primate species (sifaka, muriqui, capuchin, macaque, baboon, gorilla, chimpanzee) (see text).

for Australopiths and early *Homo*. If early *Homo* had a maximum life span in the 60–65 year range, and late Pleistocene *Homo* in the 65–75 year range going up past 100,000 years ago, then surely *maximal* lifespan in the last 10,000 years shouldn't be lower! These discrepancies further support the notion that inferences about human lifespan based on classic paleodemographic findings are wrong.

If our cross-species uniformitarian exercise is not too brazen, then we can conclude that the short maximum lifespans of our remote hominin ancestors would correspond to e_0 and e_{15} of ~25 years for the late Pleistocene *Homo*, ~20 y for early *Homo* in the early-middle Pleistocene, and a low of ~15–18 y only for our more distant great-grand-relatives, the Australopiths.

Broader Lessons

Contrary to the idea that life expectancy saw no trends prior to the Industrial Revolution, we've seen that life expectancy has fluctuated over long stretches of hominin evolutionary history, and that the whims of infection, raiding, and starvation—or plague, war, and famine, as they're called in agricultural societies—have struck down people before or during their prime.

But this backdrop of high mortality leading to short *average* lifespan should not be confused as evidence against our consistent longevity potential. Throughout, we see a long-standing potential for humans to live up to and beyond seven decades. Such an idea may not seem controversial, but old ideas die hard, and those who should know better sometimes forget. Even the great evolutionary theorist George Williams claimed that older adults were rare in the past, an "artifact of civilization." He based this on the French paleontologist Henri Vallois's study of Stone Age skeletons in 1937. The esteemed Vallois had concluded in a review of Pleistocene lifespan that "few individuals passed forty years, and it is only quite exceptionally that any passed fifty."[51]

In his classic 1957 paper on the evolution of senescence, Williams wrote, "In very primitive conditions, such as prevailed throughout almost all of man's evolution, post-reproductive individuals were extremely rare."[52] He might be forgiven, given the state of knowledge then. But he doubled down on this belief when reflecting on human lifespan four decades later. Instead, we've seen that when conditions are good enough, life expectancy by age 15 can exceed four decades, and by age 40 can exceed two decades. High fertility, combined with high mortality, results in populations in which the total number of older adults may not dominate the scene. But they're consistently there. And the ancestral inhabitants of at least one cave would agree.

In 1937, the same year of Vallois's study, a rare treasure was discovered by the French scientist León Péricard and archaeologist Stéphane Lwoff in the Vienne department of western France. Hundreds of etchings carved into limestone slabs dated to about 15,000 years ago were uncovered in a cave that was once a deep rock shelter. France is littered with

FIGURE 4.8. **Cave art from La Marche, France.** From among the many lifelike human figures carved into limestone slabs, evidence of older adults, dated to around 15,000 ya. As published in *Les gravures de la Marche. II: Les humains* by Léon Pales and Marie Tassin de Saint-Péreuse, published by Orphrys, 1976.

cave paintings, so what's special about this one? Not only were more than 150 human figures depicted on the slabs, but these figures were strikingly detailed. Among the figures are what clearly look like older adults, with wrinkles and balding pates (figure 4.8). The engravings are so vivid that some have questioned their authenticity, but there is no solid evidence that these were fakes.[53]

Postreproductive-age elders appear to be absent from the data only when researchers rely wholly on skeletal assemblages. There are no "magic bullets" to distill what we would ideally want from the skeletal and fossil record, though improvements in age estimation, assumptions, and modeling techniques have made a difference—all in the direction of prehistoric folks looking more like the ethnographic populations of chapter 3.[54] Where written records and skeletal data exist for the same period, lifespans coming from the written records look more like those of ethnographic foragers and farmers, whereas from the skeletal records they look closer to chimpanzees. What's critical here is the method, not how ancient or recent are the populations. As Nick Blurton Jones astutely summarized when reviewing the pitfalls of paleodemography, "written people look like contemporary people, bone collections look like something else."[55] Unfortunately, recordkeeping wasn't in fashion over two thousand years ago.

No one is arguing that there were a bunch of 90-year-old hunter-gatherers roaming Eurasia 300,000 years ago. Even living hunter-gatherers and subsistence farmers beyond age 85 are rare. It is the number of adults in the 50–75 year age range (what gerontologists call the "young-old") that is most revealing, and which I'll talk more about in forthcoming chapters. People were living that long in every documented population in the written record, but inconsistently in bone populations. Again, Blurton Jones invites us to wonder: "Why were ancient Chinese and the population of Roman Egypt able to live as long as Hadza, while the bone population of Athenian Greeks could not?"[56]

As we'll come to see, elders aged 50–75 in small-scale societies are still pretty vigorous, and important group members. This contrasts dramatically with observations of our closest primate relatives in the wild—chimpanzees. Famed primatologist Jane Goodall considered chimpanzees at Gombe as "old" when they reached about age 33 years. Other observers have remarked that chimpanzees by their mid-30s show evidence of wasting and frailty, combined with the "external indications of senescence include[ing] sagging skin, slowed movements, and worn teeth."[57] If our long-gone Australopith ancestors were chimpanzee-like, as believed by many experts, then it's safe to say that they aged more rapidly and died earlier than both ancient and modern *Homo*.

One Final Protest: Why Aren't We All Hadza?

The lower life expectancies once commonly reported for the post-Neolithic Age of Pestilence and Famine may be wonky outliers, unlike what I showed for contemporary hunter-gatherers like the Hadza, Ache, and Ju/'hoansi. Yet there's still a problem. Most well-studied hunter-gatherers and small-scale farmers show clear signs of booming population growth, on average about 1% per year. Some groups, like the Ju/'hoansi or the Hiwi, were at near-zero growth when studied, but others were growing faster than the price of US postage stamps.

A 1% growth rate sounds trivial, so what's the big deal? It means that regardless of its initial size, a population will double in just sixty-nine years.[58] The Ache growth rate of 2.6% suggests a doubling time of just twenty-seven years, and the super-fast-growing Tsimane, at 3.8%, should double in under two decades. When I first started working with Tsimane, their numbers were estimated to be about 7,000. Every time I present a talk or write a paper describing the Tsimane, I need to update that number. At their high rate of growth, the current ~17,000 Tsimane population will overtake the population of a country like France by the year 2252, and the 333 million-strong United States by the year 2329!

Such rapid growth couldn't have been sustained over long stretches of our species' history. For that matter, the negative population growth implied by very short lives from many skeletal assemblages wouldn't be viable, either. The long-term global population growth rate was largely near zero for much of our species' history, with accelerations evident only in the last half millennia. Most population growth worldwide has occurred just in the last century. While it's not true that there are more people alive today than have ever lived and died,[59] it is nonetheless true that the long-term average population growth rate was near zero for most of our evolutionary history.

In order to have achieved zero growth, populations in the past must have differed in key ways from most of the groups we focused on in chapter 3. Either fertility was well below that ever observed in most natural fertility populations, to just four births over a woman's lifetime, or survivorship was well below that ever observed among contemporary subsistence populations (down to 41% survival to age 15). While reducing both fertility and survivorship at the same time may be a more realistic proposition for populations in the past, doing so still wouldn't bring most fast-growing subsistence populations close to zero growth.

As I mentioned in chapter 3, it's possible that recent conditions may be unrepresentative of the past. Warfare may have been more common, or nutritional status and fertility may have been lower. But another possibility is that groups periodically experience booms and busts.

One version of this we're sadly too familiar with: virgin soil epidemics and targeted homicides against Indigenous natives throughout the New World during periods of European contact and conquest. An analysis of 117 contact-related epidemics in Greater Amazonia over the past century shows that roughly a third of native groups perished within a few years of the epidemics spreading throughout their territories. On a positive note, fewer had perished during the epidemics that occurred more recently than in those that happened much earlier.[60] What's extraordinary is that, within a decade following these catastrophes, many Indigenous populations throughout Brazil, Venezuela, and Colombia showed positive growth again, and many are growing rapidly, even rebounding to surpass their pre-epidemic population numbers.

I worked with my quantitatively gifted postdoc, Raziel Davison, to evaluate whether the different demographic scenarios above could reconcile the positive growth observed in many foraging groups today, with the requirement that long-term growth must have been near zero. In other words, we set out to resolve what Nick Blurton Jones called the "forager population paradox." Our analyses suggested that only slight altering of fertility and mortality is needed, *if population crashes occur periodically.*[61] Evidence suggests that climate varied widely throughout the Pleistocene and into the Holocene epoch,[62] but cultures varied in the degree to which they were buffered. The extent to which past foragers typically experienced increasing, declining, or zero growth in past environments is just not known. Short-term booms and busts are entirely consistent with long-term near-zero growth.

The important lesson for us is that variability in the frequency and extent of devastation from occasional catastrophes, and potentially lower fertility in the past, should not affect the inferences we've made about the potential lifespan of subsistence populations and, by extension, our ancestors. For example, in many countries, the COVID-19 pandemic caused the biggest drop in life expectancy since World War II.[63] In one dreadful year from 2019 to 2020, male e_0 dropped by a whopping 2.2 years in the US, offsetting all the gains from 2000 to 2014. Despite this devastating drop under devastating circumstances, no one would argue that aging has accelerated, or that US lifespans have been fundamentally altered. Similarly, a higher occurrence of catastrophes in the past would lower e_0 but shouldn't fundamentally alter the potential for longevity. More catastrophes in the past, however, affect how and under what circumstances skeletal materials are found.[64]

Two European examples illustrate how catastrophes can affect some metrics of lifespan, but not the critical ones. In 1882, there was a highly contagious deadly measles epidemic in Iceland, courtesy of an infected carpenter visiting from Copenhagen.[65] In 1773, Sweden also fell victim to infectious disease—mostly smallpox, but also measles and dysentery, combined with malnutrition from famine. In both countries, life expectancy e_0 dropped in half! In Iceland, it went from 35 years in 1881 to a mere 18 during the epidemic, before climbing back up to 30 in 1883. In Sweden, it went from 35 in 1771 to a paltry 18 years, before rising to 41 the following year.

Despite the large effects of catastrophe on e_0, the telltale signatures of potential longevity remain. In Iceland, adult life expectancy varied relatively little: e_{30} was twenty-seven during the measles year, but thirty-three one year before and thirty one year after. Life expectancy at age 50 (e_{50}) varied even less: nineteen plus or minus a couple years. The modal age of death was just a few years younger during the measles year.[66] In Sweden, adults had a 60%–70% chance of reaching age 50 in years just before and after the calamity, and adults over 50 made up over a third of the living population. But in the catastrophe year, less than half of adults survived to age 50, and postreproductive-aged adults would've made up ~16% of the living population—low, but still higher than described in prehistoric lifetables with similar life expectancy at birth, where adults over 50 are said to make up <5%.[67]

Even if disasters were more common in our past, the overall impact of these catastrophes on actual cohorts is quite small. Despite increasing period mortality rates and lowering life expectancies, catastrophes have almost no effect on the modal adult lifespan of about seven decades.

5

Why Long Lives?

TIKUARANGI STRETCHED his arms to the forest canopy. He was pulling an arrow fletched across twine, its sharp staircase-like head peeking out from across a long, arched black palm wood bow. Then, in a flash, the arrow soared into the tangled branches of the canopy, followed by another arrow, and another. Seconds later, I heard the hollow thud of two capuchin monkeys falling to the ground.

This was my first forest trek with the Ache, while still a green 23-year-old graduate student. We had been trudging through patches of rough secondary forest and marshes for two hours. I had never been a hunter, but I was hungry, delirious, and experiencing what I imagined to be the thrill of the food quest. Excitement surged through my tired muscles. Here was hunting and gathering for real, as a way of life, I thought. I even made the stupid mistake of bringing a cumbersome bow and arrow just "to see what it was like." (Did I mention that Ache bows are about 2 meters long?)

Four years later, I'm coming back to camp from a hunt with Fidel, this time in the Bolivian Amazon with the Tsimane. There are more than ninety Tsimane villages, but at the time, I was living with the Tsimane who inhabit the deep forest away from major rivers, and hunt more than they fish. It was the rainy season, when constant downpours knock out all the makeshift bridges over the streams that cross the territory, rendering these villages inaccessible even by logging road. My pants were ripped to shreds, and my back was soaked with sweat and blood. For the last two hours I had been carrying a large twenty-pound rodent with whitish spots running down its red-brown fur. It was a *paca*, or *naca'* in Tsimane language. Exhausted, I dropped the paca by Fidel's kitchen, and stumbled back home, collapsing in a hammock. Minutes later, Fidel came to me looking disturbed.

"Why don't you want it?" he asked. I was confused. "Want what?" He stared at me, irritated and bewildered, and then left. I didn't understand how I had given offense. I found out later, while drinking *shocdye'* with Fidel's brother. In addition to the paca, Fidel had killed and carried two ring-tailed coatimundis (like long-snouted raccoons). Because I had carried the paca back, I was entitled to take at least half of it. If I didn't want to take it with me, proper etiquette required that I promise to come back and eat communally with Fidel's family once it was cooked. In fact, I was secretly hoping to later share in the feast, but I didn't want to be presumptuous. What I thought was polite courtesy was perceived as a distaste for the local cuisine and a disregard for sharing norms. Simple rules detailing who-is-entitled-to-what encourage group members to assist others during difficult and effortful activities. Many essential tasks in daily life cannot be done, or done well, without helping hands.

Years of training in human ecology (let alone common sense) have taught me that a simple question is key to understanding human adaptations: How do we find food and comfort? It's an immediate concern whenever we find ourselves lost or in a new place. Once our ancestors were no longer confined to the trees, but instead roaming the drying African savannas, they needed new ways to make a different kind of living. This transition is very much linked to our capacity for longevity.

So far, I've hopefully convinced you that humans have had the potential for long life since the origin of our species. In addition, middle-aged and older adults have been ever-present and sufficiently well represented. More to the point, they were vital and necessary, not just luxuries bestowed by good fortune. A postreproductive life stage, I argue, is an essential part of the human life cycle that evolved because of the many contributions older adults make to the fitness of their kith and kin. The story of our longevity is a fun way to explore what it means to be human, and to understand *why* we live long lives, we have to focus on *how* we survive and thrive through multigenerational cooperation.

Dollars and Sense

In the wealthier parts of the industrialized world, we forage in the factories, office buildings, restaurants, and other businesses where we exchange our labor for the income to buy the things we need to provide for ourselves and our family. And even after the early years of diapers, little league, Girl Scout cookies, and junior high proms, kids are still considered "dependents" until about 18 years of age, around the time they graduate high school. Add in extra schooling, housing crunches, and underemployment, and the transition to adulthood can extend even further—"25 is the new 18."[1] This delayed entrance to functional adulthood contrasts with images of Oliver Twist–like children toiling miserably in the coal factory at age seven. Or of the many small hands contributing to the family farm.

It turns out that a long, slow delay into economic and social adulthood is pretty standard, even for hunter-gatherers and horticulturalists. In all populations ever studied, children don't fully feed or care for themselves. Sure, Ache kids adeptly snatch ripe fruit even high up in trees, collecting up to five times as many calories during the fruiting season as during other times of the year. Same goes for Hadza. Tasty and refreshing as they are, fruits don't make up a huge part of any group's diet, and so children still have to rely on adults to fill their bellies with protein and fat.

When infant Maggie is scanned at the supermarket checkout in the opening credits of *The Simpsons*, the flash of the cash register reveals the high cost of raising a child from birth to age 18, at least in 1989.[2] Similarly, the total food consumption of hunter-gatherer children from birth to independence is huge, ranging from 8 to 12 gigacalories.[3] Even if adolescents could cover half their daily caloric needs, that's still a level of need more easily described as more than 29,000 cans of Coca-Cola, or 21,000 Krispy Kreme glazed donuts. Calories aren't cheap, and children need help to get them.

What's for Dinner?

How hard can it really be to forage for food if you're a bona fide forager? Let's look again to chimpanzees to speculate as to what life and life history among our common ancestors might have been like. It turns out that juvenile chimpanzees can find enough fruit, seeds, flowers, insects, and other foods to meet their daily needs within just a few years of being weaned from their mothers. Imagine if your 5-year-old was as independent as our chimp-like ancestors at that age! My son, Ollie, is now 5 years old. He's decent at finding the hidden ripe wild blackberries behind the fence near the perimeter of his school, and by now could probably find his way back home if he had to walk alone. But he also recently got a tiny plastic flower stuck way up his nose that I had to extract with tweezers, and he has a hard time reaching food in the back of the fridge. Hardly independent.

The implication is that our prehuman ancestors may have had an easier time making a living, given the simpler diet we observe in chimpanzees. Both chimpanzees and hunter-gatherers are foragers, but much of a chimpanzee's diet consists of ripe fruit. Humans, on the other hand, are really good at finding the most nutrient-dense calorically rich foods that their environment has to offer, even if rare or hard to find. This is true whether food comes as mobile packages swinging from the trees, like a troop of capuchin monkeys, or is hidden underground, like starchy roots and tubers. Ripe fruits are harder to acquire than unripe fruits, since you need to know when they're available and when they're ready to eat, and you need to compete with others hungering for those same bright jewels sparkling in the trees. But when it comes to targeting mobile prey that can think, hide, evade, maneuver, and sometimes attack—humans do this more than any other primate.

There are many mistaken beliefs about what hunter-gatherers eat, and legitimate questions over whether they might be more appropriately labeled gatherer-hunters. But among nine hunter-gatherer societies where all food coming in to camp was carefully recorded and weighed, meat from hunting wild animals accounted for 59% of the total calories produced, ranging from a low of 26% among the Gwi, a San group of the Kalahari Desert similar to the Ju/'hoansi, to a high of 79% among the Onge of the Andaman Islands (figure 5.1).[4] Certainly nine groups can't be representative of all contemporary hunter-gatherers, much less our past ancestors, but it's clear that our omnivorous predilections lead us to crave a broad, eclectic menu—and one that includes meat and fish when possible.

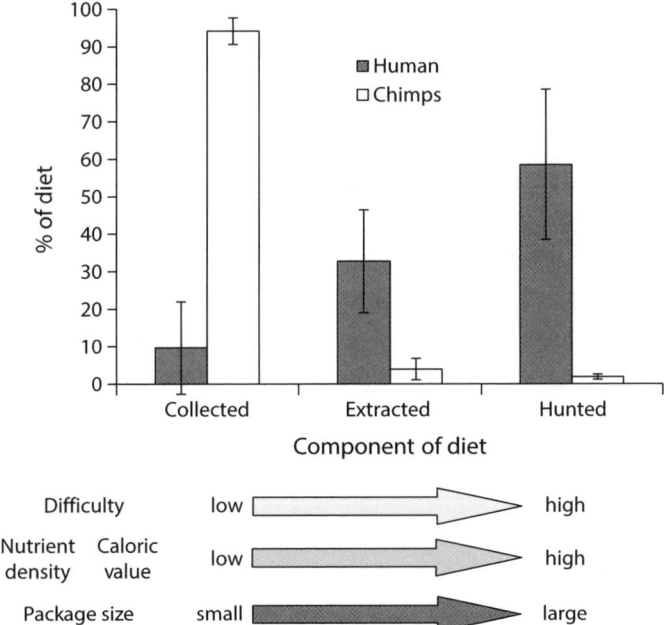

FIGURE 5.1. **The diets of hunter-gatherers and chimpanzees.**
Human diets emphasize hunted and extracted foods, while
chimpanzee diets emphasize collected foods. The foods humans
target tend to be difficult to obtain, nutrient-dense, and more
amenable to sharing. *Data source:* Kaplan et al. 2000.

A cruder analysis of more than two hundred hunter-gatherer societies shows a similar pattern: three-fourths of these groups derived over
half of their daily menu from animal foods. One of the biggest factors
affecting what you eat is not personal whim, but just what happens to
be locally available: animals and fish make up the vast majority of foods
consumed in higher latitudes, and plant foods are more important
among those living closer to the equator.[5] Figure 5.2 shows some of the
delicious foods I've been fortunate to eat while living among the Tsimane. My point here is not to argue that meat is more important than
plants, or that one diet is inherently more natural or paleo than another.
No single charmed paleo-diet (or even PaleoDiet™) will extend human
lives (see chapter 9 for more juicy details). We're omnivores, and importance can't be reduced to percentage of calories.

FIGURE 5.2. **Common Tsimane foods.** (*top*) A teenager and two children smoke sabalo fish (*vonej*). (*bottom*) Palm fruits (*vej*) at the height of their season. *Photo credits:* Michael Gurven.

Hunting, Gathering, and the Origins of Science

Much ink has been spilled describing the vastly creative and intricate ways in which hunter-gatherers make a living. Classic ethnographic films help us visualize parts of the process. In the classic John Marshall film, *The Hunters*, a Ju/'hoansi hunter, Toma, tracks a giraffe for four days and must "think" his way to find the right path. He puts himself in the mind of the injured giraffe. I once watched Bepurangi, one of the fastest-moving Ache hunters, digging a ditch for two hours, covered in dirt, just to drag out a burrowed armadillo. Another time, I watched Leóncio, a 54-year-old Tsimane hunter, carefully block off all possible escape routes with logs, then use fire-drenched leaves and smoke to panic an anteater right into his machete-wielding path.

Skills apply equally to gathering. Anthropologist friends Brian Wood and Alyssa Crittenden write about the feats involved with finding honey among the Hadza. Honey isn't just an occasional treat but, when in season, can account for half of all calories eaten. Humans are wise to exploit the meticulous labor of bees, who collect and process the nectar from flowering plants into glorious sweet honey. It's one of the most energy-dense foods around, ranging from 3,050–6,680 kcals per kilogram. Sure, you can get 7,000 kcals if you eat a kilogram of butter, but there's probably a sound evolutionary reason why that sounds disgusting.[6] And not just Hadza but—whether desert-dwelling, savanna-sauntering, or forest-foraging—many subsistence groups around the world rely on honey. Combs are also rich in fat and protein when you eat the grublike larvae inside, as the Ache and many other groups do. Honey is usually located high up in trees. Despite losing our arboreal tree-limb dexterity, humans are surprisingly adept at climbing trees.[7] Or in the Ache case, chopping them down. Hadza climb large baobab trees and weather the onslaught of stings (figure 5.3). They also smartly follow the honeyguide bird to lead them to their next score of rich honey.[8]

To accommodate the broad human diet, hunter-gatherers employ a wealth of knowledge and strategies, as the examples above nicely illustrate. They make and use tools, from digging sticks, fishing nets, and hunting snares to bows and arrows, baskets, and so much more. They

FIGURE 5.3. **A Hadza man climbs a baobab tree to get honey.** *Photo source:* Brian Wood.

beat, mash, slice, and sift to process and cook foods to reduce fiber, eliminate toxins, or extract the nutritious morsels from some inedible packaging. All that processing means we don't need to have the same huge guts that chimpanzees have for processing bulky fibrous foods. Hunter-gatherers also cover huge distances daily to get all the goods, but without a shopping cart. For example, Ache and Hadza on trek cover an average of 5–9 miles per day on foot, roughly two to three times the distance covered by chimpanzees[9] (and up to four times what the average American adult walks per day). Pooling their collective knowledge, Ache are privileged to dine from an expansive menu of seventy-eight different mammals, fourteen fish species, twenty-one reptile species, and over 150 species of birds.

Too Hard to Handle?

When humanity shifted from an ancestral chimpanzee-like diet of ripe fruit, leaves, insects, and the occasional colobus monkey to one covering a broader range of plants, animals, and eclectic etcetera, it was a total game changer—and not only because of the reasons I've mentioned. Adopting the human diet also completely changed how people interacted with each other. While the human ecological niche features lovely food bonanzas, the unpredictability of the food quest often means coming home empty-handed after a long day's toil. This is most common for hunting. Ache hunters return to camp empty-handed 40% of the time, Hiwi hunters 65% of the time, and Hadza, when they are targeting big game, 96% of the time. Imagine working seven tiring hours with no paycheck. Tsimane, though sometimes armed with shotguns and bringing with them capable hunting dogs, are still unsuccessful almost half the time. Even fishing, usually considered an easier activity, whether by bow and arrow or by hook-and-line, is a bust a quarter of the time.[10]

The high failure rates of hunting have led some to argue that hunting is not a reliable way to provision children and others. As it's largely men who hunt, we might question their intentions. Hunting is a challenging and demanding contest—a gamble—where success might signal how awesome you are, increase your status, and help you attract mates and business partners. In my graduate school work, I became entangled in debates about what motivates hunter-gatherer men in their choice of subsistence activities, and what motivates the hunter-gatherer women who prefer husbands or mates who are good hunters. Controversies like these are exciting, as they often spur a flurry of studies to figure out which way is up.

The bottom line is that meat is a valuable commodity, and the pursuit of meat is worthwhile, regardless of the difficulties just mentioned, and no matter whether it is consumed alone, by family and friends, or offered up on the informal community market. It should come as no surprise then that meat makes up a substantial part of the diet of most hunter-gatherers. There are no vegetarian hunter-gatherers by choice. And meat is never eaten begrudgingly. On the contrary—meat is highly desired everywhere. Many groups even use separate vocabulary to voice their "meat hunger," like the Mbuti of central Africa (*ekbelu*). I've heard many Tsimane complain of hunger despite being surrounded by tasty plantains hanging from the rafters of their house. A few days without meat is enough to motivate complaints. In the film *The Hunters* that I mentioned earlier, Toma is encouraged by his wife to hunt because her "breasts are lacking milk." Among the Maori, when hunters brought in large quantities of fish, birds, or rats, New Zealand anthropologist Raymond Firth wrote that women would "dance, caper about, and chant an *umere* or song of joy."[11]

The desirability of meat isn't just chalked up to its rarity. It was once believed that protein and fat were in such short supply in tropical rain forests that the main reason for groups raiding each other was to gain access to territory where the hunting might be better. Instead, it's more carbohydrate-rich foods that are in short supply in forests. Many forest groups, like the Mbuti or the Agta, will trade meat for rice or other starchy foods from neighbors to help balance out their diet. On one Ache forest trek, I was eating so much meat that I felt feverish, staggering along daydreaming of french fries, mashed potatoes, and Thanksgiving yams.

Given meat's delectable value and its cross-cultural desirability, who in society should expend considerable effort hunting? Despite a penchant for eating everything, humans aren't exceptional generalists. As foods vary in how they're acquired, divisions of labor by gender, by age, and by skill are universal. Not to say that we couldn't squeak by being Jacks-and-Jills-of-all-trades, masters-of-one-or-even-some. The big reason for divvying up tasks according to ability and other considerations is to *improve efficiency*. That reason may sound more boring than men hunting to impress women or women gathering because roots and berries are predictable and profitable, but sometimes the boring explanation is the correct one. For omnivorous hunter-gatherers, preferred foods come in many shapes and sizes; but obtaining different foods requires different tactics and skills. And skill mastery can sometimes take years of practice. If everything were easy, being a generalist might make more sense. But many roles—not just hunting, honey gathering, or being a shaman—see improvements over time beyond initial proficiency, taking up to two decades in some cases to master. Under these conditions, specialization is practically inescapable. No effective sports team, business, school, party planning committee, or family ever has everyone do exactly the same things.

With the need to specialize, it's usually men who participate in riskier hunts, especially when treks are long and far from camp. Men, though physically stronger on average than women, are not "built" to hunt, nor are women better suited to gathering. Rather, women face trade-offs due to on-demand breastfeeding: active pursuits of game aren't usually compatible with keeping nursing infants safe. Childcare constraints on the part of women, therefore, give men a comparative advantage, leading to the widely observed sexual division of labor in hunter-gatherers.[12]

Multigenerational Cooperation

The solution that helps everything work out—unpredictable success rates, hard-to-learn foraging strategies, and efficiency in divisions of labor—is multigenerational cooperation. Pooling food and other resources is a critical way for hunter-gatherers to ensure there's still food available when they themselves have been unlucky. Specializing on hunting high-risk game that provides fat and protein only makes sense when others in the sharing network target other, more predictable, foods. If enough hunters give it a shot, and pool their collective returns, they can greatly reduce the chance they go without meat for long.[13]

The human life course couldn't exist without sharing. It is especially critical during infancy, childhood, and adolescence when daily production often falls short of the nutritional needs of growing bodies. Even when adults are in their prime productive years, they still need to share. The evolutionary legacy of pervasive sharing hangs with us still, even when all the food we need is available at local stores. We still feel a warm glow when sharing food with others, and when others share with us. We avoid people who appear stingy, and praise those who are generous.

Hunter-gatherer sharing shows a consistent pattern. People favor spouses, their children, close kin, friends, and reliable social partners that they tend to travel with (see figure 5.4). From an evolutionary perspective, sharing with family makes sense: kin share genes by descent and so helping kin increases our inclusive fitness. But social relationships between nonrelatives can also intertwine the fates and fortunes of people in ways that foster sustained affiliation and cooperation. For example, marriage introduces new "fictive" kin relationships (in-laws or *affines* in anthropological parlance), who share common interests in the welfare of grandchildren.[14] Consider *compadrazgo*, or the godparenting relationship. It's a pervasive fictive kin system in much of Mesoamerica among children, parents, and godparents. Borrowed from Roman Catholicism in medieval Spain, its spread in central Mexico after the sixteenth century may have been a postconquest aid to buffer against a variety of stresses and disruption.[15] Formal fictive bonds also tie together foragers with their farmer neighbors with whom they share exchange relationships. Nomadic Raute foragers of western Nepal cement fictive bonds, called *mit*, with nearby sedentary villagers, through blessings, red *tika* powder on their foreheads, and referring to each other with fictive kin names.[16]

Other reasons have been proposed to explain why hunter-gatherers share food from an evolutionary standpoint. These are usually based on the logic of strategic donors who stand to benefit from their generosity, either in the short term or the future. When there's little storable wealth, food is the main game in town. Revealing the benefits and costs of sharing is a bit of an industry in my field. It's a case study in the behavioral biology of apparent altruism. Food is hard to come by, so why ever give it away? While kinship, reciprocity, and other strategic means to fitness ends have been successful in explaining some of the sharing patterns, they don't quite capture the on-the-ground flavors common to most, if not all, hunter-gatherers: a moral obligation to give when you can, the socially painful sting of being considered stingy or lazy, and the keen attention placed on "need." Children, pregnant women, injured or sick,

FIGURE 5.4. **Dinner on the Maniqui River.** An extended Tsimane household eats fish stew, or *jo'na*, out of a common pot. *Photo credit:* Michael Gurven.

and sometimes elderly are all categories of people who may produce and share little, but they are rarely excluded from group feasts. Instead, they're often given equal, if not privileged, access to food.

Sharing is a serious matter. It's a joyous occasion to eat communally, but it's also rife with tension. In one telling tale, a greedy Ache hunter, hoarding the fat from some kills and not sharing it with his thin wife, angered so many people in camp that a group of men killed him "by spearing him and then clubbing him to death."[17] Cultural norms help folks navigate those delicate decisions by zeroing in on preapproved targets when there's not enough to go around. Those norms, and enforcement against their violation, exist to favor long-term interests from a broad life course perspective.

When the forest, savanna, or desert is your half-stocked corner market, there's often not enough food to go around. Asking, or begging, can be stressful for all involved. But regardless of the reasons for sharing in any particular instance, the big picture is that there's a consistent net flow of food from the haves to the have-nots, the lucky to the unlucky, and from older to younger. For example, among Tsimane, adults in their 40s show a peak net flow of food to their children, whereas adults in their 60s show a peak surplus flow to their grandchildren. By the time Tsimane elders reach their 70s, there is little surplus food left to give.[18]

With sharing, high failure rates due to bad luck or a broken arm are buffered. That means having something to eat. Consumption is smoothed, risk is reduced. This is the miraculous result of sharing and cooperation: changing an uncertain world into a more predictable, more certain, high-yield return. Six foragers each with a 60% failure rate pooling their catches at the end of the day can reduce the probability of going without food on any given day to 5%. People are really good at figuring out that high-variance strategies require sharing to thrive and survive. People even adopt sharing practices akin to reciprocity when immersed in an artificial virtual world of computer games and experiments. But they don't resort to sharing when there's little gain, when items can be acquired predictably and synchronously.[19]

A Life Course Worth Living

Remember that hunter-gatherer women tend to have between five and eight children over the course of their reproductive lives. The average spacing between births can be as long as four years, as among the Ju/'hoansi, or as short as just over two years, as among the Tsimane. Once you start building a family, you might have three or four dependent children at the same time. Having multiple dependents at the same time places a huge demand on parents, even if they're otherwise making a decent living. Whether you're a low-fertility Ju/'hoansi or a high-fertility Ache, the net caloric burden of feeding your family can be huge. It peaks by the fourth decade of life, when an Ache woman has about four living "dependents": one infant, one child, one juvenile, and one adolescent.

In order to see the profound impact of sharing on shaping our evolved human life course, let's look closely at patterns of food production and consumption. I've compiled updated information from the Ju/'hoansi, Ache, Hiwi, and Hadza and combined them to give a composite picture of subsistence economics over the life course.[20] Figure 5.5 (top) shows

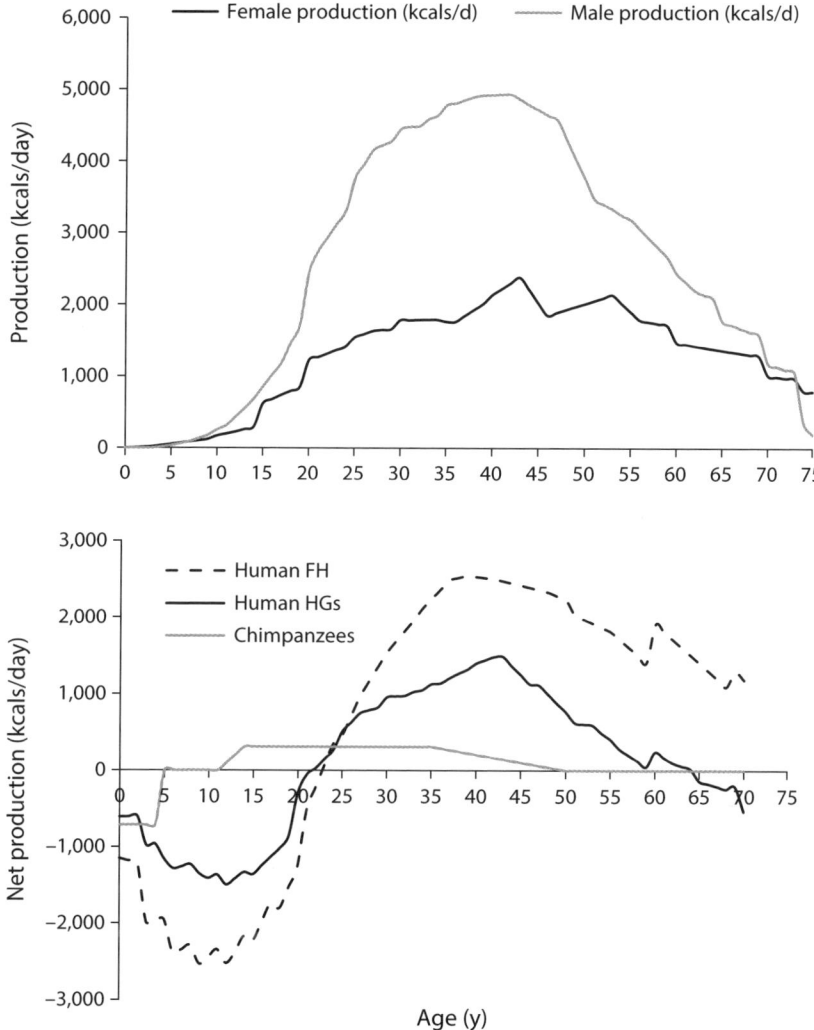

FIGURE 5.5. **Productivity for hunter-gatherers and horticulturalists.**
Humans eat more than they produce until age ~20 and generate large surpluses
from ages 30 to 50. (*top*) Daily food production (kcals/day) for females and
males, combining hunter-gatherer groups (Ache, Hiwi, Ju/'hoansi, and Hadza).
Solid lines indicate daily caloric production for females (black) and males
(gray). (*bottom*) Net food production (after subtracting daily consumption
from production), combining sexes, comparing hunter-gatherers (HGs),
forager-horticulturalists (FH), and wild chimpanzees. Horticulturalists include
Machiguenga, Piro, and Tsimane.

that caloric production is low for the first decade of life. Even by age 20, when people are practically adults, the amount of food they produce just accounts for how much they eat. No net surplus. Upon reaching age 20, which as we saw in chapter 4 is the lower bound estimate of life expectancy for prehistoric humans, women and men reach only about *half* of their maximum production and are only just starting to be able to pay back their cumulative debt from prior dependency.

Foragers reach peak "income" generation around age 40. Adults maintain a high level of productivity for a few decades, returning back to the lower level of their early twenties by the time women are age 70, and men are age 60. The bulk of surplus production over the human life course occurs in middle adulthood.

For groups mixing foraging with horticulture, such as the Tsimane, Machiguenga, and Piro (of the Peruvian Amazon), production surplus is even higher than among hunter-gatherers (figure 5.5 bottom). And the surplus remains higher until even older ages. This is because older adults are able to manage and maintain many of the skills-intensive but less physically involved aspects of farming, more so than in a pure foraging context. The difference is more obvious in men. Among horticultural-ists, the sum total of lifetime production attributed to age 50 and beyond is a whopping 38% for women and 43% for men! For hunter-gatherers, though the peak surpluses are lower, the percentage of lifetime production from age 50 onward for hunter-gatherers is an impressive 39% and 29%, for women and men, respectively.

This income profile is extraordinary. Humans basically take a hit for close to two decades before they start breaking even on a daily basis. But then they generate large surpluses and pay it forward. Delayed financial independence and midlife surpluses may be familiar to those of us used to fledging our 18-year-olds into the workforce, or 20-somethings graduating from college with a competitive advantage when it comes to wages. But even in the absence of formal schooling, it takes a long time to start "making money" as a forager, and an even longer time to really hit your stride. In marked contrast, based on chimpanzees, it's likely that our most common ancestor started paying their own way shortly after weaning, and showed only a modest surplus in adulthood. There's no need for surplus when most can cover their own needs. Few chimpan-zees are even alive by the age of peak human productivity.

This age profile for subsistence populations is fundamentally human. It appears in the absence of schooling and generalizes to our more familiar context of modern industry. There's a long tradition of docu-menting age patterns of productivity, from George Miller Beard's 1874 study of scientific achievements through Harvey Lehman's 1953 classic, *Age and Achievement*. It reveals a robust pattern of human productivity across the life course: an inverted-U shaped curve, with similar peaks in early to mid-adulthood, between the 30s and 40s, depending on the field, industry, and how productivity is measured. Learning and expe-rience matter, delaying the trajectory toward the peak. Whether con-sidering manager-assessed widget production in a factory, furniture assembly, scientific discoveries, or artistic works, even when Nobel Prize–winning work was done, the peak is in mid-adulthood. When physical strength, manual dexterity, and cognitive processing speed matter for productivity, efficiency falls from its peak more steeply with age. Elite athletes reach their peaks early, from female gymnasts in their teens to sprinters in their 20s and marathoners in their 30s, and it's usu-ally downhill from there.[21] That fall is less dramatic when productivity is more about experience-based cognition (better known as wisdom). The same patterns apply to subsistence populations.

You Are What You Eat (and Produce)

Imagine you're in your senior year of high school and aliens finally invade Earth. They proclaim that, in five years, the planet will be con-sumed as part of Season 4 of a popular intergalactic TV show. Any plans you may have had for attending university now seem a bit mis-guided. What's the point of delaying adulthood for at least four more years, and at great financial cost? Even if you could land a high-paying job immediately after graduation, you'd have it for only one year until planetary obliteration. Maybe hedonism is a better option, or joining a doomsday cult. But you'd still need money, so better to take a job when you graduate high school—it might mean a salary that's lower than what you'd get if you went to university, but at least you'd start earning right away.

If the time horizon of doom is looming, it pays to be present-oriented. The future is uncertain, but expected to be short. If the future looks *hopeful,* with a distant time horizon, it may pay to incur short-term costs—like going to college, or learning difficult foraging skills—for long-term gain. This argument is simple, but powerful in its implication. It suggests that we'd never have taken the yellow brick road toward a hunting-and-gathering career without expecting a reasonably long lifespan. Following the economic logic of figure 5.5, you'd have to live to at least age 40 just to break even. But it would be great to do better than breaking even: living longer to bask in the surplus glow. And conversely, the long lifespan wouldn't exist without a hunting-and-gathering livelihood, which itself could never work without multigenerational sharing and cooperation.

This proposal suggests a coevolution of our very human social worlds and human life history, a tantalizing idea first developed in 2000 by one of my graduate school mentors and longtime collaborator and friend on the Tsimane project, Hilly Kaplan, in one of my favorite academic papers—one so long, and so important, that it took up the entire issue of a journal. Similar ideas linking economic productivity and lifespan were also made around the same time by the economist Ron Lee, who had long studied intergenerational resource flows in different nations.[22] Much that follows here builds on these ideas.

Live Long and Prosper

The critical implication needs emphasizing: *long life is a fundamental part of the adaptive package that defines our humanity*—not a post hoc adornment layered on top of our success as a species. Our longer adult life is intertwined with the extended period of life before we're physically or socially adults. Human children grow slowly, in large part to help build and fuel our large brain that helps us absorb and learn. That makes them effective energy-vacuums early in life, as they develop and practice a rich repertoire of skills.

The human brain is central here. Remember when it figured so prominently in our estimates of primate and hominin longevity in chapter 4? There's a good reason for that. The human brain is three times the size of a chimpanzee's. Natural selection modified the human brain in a way that likely departs from that of other apes, and not just in size. A big change is in the prefrontal cortex, the area associated with symbolic thought, knowledge of appropriate social behavior, decision-making, planning, cognitive control, and working memory.[23] The difference between human and ape brain size is not just a matter of more of the same, but potentially new ways of thinking. The ability to think about things not directly observed by the senses. To make (imperfect) inferences about causation, to pay close attention to emotions, thoughts, beliefs, and intentions of others.[24] Our need to explain causes is so great that we make up motives for the wind in the willows and ascribe emotions to weary tumbleweeds.

Even the cerebellum expanded in apes and humans over evolutionary time, though its relevance has been largely ignored. The cerebellum actually has more neurons than the neocortex and is critical for sensory-motor control and the learning of complicated action sequences. While the larger cortex affects reasoning and conscious thought, the cerebellum may contribute to our "technical intelligence," something vital to us avid tool-users. Human brains also show greater connectivity across regions, leading to higher-order functional integration—hence it's not just the size but the wiring that has been shaped by selection since we veered from our common ancestor.[25] Taken together, the changes in brain size and organization helped cement the "cognitive niche" of human foraging and social behavior, a term first coined by my late colleague, John Tooby.

While an individual big brain-holder may be better equipped to figure things out than a smaller-brained ancestor, much of what we learn and achieve comes about from our collective brains working and learning from each other. Language helps ensure high-fidelity transmission of information in social groups, and for knowledge and traditions to build and accumulate. Such cumulative culture is another human hallmark, enabling us to form complex solutions to common problems.

This is what the anthropologist Joe Henrich calls the "secret to our success."[26]

There's an irony to having a large, powerful, and well-connected brain but yet remaining relatively helpless for so much longer than our primate cousins. Does it really take such a long time to be a good forager?

Yes. Measuring success rates and calculating wage rates, like we did in figure 5.5, showing how many calories are gained per hour spent in the food quest, tells us that it takes decades to become really proficient. Better foragers earn higher caloric salaries. Ache, Tsimane, Hadza, Machiguenga, and Gidra hunters of Papua New Guinea reach peak return rates in their late-30s through late 40s. Another study including many more groups confirmed a lag of up to two decades between reaching adult body size and peak performance in hunting.[27] This lag is not due to lack of motivation, or laziness. So why then such a delay in becoming a good hunter?

It's difficult for inexperienced outsiders (like me) to develop the kinds of tests that could best differentiate people of different ages by intricate skill and knowledge. I once set up "Forager Olympics" competitions with my colleagues to figure out which components may be more or less difficult. For one contest, we shot arrows at plantain hearts that were hanging from a tree, and also positioned on the ground, some distance away (figure 5.6). But even bored anthropologists can succeed at these tasks with enough practice. Among Tsimane, I found that the ability to walk effectively in the forest and to identify signs of prey improved throughout late adolescence and early adulthood—in sync with becoming taller, stronger, and more agile with age. Once adults, hunters didn't differ much in how many sounds, smells, tracks, and scat they identified that could be linked to potential prey. What separated the wheat from the chaff was the process of integrating that information, to eventually finding yourself close enough to the animal to make a successful shot.[28]

Most attention by scholars to date has focused on hunting, partly because of its prominence in hunter-gatherer diets and its historical legacy, and partly because of the male bias. But other areas of foraging food production can also take a while to attain proficiency. Hadza and

FIGURE 5.6. **Tsimane Olympics**. (*above*) A Tsimane man aims for a
plantain flower bud placed a distance of 16 m away and 10 m off the ground.
(*next page*) A Tsimane teenager is timed to pound and sift 1 kg of rice.
Photo credits: Michael Gurven.

Hiwi women are great at finding roots and knowing how to dig for them.
Younger girls can dig (maybe not as effectively as older women), but
they don't know where to look for the roots. Finding and collecting
shellfish among Gidjingali foragers of Australia also shows delayed proficiency. Same goes with the difficult processing of mongongo fruits and
nuts after collecting them en masse in the Okavango Delta.[29] Mongongo
is one of the most important and nutritious food staples of the Ju/'hoansi
and other San peoples of the Kalahari Desert. Mongongos have a fruity
pith surrounding a hard shell that itself surrounds another shell, below
which is the tasty nutmeat. One ingenious method involves collecting
the nuts from elephant dung. After going through the elephant's diges-
tive system intact, the outer shell is easy to break apart. Deer-like kudu
apparently also feast on the sweet fruit, but regurgitate the nut out their
mouth, ready to be collected by patient foragers.

FIGURE 5.6. (*continued*)

Expertise and Long Life

So, it takes up to two decades from initial basic proficiency before adults are really productive, into their 30s and beyond. Many other skills that hunter-gatherers and other subsistence populations must learn to successfully navigate their physical and social environments also may take a long while to master. But how to compare the "efficiency" or skill mastery for activities like making and fixing tools, raising fair-minded children, telling captivating stories, or just being viewed as trustworthy?

One roundabout way is to avoid assessing the ability altogether, and just ask people who they think are the experts! My former graduate student, Eric Schniter, and I did just this among Tsimane. By age 20, most Tsimane are competent in most tasks, such that people were comfortable saying "I know how to do this." Self-reported proficiency at

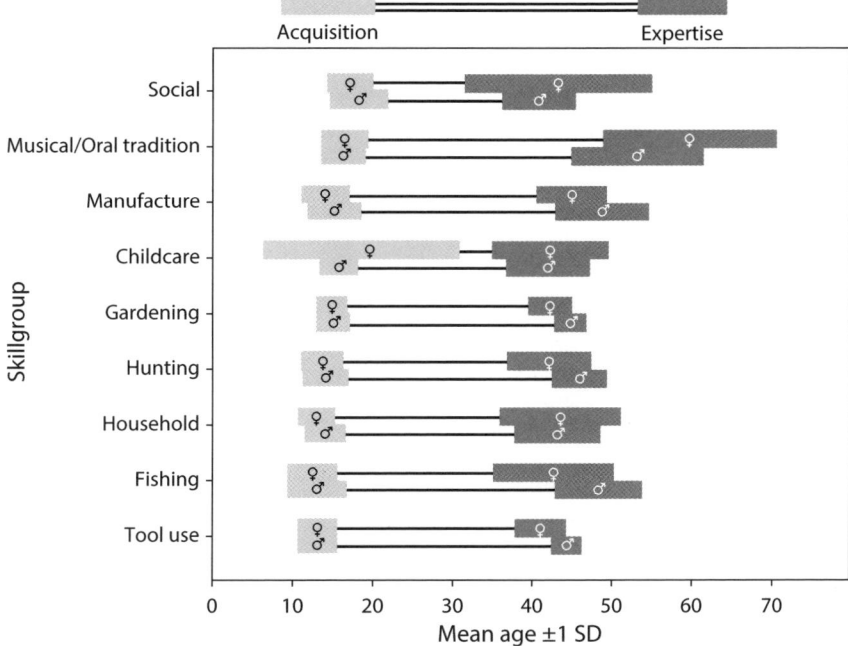

FIGURE 5.7. **Skill development over the Tsimane life course.** Ages of skill "acquisition" are based on self-report, whereas expertise reflects ages of those nominated as experts across ninety-six skills grouped into eight categories, like childcare, gardening, and craft and tool manufacture. Expertise comes two to four decades past the age Tsimane report first learning how to perform the skills needed in their environment. *Source:* Schniter 2009.

a more substantial level, however, occurred about a decade later. Even more extraordinary are the ages of those nominated as "experts" in different skill domains. Across *all* domains, experts tend to be over age 40 (figure 5.7). It could be that older adults are held in high esteem, and so a form of respect is to name older adults as the experts, what psychologists call a "halo effect." This may be the case, but the more an activity was rated by Tsimane as difficult, the more that older adults were nominated for showing proficiency and expertise. Of all skill categories, expertise was most concentrated among postreproductive-aged elders—for storytelling and playing music. In the absence of formal schools, these genres form the foundation for traditional pedagogy and socialization (which I explore in chapter 6).

Certainly the longer people are alive, the better they can get at many activities, unless physical decline compromises performance. Older adults can cultivate new hobbies and skills, too. We saw how food production declines in later adulthood, which reduces the potential contributions older adults could make to their kin. Do other activities that older adults engage in make up for their lower productivity? Short answer: Yes, but only for another decade or so.

Mothers and Grandmas Are Warm Hugs and Sweet Memories

By now, I hope I've convinced you that postreproductive longevity represents a true life stage universal to our species. But how did such a novel life stage evolve in the first place? The ability to continue reproducing beyond the mid-40s was not extended along with life, suggesting some constraints are at play. Like all mammals, human females are born with a fixed supply of follicles that are subject to gamete selection and decay, in a process oriented toward preserving embryo quality called follicular atresia. By birth alone, the number of viable eggs reduces from 6–8 million to 1–3 million. By menarche, it's down to about 300,000–400,000, and by age 40 down to roughly 5,000–10,000. Most eggs never make it through the gauntlet. Along with the exponential decline in viable egg supply, an older mother experiences a higher risk of miscarriage, and a greater chance that her children will become motherless if she dies. By "stopping early," mothers could avoid these costs to their fitness by ceasing reproductive function altogether with menopause. Whatever the reason, living beyond the ability to reproduce suggests that older individuals still can have "reproductive value" if they increase fitness through nonreproductive means.

George Williams was the first to propose back in 1957 that beginning at ages 45 to 50, mothers might benefit more, evolutionarily speaking, from investing energy and resources in existing children rather than from giving birth to new ones. Thirty years later, anthropologist (and, in an ironic twist, my academic grandmother) Kristen Hawkes coined this idea as the "Grandmother Hypothesis." Inspired by observations of "hardworking" older Hadza women producing bundles of starchy roots buried underground, Hawkes and colleagues proposed that older women increase their inclusive fitness by enhancing the fertility of their daughters and the survival of their grandchildren. She argued that grandmothers do this by provisioning small children and providing support to younger generations.[30] Putting aside for a moment that the total amount of food production by older women in hunter-gatherers, including the Hadza, is relatively low, the revelation that grandmothers matter elicits one of those "Duh, of course they do!" eye rolls.[31] But remember, the argument is that grandmothers don't help just because they're sitting around with nothing better to do. The proposition is that they live long lives as grandmothers precisely *because* of the assistance they provide—acting, in the words of the primatologist Sarah Hrdy, as the "ace in the hole." Among foragers, the resources acquired by women often involve strength and stamina, disadvantaging young children and thereby increasing the value of older women's labor contributions.

The limited evidence illustrates myriad benefits of grandmothers. Among the Ju/'hoansi and Hadza, grandmothers are productive foragers until their 70s. As Lorna Marshall, a long-time ethnographer of the Nyae Nyae !Kung once wrote, "Women continue their gathering into old age, as long as they are able. Most of the old women we knew, . . . spry and wiry as they were, gathered regularly."[32] Ache grandmothers are also hard workers and effective babysitters while their adult children are out working. Hadza grandmothers who worked more hours per day saw their young grandchildren experience positive weight gain, a result replicated more recently by Nick Blurton Jones with a larger sample.[33] Hadza grandchildren also were more likely to survive if their grandmother was alive, even when accounting for the possibility that long-lived grandmothers may simply have more robust children and grandchildren regardless of their help. Among the Ache, however, the effect is less clear. Ache grandmothers had little to no effect on offspring fertility or on grand-offspring survival.[34]

Having a living grandmother also makes a big difference for a daughter's or daughter-in-law's fertility: the gap between Hadza births is 22% shorter if at least one grandmother is around. That's a big difference in how many children a woman can have over her lifetime. On average, a woman by age 45 had about 1.5 more children if she had a living mother or mother-in-law.

As expected, the advantages of having grandmothers around extends well beyond hunter-gatherers. Among rural farmers in the Gambia, the anthropologist Rebecca Sear and colleagues found that maternal, but not paternal, grandmothers had a positive effect on grandchild nutrition and on early child survival. But only paternal grandparents had a positive effect on a daughter's fertility.[35] In a Japanese village from the seventeenth to nineteenth centuries, having a maternal grandmother around was associated with higher grand-offspring survival.[36] In eighteenth- and nineteenth-century northern Germany, having living maternal grandmothers meant higher grandchild survival, but more paternal grandmothers had the opposite effect.[37]

But does all the benefit really come from grandparenting? The long-term dependence of children in a natural fertility context without effective birth control means that a mother who births her last child, say at age 40, won't finish parenting until she is well past the age of menopause. The notion that most of the benefits to longevity come just from helping children rather than grandchildren has been called the "Mother Hypothesis" by anthropologist Jocelyn Peccei.[38]

According to the Grandmother and Mother Hypotheses, longevity is driven by natural selection acting on women, and the impact of transfers on biological fitness, from grandmothers to grandchildren, or mothers to children, respectively.

Where Are the Grandparents?

You might be scratching your head wondering how parenting adult children or helping grandchildren is even relevant among hunter-gatherers, who, as we saw in chapter 3, experience much higher mortality than what we're accustomed to seeing in all contemporary nations. We learned that many do survive until age 50, but overall, how many kids have living grandparents?

Figure 5.8 compares the extent to which hunter-gatherers and contemporary Americans at each age could expect to have living parent(s)

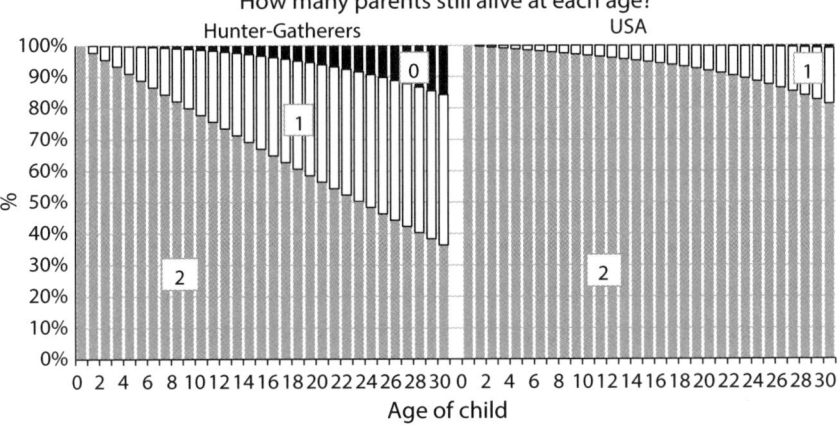

How many parents still alive at each age?

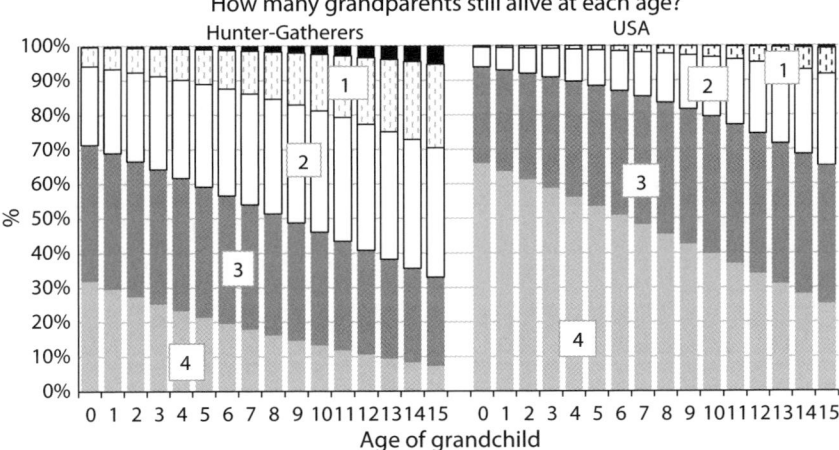

How many grandparents still alive at each age?

FIGURE 5.8. **Close kin availability by child age, for hunter-gatherers and contemporary Americans.** I assume hunter-gatherers have their first child at age 20, and Americans at age 30. (*top*) Number of parents who will be alive at each child's age. (*bottom*) Number of grandparents alive at each grandchild's age. Numbers given are from the perspective of the firstborn. While their mortality is high compared to contemporary Americans, hunter-gatherers have higher fertility at younger ages, which ensures that many hunter-gatherer children will have living parents and grandparents. *Data source for US:* National Center for Health Statistics 2017.

and grandparent(s). Hunter-gatherers tend to have at least one parent alive, for the most part, up through early adulthood (figure 5.8 top), and at least one grandparent alive up through adolescence (figure 5.8 bottom). Over half of hunter-gatherers have two living parents up through age 20, and at least two grandparents up through age 15. Though mortality is high, hunter-gatherers first become a parent about a decade earlier than in the United States or Europe, like age 20 versus age 30. The difference is even more striking for grandparents. A hunter-gatherer will be a Nana or Papi by age 40, instead of around age 60 as in most countries in the Global North. If I were Tsimane, I'd already have six kids and almost three grandkids, instead of just two young kids.

What about Men?

For the majority of mammals, fatherhood ends with insemination, and so mammalian life history evolution is often understandably framed in a female-centric way. A key difference between men and women is that men don't experience menopause and can therefore impregnate a willing partner well into their seventh and eighth decades of life. Tony Randall, of *The Odd Couple* fame, fathered children when he was 77 and 78 with his 28-year-old wife. The father of singer Julio Iglesias Jr. fathered at age 90, but sadly passed away before the birth of his daughter Ruth. These, and other stories of prominent men fathering children late in life, are examples of what my late friend Frank Marlowe called the "patriarch hypothesis." Marlowe, who like Kristen Hawkes also worked extensively with the Hadza, argued that, as men age, they accrue status and power that they use to obtain reproductive benefits. These fitness benefits at later ages could, in theory, select for their greater longevity.

While men may be capable of reproducing well into their final years of life, is this really relevant to human evolution? How many of our ancestors were Mick Jaggers or even Alec Baldwins? In subsistence populations, men are usually older than their wives, and divorced men often remarry and have children with younger wives. On average, however, hunter-gatherer men have children at older ages than their wives, but their fertility rates decline too, offset by just a decade or so from women's.

In more stratified societies with inherited wealth passed along male pat-
rilines, like many pastoralists and some gerontocratic Australian aboriginal
groups, men do have kids at much older ages. (Despite the variability,
these human patterns contrast strikingly with what has been reported
among chimpanzees. Chimpanzee male fertility at two famous sites,
Gombe and Kanyawara, peaks squarely in adulthood; males and females
show a roughly similar share of total fertility at age 50+ [~9%].[39])

A fertility rate is the average number of children born to a person per
year. Comparing male and female fertility rates for six groups, I find that
male fertility after age 50 accounts for as low as 5% of total lifetime fertility among Ju/'hoansi and Tsimane, 15% for the Hadza, and up to 19%
for the Ache and Yanomamo. The agricultural Mandinka, Wollofs, and
Fulas of Gambia were highly polygynous—more senior men often have
multiple wives. So over a third of their total fertility occurred after age
50. How much extra fertility beyond age 50 is enough to push against
the proverbial "wall of death" mentioned in chapter 2? Mathematical
modeling suggests observed levels of later-age male fertility may be
enough to extend lifespan beyond age 50.[40]

But why would a young woman choose to have children with older
men in the first place? One possibility is that it's not her choice. That
would assume that older men hold much influence and power. Widespread coercive matings or long-term marriages with older men are
inconsistent with the greater gender equality described in many hunter-
gatherer societies. But arranged marriages are not uncommon. A third
of Tsimane women I interviewed told me they were forced to marry
against their will by relatives, that they were too young, or didn't want
to marry the particular suitor. In contrast, only 11% of men said they also
married against their will.[41] These arranged marriages, like those among
many hunter-gatherer groups, often take the form of reciprocal exchanges
of marital partners across families.[42] But, like arranged marriages elsewhere, they're not organized just by men. In any case, Tsimane spouses
in these arrangements usually grow to love each other, or they get
divorced.

As suggested by the food production profile of figure 5.5, older men
may be preferred not because of undue influence or status per se, but

because they're the ones with all the cash. Whatever the form of wealth, be it food, livestock, or endless supply of dad jokes, if older men have more of it, they may be viewed as better marriage material (by available women but also by their families) than younger guys still coming in to their own. Cross-culturally, men tend to be older than the women with whom they have children. The age gap is greater when women marry at younger ages.[43] But the reason for this discrepancy takes us back to our distinctly human delayed productivity. And with delayed production comes an additional form of men's contributions to fitness. Men can channel surplus production to their partners, children, relatives, and others, just as women do.

It Takes Two to Tango (and a Village to Square Dance)

The shift to the difficult feeding niche, where brain growth and development are privileged at the expense of somatic growth, but where such a sacrifice early in life leads to big surpluses later in life, has been called the Embodied Capital Model by Hilly Kaplan, my longtime collaborator who I mentioned earlier. According to the model, the big sacrifice prior to adulthood requires a long adult life to make the human enterprise worthwhile. It built on earlier ideas in anthropology, and from rich ethnographic observations from working with the Ache, Machinguenga, and Piro, together with the brilliant minds of Kim Hill, Magdalena Hurtado, and Jane Lancaster during the halcyon days of the Anthropology Department at the University of New Mexico when I was a graduate student.

The gains that come after long delay come as a result of all the skills, knowledge, and abilities accruing during development that increase biological fitness. These inputs are like investments in a capital stock, not in Gringotts Wizarding Bank, but embodied in your person. It's an evolutionary take on a familiar idea, that schooling, training, and other forms of experience are the "human capital" that increases our economic value in a competitive labor market. The idea borrows elements from the Grandmother and Mother Hypotheses in expecting older women to make valuable contributions, but also includes men's contributions. It's a fairer two-sex model, where the specific contributions of men versus women will depend on the particular divisions of labor. In the embodied capital model, the gains in adult productivity due to prior investments in skill acquisition help select for lowered mortality rates and greater longevity. And lower mortality makes it more lucrative to pitch for the delayed return strategy earlier in life. In other words, the long human lifespan coevolved with the lengthening of the period leading up to adulthood and increased brain capacities for information processing and memory storage, and the flow of food, information, and other resources within and among generations.

This approach dovetails nicely with ideas introduced in chapter 2. Consistent with fundamental life history trade-offs, lowered exogenous mortality from the productive social niche of humans selects for delayed entrance into adulthood and longer lifespan potential via greater investments in somatic maintenance and repair—that is, a slower life history. The kick-start, owing to hominin proto-sociality, could have been reduced mortality from predation (thanks to fire and weapons), and improved survival from extra support when sick or injured, as we discussed in chapter 3. Once started, the stage was set for further longevity gains like those envisioned by the embodied capital model. Consistent with the disposable soma idea, mortality increases rapidly after seven decades because reproduction and other productivity declines. Productivity declines because of physiological aging, which itself is the result of an evolutionary process. Natural selection doesn't favor unlimited maintenance of our bodies faced with cruel entropic forces that deteriorate how our cells, organs, and body functions. Bodies are maintained only up to a certain point. It's cheaper instead to maintain integrity of our smaller germ cells than further preserve our bodies, given the backdrop of exogenous mortality.[44]

Enter: The Extended Human Family

A supportive social system enables mothers to wean their children early, and kids to take their time growing and actively soak up everything there is to learn. Such a system where many hands on deck help raise kids and buffer against the hard times has been called "cooperative breeding."[45] Among cooperative breeders, non-breeders are "helpers" at the proverbial breeder nest. When you're no longer exclusively paying your own way, energy budgets in effect become "pooled" with others.[46] Pooled budgets, where you can still eat and thrive despite barely being able to tie your own shoes, fosters interdependencies among families and folks looking out for each other.

Consistent with these ideas, human families and social systems are organized in ways that facilitate helping and mutual aid. Co-residence enables the kind of coordination that makes the benefits of group living outweigh potential costs, like fork-stabbing at the dinner table, or cheating with your brother's wife. The stereotypical nuclear family with two married heterosexual partners and up to two kids that has come to characterize the post-mid-twentieth century household of the Western middle-upper class is but one of many possible forms of the human family. From the broader perspective of human experience, it's a complete anomaly.[47]

If cooperation and group living in the context of a difficult but rewarding foraging niche helped foster longevity evolution in humans, then we'd expect some glimpse of these connections beyond humans. The biggest study to date of almost a thousand mammalian species shows that group-living leads to higher longevity compared to solitary living, and not the other way around.[48] Group living helps reduce mortality from predation and starvation, and strong social bonds also help improve survival. Average group size doesn't seem to correlate with longer maximum lifespan. But being in a larger group does not necessarily indicate greater cooperation or coordination than smaller groups.

Among group-living animals, "cooperative breeders" take us a little closer to humanlike social systems. Cooperative breeding is rare, accounting for only 1% of all mammals. Current evidence suggests cooperative breeding does not consistently associate with longer lives in mammals. But it certainly leads to a greater number of babies, and they're born at a comparably fast rate. One clear example where cooperative breeding makes a big difference is among African mole-rats. They're relatively long-lived, given their subterranean lifestyle protecting them from terrestrial and avian predators and inclement weather. What's amazing is that the highly social cooperatively breeding mole-rats are much longer-lived than their solitary sister species. Everyone's favorite furless critter, the naked mole rat (*Heterocephalus glaber*) of the Horn of Africa lives up to three decades in captivity, and their mortality rates do not increase with age. Consistent with this absence of "actuarial" aging, they also show little evidence of physiological decline with age. The icing on top is that they're miraculously almost free of cancer.[49] That remarkably places the naked mole rat as a "non-aging" animal, and hence a new non-wormy model ideal for studying how to slow aging. Naked mole rats are also coprophagic, a fancy way to say they eat their own feces. If that's their secret to long life, I'll take mine short.

To assess cooperative breeding in humans, a flurry of studies evaluated who helps most, in both historical and contemporary populations. Compiling forty-five studies that investigated the effects of kin on child mortality in natural fertility and high mortality populations, Rebecca Sear and Ruth Mace found wide variability in whose help matters.[50] As might be expected, moms matter everywhere, and the effects are strong. Dads matter too, but less consistently than mothers. Among Ache, for example, a mother's death increases the chance her young child dies fivefold, compared with threefold when fathers die. Tsimane fathers protect their children against infanticide, but their presence is unrelated to survival otherwise. Grandmothers seem to matter in over half of the studies, and mom's parents matter more than dad's parents. Grandfathers also make a difference in the survival of their grandkids in some studies, though they often get ignored, lost in the shadows of the grandmother and patriarch hypotheses.[51] Above and beyond the effects of adult kin, older siblings seem to matter a lot, too.

Taking in the big picture—close kin are important helpers influencing fitness, though who helps differs from place to place. This is to be expected. Local circumstances affect who is alive, available, nearby, and capable. Sometimes only your second cousin lives nearby. Or you may have two older brothers, but no living grandparents. For example, Aka grandmothers of the Congo Basin help buffer the nutritional status of young children. Aka are highly cooperative and egalitarian foragers who spend much of the year in small forest camps hunting small game with nets and collecting fruits and honey. When grandmothers are absent, other kin help fill in the gap. Except buffering is stronger when residing with the kin of a child's mother than with the kin of the child's father. On the other hand, a paternal grandmother's help is hard to replace.[52]

These types of studies are also indirect, linking whether a certain category of kin was alive or not to some outcome like child survival. In other words, these are correlations. The reason for the correlation might not be due to the help given and received. Families living in more productive areas, or who won the genetic lottery, may both include long-lived members who live locally and show higher fertility and child survival. When attempts are made to adjust for these factors statistically, the positive impacts of kin usually remain, though effects may be smaller.

Unlike women, men can never be certain about their paternity. From an amoral evolutionary standpoint, investing in kids who are not your own is not usually ideal. For this reason, some believe that maternal grandparents, and grandmothers in particular, should be the most invested in their grandkids (and paternal grandparents less so). There does appear to be general support for this across studies, where maternal grandparents seem to more consistently improve grandchild survival than do paternal grandparents. Even in urban Western settings, maternal grandparents are reported to be more emotionally close to, spend more time with, and spend more resources on grandchildren than paternal grandparents do.[53]

Living in close proximity to each other would normally be expected to facilitate help, but it could also promote competition for limited resources. Kin competition is underappreciated in models of cooperative breeding where the focus is more on the warm fuzzies of group living than on the turbulence of familial discord. But when resources are limited, and there aren't many good options for folks, anything goes. Among polygynous Kipsigis agropastoralists of Kenya, limited land access leads to severe competition for inheritance among co-wives and their children, and particularly among siblings. The anthropologist Monique Borgerhoff Mulder tells ghastly tales of conflicts between siblings over land: a man burning down his brother's house with his brother's family inside, and a woman trying to poison her husband's brother's children. The Kipsigis are also patrilineal—land and wealth inheritance passes from fathers to sons. A father's kin, especially uncles, therefore plays a more important role than a mother's kin in protecting Kipsigis children, and especially in richer households. In poorer households, paternal grandmothers and some maternal kin seem to buffer young children from mortality. But overall, maternal grandmothers showed no effect on child welfare.[54]

The bottom line is that the arduous task of raising multiple children does not fall on parental shoulders alone. It's aided by broader networks, where the specifics of who and how much vary by local conditions. Flexibility is key.[55]

May the Force of Selection Be with You

In chapter 2, I mentioned that the force of selection declines with age, and how deleterious alleles that shriek "BASTA!" can accumulate at late ages because they're believed to be invisible to selection, hiding within its dark shadow. But given everything we've just covered, we revisit the traditional dogma. Social species with multigenerational cooperation may show a fundamentally different pattern in how the force of selection operates with age. If older adults can "indirectly" reproduce by increasing the survival or fertility of close kin, they can potentially peek out of the selection shadow and cast new light.

In another adventure with my postdoc Raziel Davison (a talent in the fields of population and evolutionary ecology), we set out to estimate the strength of selection in a way that considers the fitness value of food transfers made to other group members.[56] Other forms of aid are possible, but food is a sensible start. Using age profiles of food production like those in figure 5.5, we estimated the effects that sharing food could have on lineage fitness, putting everything into a useful currency to compare against new babies, the gold standard of fitness calculations. When there's surplus production and hungry bellies to fill, the potential impact of transfers on biological fitness can be strong. By the 40s, even before menopause, the fitness value of transfers can be "worth" more

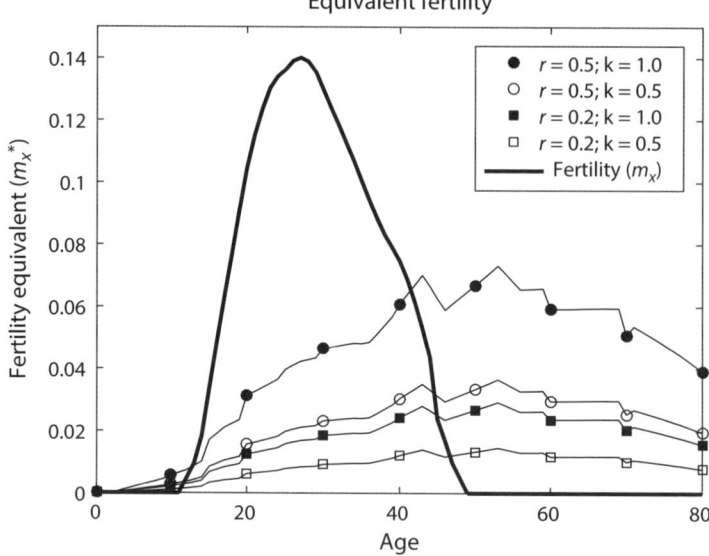

FIGURE 5.9. **Force of selection in postreproductive life.** High productivity throughout adulthood renders middle-aged and older adults more advantageous from an evolutionary perspective. For hunter-gatherers, the fitness impact of food transfers made by adults at each age is expressed as babies, or "fertility equivalent," for easy comparison to fertility rates. By the 40s, the food contributions to fitness are greater than fertility. Fitness impacts of food transfers are lower if recipients are less genetically related (lower kinship, r), or if others don't adhere to norms of sharing (lower cooperation, k). Under best of conditions, extra fertility is equivalent to three extra children. *Source:* Davison and Gurven 2022.

than fitness gained from reproduction (figure 5.9). Even when you take into account that some recipients may not be close kin, or that some group members may be defectors not buying into the food-pooling norm, the force of selection at postreproductive ages can still be sufficiently positive. How sufficient? A hunter-gatherer living from age 50 to their 70s "gives birth" to the equivalent of up to a few extra kids after the age of menopause.

As you'll recall, less than a third of hunter-gatherers live to see age 50. But when viewed through the lens of fitness contributions due to transfers, we see a clear benefit to being alive beyond the age of menopause. But only for a couple decades. Past seven decades of life, there is a remarkable

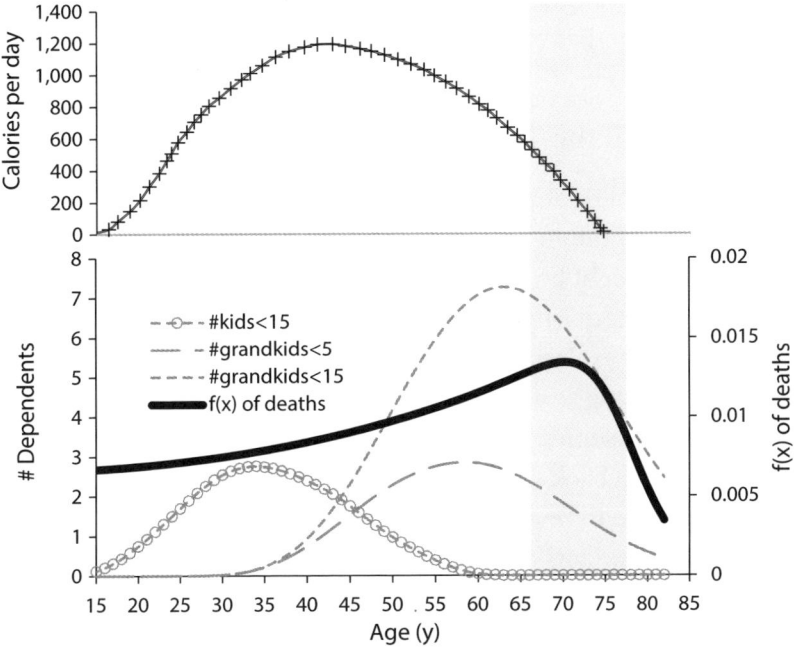

FIGURE 5.10. **Dependency, transfers, and modal lifespans.** (*top*) Net food transfers to Tsimane children and grandchildren. (*bottom*) Number of direct descendants (total children and grandchildren younger than 5 y, or younger than 15 y). Overlaid in bold black is the frequency of adult deaths. Modal lifespans (shaded region) peak when dependency is declining and net transfers to descendants is small.

convergence: productivity falters, net food transfers to children and grandchildren slide toward zero, and there are fewer young direct descendants who stand to benefit from transfers. This tight relationship between declining production, transfers, and number of close kin beneficiaries is shown in figure 5.10. After seven decades, the value of being alive, at least from the cruel, narrow view of natural selection, diminishes rapidly.

Herein lies the primary reason behind the book's title: the net fitness benefits of being alive keep hunter-gatherers vigorous long enough to fledge their children and ensure that many of their grandchildren survive childhood. This is about seven decades. Sure, there are benefits to being alive even longer, but those seem small relative to the high costs of keeping our bodies breathing, capable, and competent.

Will the True Model Please Stand Up?

These models were developed to explain the extended lifespan that evolved in the hominin lineage. And while mathematical models demonstrate what might be plausible, it's hard to decisively pit all of these models against each other. The evolutionary scenarios I've mentioned are not mutually exclusive. They differ in their focus on women (grandmother, mothering), men (patriarch), or both sexes (embodied capital). It is often assumed that fitness advantages leading to longer life in one sex can piggyback and gift the other sex with similar longevity. But this doesn't have to be the case. We're used to women outliving their partners. The residents of senior centers are disproportionately female. Female longevity exceeds male longevity in most human populations. And the sex difference is greatest in postindustrialized countries with the lowest mortality. Even among anthropoid primates, female lifespan often exceeds male lifespan, especially in social systems where females do the majority of parental care.[57]

A fascinating study by the evolutionary demographer Sam Pavard and mathematical ecologist Christophe Coste does the best job of combining elements of the models I just discussed into one overarching framework, to determine which model best maintains the Jedi-level force of selection at postreproductive ages. Pavard and Coste cleverly compare effects of maternal care, grandmothers (sorry, grandfathers!), paternal care, and the fitness gains of male fertility beyond the age of menopause, on the force of selection with age.[58] Compiling estimates from the literature, they show that mothers are critical to their children's survival, but that benefit extends only to about ten to fifteen years after menopause. A grandmother increases survival of a grandchild by about 10% on average—a small amount per child, but a big effect once summed across all her grandchildren. Adding older male fertility and paternal care above and beyond parents and grandmothers doesn't make a big impact on the effect of selection. The important point is that sociality in all its myriad forms has positive effects leading to longevity, as long as the cumulative impact on kin fertility or survival is strong enough. Our sociality has helped push back the harmful effects of late-age diseases and enabled us to live longer lives.

One problem with applying these models to living populations is that the conditions that maintain older age survival today may differ from those that selected for it in the first place. Most attempts to test the models are really tests of the former.[59] So while there's no definitive single reason why longer life evolved in our species, more tests across populations and environmental and cultural contexts will be needed to better assess the applicability of these models.

Human longevity may have a specific, unique history, but we've seen that sociality and longevity are linked across mammals. As lightning rarely strikes just once, surely the potential benefits of longer life in a complex foraging niche must appear in other species. Generalizability and parsimony demands that a good model should help explain life history trait evolution beyond just humans. And indeed it might.

Aspects of the embodied capital approach that tie in to longevity—like slow growth and delayed productivity with greater allocare of juveniles, and larger brain size—have found some promising support in other animals. Mammals whose feeding niche is best described as "complex," in terms of both the motor skills and cumulative knowledge required for proficiency, take longer to show adult-level competence in foraging skills, close to the age at which they typically first reproduce. During that longer period of training, they're provisioned by others, just like in humans. Among some species of cooperative hunters ("social carnivores"), such as gray wolves, bottlenose dolphins, and spotted hyenas, resource sharing is common, and foraging skills seem to peak even later, after the age of first reproduction.[60] Slow development in primates, and active provisioning in social carnivores, serve to buffer the low productivity of juveniles during periods of learning in these species with complex feeding niches. Humans, of course, use both strategies. Among primates, a complex foraging niche is also associated with having a large brain.[61]

Other comparative work highlights how help can strategically subsidize growth and reproduction.[62] In a study of 478 mammals, a team of

primatologists found that who helps makes a difference in the evolution
of key life history traits. More reliable male care was associated with
the evolution of larger brains, while greater help from others was linked
to higher fertility. When help is available and needed, it surely makes a
difference.[63]

The Drunk History of Human Evolution (or, a Brief Story of Food, Fire, Brains, Teeth, Guts, and Longevity)

Even if we have a good sense of *why* longevity may have evolved in
humans, that doesn't lay out the story of *how* our long, seven-decade
lifespan came to be. That of course is a work in progress, with key pieces
from previous chapters that we can now assemble.

Though it's tempting to believe that a giant black monolith signaled
the dawn of a new age for our species, the real prime mover in the
environment of our ancestors was the emergence of the open savannas
and its mosaic woodland ecosystem throughout eastern Africa during
the Pleistocene several million years ago. (And probably other parts
of Africa too, like in the South and possibly even as far west as Morocco, where recent findings show the earliest human occupation back
300,000 ya.) As the old adage goes, we didn't come down from the
trees, as much as the trees left us, with the drying of the landscape.
The hominins that explored these new environs had a whole new
menu of goodies to eat. It's a splendid thing, too, because climate fluc-
tuations increased habitat unpredictability in a way that put a big premium on dietary flexibility. Plants hold many creative solutions for
surviving seasonal and unpredictable environments that are great for
human palates. Underground storage organs, like roots, tubers, corms
and bulbs, are excellent sources of energy, vitamins, and minerals.
Nuts and seeds supply much-needed protein and unsaturated fats.
And, of course, open areas are filled with animals, especially dry-adapted
grazers.[64] The shift in the diet from a reliance on plants to more extractive foraging and greater dependence on animals by at least 2 million

years ago is the prime mover for at least one group of hominins, the ones leading to us.

Our early *Homo* ancestors had other things going for them too. They were already bipedal. No longer needing your hands for knuckle walking is a nice perk of being upright. More importantly, bipedality helps conserve energy when you need to walk longer distances than our arboreal plant-loving cousins. Traveling prolonged distances bipedally can save human foragers over 480 kilocalories per day.[65]

A shift to humanlike hunting and gathering blows away what our ape cousins (and our presumed ancestors) could accomplish per unit time. Anthropologist Tom Kraft, while working with me as a postdoc, painstakingly assembled the relevant numbers from all prior studies of apes and humans to assess efficiency and wage rates of primates. We showed that humans are not much more efficient than the great apes, in terms of calories gained relative to calories spent. That's because scoring a palatable bonanza can soak up a lot of energy. But humans are far more efficient with their *time* compared to apes. Human forager salaries, or calories gained per unit time spent working, are much higher than what apes procure: ~730 kcals per hour by hunter-gatherers versus ~250 kcals per hour among great apes, on average. When foragers add domestication to their repertoire, like the Tsimane and other horticulturalists, they do even better: about 2,160 kcals per hour.[66]

That's an impressive haul, a pattern typical of human food production systems up to and beyond the Industrial Revolution. Mechanization and other ways of "externalizing" production have vastly increased the surplus energetic gains we have achieved per unit time of labor. Over the last century, the proportion of income Americans spent on "subsistence" declined considerably. In 1900, Americans spent over 40% of their income on food, down to under a third by 1950, and to 10% by 2000. Assuming a 40-hour work week, that could yield us 2.4 extra hours per day, if housing, education, and other costs weren't so hefty. Even though modern societies externalize most production through advanced technology, supply chains, and industrial organization, hunter-gatherers were the original economizers, converting labor to energy far more effectively than our ape cousins.

One of our earliest forms of technology is the strategic use of fire. Traces of wood ash suggest the controlled use of fire by the time of early *Homo* < 1 million years ago, though unequivocal widespread use of fire is not evident until roughly 200,000 ya, when anatomically modern humans arrive at the scene. As mentioned earlier, fires are effective deterrents against scary carnivorous predators. They keep us warm in the cold, and as the anthropologist Richard Wrangham cogently argues, cooking helped make us human.[67] If even thinking about the aroma and taste of toasted bread, charred burgers, or dark roast coffee makes you salivate with delight, you have the transformative chemical processes of cooking known as Maillard reactions to thank. Those aromas convince our brains that cooked foods are both nutritious and safe to eat.[68] Indeed, the use of fire for cooking makes provisioning more efficient by improving the digestibility of food and making its energy more accessible. Cooking also increases the number of foods available for weaning infants. As we've seen, early weaning and fast reproduction are landmark human traits.

Lastly, cooking also reduces mortality. It detoxifies certain foods and helps eliminate food-borne pathogens. And certainly, as appreciated by anyone spotting hyenas or lions in the periphery, fire is an effective deterrent to predators. Not just on the open savannas, but also beneath the tropical rain forest canopy. I remember waking up to Ache stoking the campfire, trying to keep it lit throughout the night while everyone slept on or near each other for warmth and safety.

The hunter-gatherer foraging niche forced our ancestors to rely on foods that were more difficult to acquire, but that came in large energy-dense packets amenable to sharing. The scale of human hunting and foraging is unique among primates because of this widespread deliberate sharing of spoils and the long-distance transport of kills, honey, and other big food bundles to a central home base for processing and feasting with others. As I discussed earlier, sharing is ubiquitous, without which most hunter-gatherer livelihoods wouldn't be viable. We slow down body growth to build and maintain a large brain, not just to be skilled foragers but to be skilled *social* foragers. And to reap the gains of cooperation. Debates continue about the details, like how and when meat was first reliably obtained, whether through scavenging spoils from the efforts of other top predators, or by deliberate hunting. Whatever the case, by 2 million years ago, *Homo* was regularly hunting, with the use of flakes and other tools to bring down and process animals and plants.[69]

According to skeptics, the lack of definitive evidence of large game hunting among early *Homo* is grounds for rejecting ideas that long, delayed learning early in life is relevant for explaining our extended lifespan. Though big game hunting may look awe-inspiring when fleshed out in museum murals, it might not be the clincher. Many groups, like Ache and Tsimane, for which substantial delays in hunting performance have been reported, are primarily *small* game hunters. Even Hadza men hunt a lot of small game—quite extensively so, in fact, despite vigorous protests that they are, have been, and will always be big game hunters.[70] While how large game was obtained may be more open to interpretations of hunting versus scavenging, small game is likely to have been hunted by early *Homo*. Even megafaunal specialists were hunting small and medium-size game.[71] Given that all chimpanzee hunting is directed toward small game such as colobus monkeys, the first place to look for expanded hunting among early hominins is in increased frequency of small game hunting. Small game hunting has been an important component of human diets for at least 200,000 years, especially during local shortages of big game.[72] Do the cognitive skills required for targeting small versus large game hunting differ? Beats me. But there's no reason to think that they should. Small game hunters often rely on a greater variety of different prey items than large game hunters, and so must know a lot of specialized information about the behavior and ecology of many different animals.

In chapter 4 we saw that the brains of our most ancient ancestors prior to *Homo* were similar in size to those of chimpanzees. The tight correlations between brain size and life history traits suggest also a similar lifespan with chimpanzees—not just shorter lives, but earlier maturation, too. When brain size increased by the time of early *Homo* prior to 2 million years ago, that's our first guess for when longevity might have first started to increase.

In addition to having big brains, humans also have comparatively small jaws and small guts. Apes have smaller brains, larger jaws, and more guts. Teeth are amazing, not just because they preserve well, allowing folks to study them eons later.[73] Their shape, size, and other properties help tell us what was eaten, and how quickly their surrounding bodies developed. We know from the timing of molar tooth eruption that the juvenile period was longer for *Homo erectus*, and fairly prolonged by the time of *Homo ergaster*, ~1.4 million years ago, compared to prior hominins. And, given the proposals that delayed development coevolved with adult lifespan, slower development early in life links to longer lifespan in the hominin record.

Teeth have reduced in size over the course of hominin evolution. The pointed canines of apelike ancestors are a sharp (pun intended) contrast with the small incisor-like canines of modern humans. Even our wisdom teeth are small or don't fully develop in modern humans, while other hominins have huge wisdom teeth. The reduction in tooth size has been attributed to cooking with fire, and with tool use.[74]

The scant evidence so far best supports extended preadulthood, including the novel life stages of childhood and adolescence, appearing first among late *Homo erectus*, then becoming more pronounced later in the *Homo* genus. Early *Homo ergaster* shows both significant brain expansion and a prolonged development, but less so than modern humans. Early life growth patterns likely occurred around the same time as increases in brain size and improvements in tools, shelter, and use of fire—all ways to further reduce exogenous mortality and to put a premium on a longer adult life.

Other indirect signs of physical changes throughout our evolutionary history also shed light on the emergence of our recognizable humanity. Male and female body sizes tend to look more similar to current human differences, with males 10%–15% larger than women on average, by late *Homo erectus*.[75] In the animal kingdom, and especially among primates, sex differences in body size reveal insights about the mating system and social organization, a link appreciated by Darwin himself. According to sexual selection theory, Darwin's lesser-known sequel to natural selection, larger differences mean greater competition between males in more polygynous mating systems where dominant males outreproduce subordinate males. Male gorillas, with their harems, are over 50% bigger than female gorillas, whereas monogamous gibbon males and females have similar body sizes. Given the roughly similar body sizes of early humans, the types of heavy competition between males common in highly dimorphic and polygynous species was most likely minimal. This is consistent with the more heavily two-parent pair bonds and extensive investment in children we see in modern serial monogam-ish *Homo*.

To rehash, a more humanlike diet takes longer to attain proficiency in but reaps large gains later in life—a life that is longer because it needs to be. In concert with longer adult lifespan, early life extended as a critical time for learning, practice, and building a desirable reputation. Group behavior and extended sociality may have helped reduce mortality in some hominins, which then set the stage for the feedback effects I laid out in this chapter. Interdependence through enhanced sharing and cooperation became necessary for this new lifestyle. It tied all the loose strands together. All of this required enough longevity: vigor in the 40s and 50s, and sustained "value" in the 60s and 70s.

The standard approach that says our elder years are invisible to selection is wrong. Longer adult lifespan adds surplus energy and helps facilitate learning and sharing across generations. Under good conditions, this arrangement enables human populations to grow rapidly. We're a pioneer species, colonizing with babies. Population crashes from periods of war, pestilence, and famine may have kept human populations in check over most of our species' history, but when misfortune subsided, we often rebounded. This is the situation among many contemporary subsistence populations today. Thanks to multigenerational cooperation, we have an inordinate ability to ramp up fertility and boost our collective survival.

6

To Be of Use

I want to be with people who submerge
in the task, who go into the fields to harvest
and work in a row and pass the bags along,
who are not parlor generals and field deserters
but move in a common rhythm
when the food must come in or the fire be put out.

—MARGE PIERCY, TO BE OF USE (1982)

Ju n!aan n/ui /oa n//ae, kom, koara. Ju!ae!ae'm ko ku n//ae/'an e ko Hoesi
oosi, te ha n/ai /oa //kae toan te koara ka. Te ha n/ai ku //kae o kxae ka.

(The old person who doesn't tell stories just does not exist. Our
forefathers related for us the doings of the people of long ago and anyone
who doesn't know them doesn't have his head on straight. And
anyone whose head is on straight, knows them)

—OLDER JU/'HOANSI WOMAN

IN THE opening scene of the 2002 movie *About Schmidt*, actuary Warren Schmidt, played by a seasoned Jack Nicholson, stares at the clock
above his office desk, as it ticks loudly toward the final hour of his final
day of work. At 5 p.m., Warren is officially retired, and soon finds his
work-less life without purpose or meaning. Lonely, he begins writing
letters to Ndugu Umbo, a 6-year-old Tanzanian boy he has "fostered"
through a mail-order program he saw on TV.

Hunter-gatherers wouldn't understand this movie. Why does Warren have only one child? And why does that child live so far away? Then there's the more perplexing question: What is retirement?! Such a discrete division between work and post-work life doesn't exist in hunter-gatherers or most other subsistence societies. For the most part, people work until they can work no longer, and even then they strive to make themselves useful. To be *of use* is a recurrent theme. Elders may earn the right to relax a little, but there's no free ride.

A hard life may seem at odds with what you've heard about or seen in depictions of hunter-gatherer societies in films like *The Gods Must Be Crazy*. And if you think that hunter-gatherers follow the Zen path to affluence—wanting very little because they have all that they want—you wouldn't be alone. This was a popular view first propelled into the public imagination in the 1960s by the anthropologist Marshall Sahlins. Based largely on some initial observations of the Ju/'hoansi, and Arnhem Land aborigines in Australia, Sahlins argued that hunter-gatherers work just three to five hours per day. According to this view, markets and modern urban life push you off the gravy train to relentlessly pursue material goods. During the tense and dark days of the Vietnam War, the idea of communal-living, leisure-loving hunter-gatherers was enticing—an idyllic way of life we used to have, and to which we could perhaps again aspire. Too bad that view is wrong. If you take a broader, less ethnocentric view of what constitutes work, include more representative sampling of people's time, and open your study to more populations, this romanticized portrait of the "original affluent society" looks not so original nor affluent.

If we include time spent getting firewood, water, and other materials, and food processing and preparation and other work-related tasks that extend beyond the literal bringing home of the bacon, hunter-gatherer work lives are not so different from the rest of us. Sure, with less physical separation of the workplace, categories of work, family, and leisure are more mixed, and work schedules are more flexible. But all in all, hunter-gatherers like the Ache and Hadza work more than forty hours per week. Men tend to work forty-five to fifty hours per week, and women about fifty to fifty-five. Horticulturalists and pastoralists work roughly similar amounts. Based on thousands of momentary snapshots (recorded systematically by many vigilant graduate students over the years), I estimate that, in the early 2000s, Tsimane men worked about a thirty-four-hour work week, women about thirty-eight hours.[1] The modern amenities of industrialized life don't shorten our work week, nor does the rapacious capitalist quest to impress our neighbors triple the work week. Based on recorded observations of work activities in multiple societies, my colleague Joe Henrich and his students found that industrialization and commercial activity increase work time modestly for men (from 47 to 55 hours per week), and not much at all for women, who work longer days than men no matter what.[2]

But even when some groups like the Hiwi work less, it's not because they have "limited needs." As the anthropologist Bruce Winterhalder once said, "The Zen economy has an ecological master"[3] (which is a veiled way of saying that environmental and social factors often affect whether it's worth it or not to work harder). For example, if it's blistering hot out, you're liable to become dehydrated, get sick, or worse. Better to conserve energy and work strategically when you can.

All these numbers about work weeks so far reflect all working-age adults. But what about the elders? How do they spend their days? How hard are they working? What is life really like for older adults in subsistence societies? If the postreproductive life stage evolved because of the contributions that older adults make to their inclusive fitness, then older age should not just be a time to take luxury cruises, learn to make fancy wooden pens, or embark on salt-cave self-discovery retreats. We'd expect time well spent in activities that help others, especially kin, and that those activities should make a difference in people's lives. Divisions of labor in complex societies opened important roles for elders, up through their seventh decade, throughout human history. In this chapter, I'll explore the different ways that elders "belong" and have made themselves useful.

Not Richard but Leo

The public and private lives of older adults in small-scale societies have been overlooked for quite some time. Classic ethnographies from the early-to-mid twentieth century usually focused on younger and prime-aged adults, a tendency that is especially pronounced, according to the founder of medical anthropology Margaret Clark. Clark attributed this bias to how Americans ignore the aged in their own society. It has been an unfortunate oversight, because anthropologists often enthusiastically converse with the older adults in the communities where they work. They're relied on as great sources of information about the way "it used to be." Not until the 1970s was there a serious attempt to point an ethnographic lens on older adults and to ascertain their quality of life given the experiences shaping their lives. To date, however, these studies are mostly confined to specialized fields on the margins, like cross-cultural gerontology, or the "anthropology of aging."

The first real attention on older adults in subsistence societies was pioneered by the Yale sociologist, Leo Simmons. Simmons was decades ahead of his time. He sifted through hundreds of reports and ethnographies and attempted the first truly comparative, cross-cultural analysis of old age. Published in 1945, his *The Role of the Aged in Primitive Society* was a game changer.[4] He coded hundreds of variables among seventy-one preindustrial societies to capture different aspects of how older folks participate in society and how they're treated by others. Groups included some familiar folks we've already seen, like the San (!Kung) of Botswana, but also many others, like the Ainu of Japan, Lengua of Argentina, Seri of Mexico, and Yakut of Siberia.

Simmons was interested in both human universals and how the status and treatment of older folks might vary among societies. He meticulously coded traits as either missing or evident on a mysteriously sliding scale from "absent," "incipient or unimportant," and "intermediate importance" to "strong social importance." The common refrain that "correlation doesn't mean causation" makes responsible scientists pause. But Simmons was a daredevil, glibly reporting 1,146 correlations in his book.

Many have criticized Simmons for his lack of standards about who constitutes the "old," his haphazard sampling of societies, and his inclusion of related groups. He was also opaque about how he defined or scored cultural traits (how do you rate societies on "phallicism"?!) and never specified whether scores reflected cultural ideals or actual cases. Some mix of these complaints are unavoidable when trying to shoehorn subjective ethnographic information into a quantitative framework.[5] Say what you will, and the work could be better, but at least Leo Simmons tried. Remember, this was 1945. While the West was picking up the pieces after the war, Simmons was busy launching new fields—the systematic comparative study of human culture, later referred to curiously as "holocultural studies." If that wasn't enough, Simmons was also an early advocate for considering the larger sociocultural environment when making recommendations for treating disease.[6]

The overarching view from Simmons and the holocultural studies that followed is that older adults occupy many roles. Simmons coded information on thirty-two different roles older adults could have. Many sound familiar: educating and caring for the young, being chiefs or members of councils of elders, practicing shamanism and treating illness. Others are likely less familiar: "professional decorators" or "defenders of the status quo." Methodological issues aside, let's look at the big picture.

I am taking you down this obscure anthropological path because each society is like a natural experiment in how to come up with a way of life. Groups don't need to invent it all de novo. They borrow heavily from neighbors or from conquerors who enforce new traditions, and they inherit traditions from past generations. In other words, cultures evolve, and can adapt quickly to new local environments, and can generate their own dynamics. Horticulturalists across the Brazilian, Bolivian, and Ecuadorian Amazon share many cultural traits, but also differ in many ways. Cross-cultural studies are vital for telling what may be universal and what varies—and why.

I reanalyze Simmons' raw data to highlight cross-cultural observations from valuable, but imperfect, older ethnographic information.[7] Figure 6.1 compares the proportions of older men and women noted to take on each of fifteen different roles in their society. Older men participate in 89% of these roles, and older women 75%—an impressive level of involvement. These roles vary from place to place, but the tally looks similar among societies with different livelihoods (see figure 6.2). That is, elders living among hunters/fishers, collectors, farmers, and herders have multiple roles they serve in their societies.

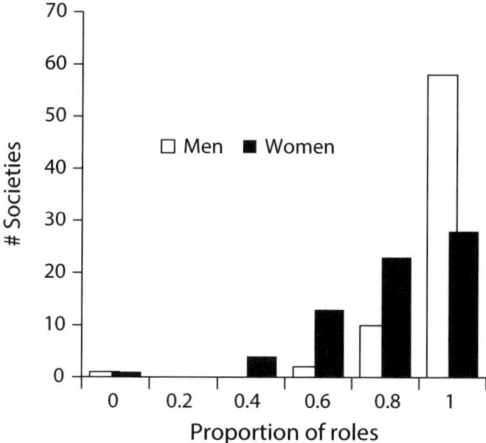

FIGURE 6.1. **Social roles of the aged in small-scale societies.** Proportion of social roles held by older men and women, across seventy-one societies coded by Leo Simmons. In most groups, older adults exhibit the vast majority of activities. Older men tend to exhibit more social roles than older women in Simmons's sample. However, where there are more roles for women, there are more roles for men (correlation $r = 0.24$, $p = 0.049$). Roles include fifteen types of participation, such as aiding in childbirth, initiating and educating young, and practicing shamanism.

What's clear is that older adults are hardly idle. There is no retirement. An idea from 1960s gerontology held that withdrawing and disengaging from others and society were "natural" universal aspects of growing old.[8] And in some versions, withdrawal was seen to even benefit the elderly by conserving their dwindling strength and "ego energy," whatever that is. Not only does withdrawal fly in the face of current thinking about healthy aging, but from surveying across small-scale societies, withdrawal is neither universal or natural, nor desirable.

Having just taken the long view from space, let's now explore in more detail the myriad ways that elders make valuable contributions. They provide, build, instruct, and heal. They nurture, care, and protect. They lead and guide, mediate and advise. They sing, narrate, and entertain. A skeptic might say that living longer gives us time to keep busy. That these roles exist because we live long. My argument spins that argument around—we have long lives *because* of these roles. The goods and services that middle-aged and older adults provide are not just leisure

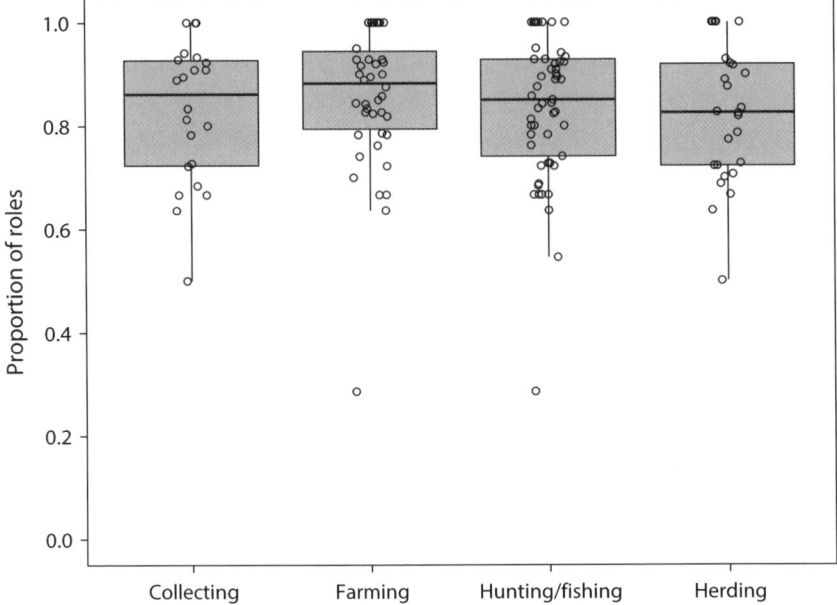

FIGURE 6.2. **Social roles held by aged men and women by economic liveli-hood.** Elders hold important roles in all societies, regardless of economic mode of subsistence. Economic mode ranked as absent or minimal (0) versus moder-ate to exclusive (1), for four types of livelihood: collecting, agriculture, hunt-ing/fishing, and herding. The box-plot shows the mean and inter-quartile range among the same populations in figure 6.1.

hobbies to top off a well-lived long life. Divisions of labor and specializa-tion privileging the experience that comes with age are central to what it means to be human. What elders provide is fundamental for gleaning the most out of human ingenuity. In chapter 5, I explained why longev-ity evolved by demonstrating that "value"—from a natural selection perspective—doesn't vanish with the ability to ovulate or give birth. Now I'll explore *how* that maintenance of value occurs.

Still Crazy (Productive) after All These Years

Let's start with the fundamentals: food. Remember, from the age produc-tion curves in chapter 5, that hunter-gatherers and horticulturalists show big surpluses in middle adulthood. By age 60, hunter-gatherer men fall back to what they earned in their early 20s, and horticulturalist men fall back to what they could produce in their early 30s. On the other hand, women's productivity doesn't budge much, on average. Once you take into account personal daily food needs, the net surplus available for others evaporates by around age 70 in hunter-gatherers, and a half decade later in farmers. In chapter 5, we saw that those extra calories are not just bonus dessert. Under reasonable conditions, the model that Raz Davison and I came up with shows that surplus production yields can increase survivorship and fertility. Summing the calories produced from age 50 through the seventh decade of life is "worth" the equivalent of at least one extra child under most conditions of our model (where much sharing is with non-kin and reciprocation is weak). Under stricter conditions, where sharing is confined to close kin who reciprocate, the extra food can be worth more than three children.[9] In biological fitness terms, that's a huge impact.

Of all subsistence activities, the steepest declines in productivity with age are in hunting, due to the physical vigor and sharp faculties needed for success. But older men can adapt their activities in ways that still make them useful, even in the domain of hunting. Among Ju/'hoansi, a man in his 60s may still accompany a younger hunter, usually his son. Together they interpret tracks, but the younger guy does the harder labor, and makes the shots. Men in their 60s may use traps and snare lines to catch guinea fowl or hares. Among G/wi, another San group like the Ju/'hoansi, older men still go on single-day hunting trips nearby camp. They target small animals, like the antelope-like duiker and steenbok, rather than the large (but more rarely acquired) megafauna prizes tracked for days, like giraffes and wildebeests.[10] Old Hadza men will hunt until their dying days, though also switch strategies to accommodate their more frail condition, like foraging for berries and baobab fruits.[11] Groups like the Tsimane can get by hunting at older ages, because they often bring well-trained dogs to help with the chase, and sometimes with the kill.

Apart from the actual hunt, older men still contribute to production by making and repairing arrows. Older Ache and Tsimane are renowned for making the best bows and arrows. Older Tsimane women make the best handbags (called *saraij*), used for carrying game, roots, and other forest treats. Toolkits, pottery, basketry, weavings—these have long been a preferred domain of elders.

Walking Libraries, Talking Televisions, and Expert Tinkerers

Sans books, television, and Google, where would you go to look something up? What settles an argument with your friends about which leaves best relieve the itch of scabies or how many rows of teeth *vonej* fish have? How do you figure out whether that darling charmer from two villages downstream is of the right marriageable lineage or not? In this regard, elders hold two major irreplaceable assets: a long life of experience that pre-dates the birth of many other group members, and (hopefully) a good memory. Lucid, reliable recall is a collective asset— like a 10 TB hard drive that others can access. Older adults are like a breathing Wikipedia rife with facts about plants, animals, land, and the like, but they know so much more. They are masters of family histories and genealogies, and of traditions and folklore. They navigate through webs of local politics and culture, and apply their decades of crystallized knowledge to helping others. (You may be excused for not remembering what you ate for breakfast, but repeating the same stories from fifty years ago shouldn't be viewed as a nuisance—it may have adaptive value!)

One attempt to capture the extent to which older adults are masters of information, and information control, scored a random sample of twenty-six societies, including hunter-gatherers, farmers, herders, and what could be best described as "rural peasants." Controlling information in different spheres of life comes in the form of teaching, consulting, entertaining, arbitrating, making decisions, and participating in ceremonies, feasts, and the like. In over half (14) of these societies elders engaged in four or more of these activities. Only in six did they participate in just one or two. Civic engagement, consulting, and making decisions almost universally involved the aged.[12]

When older adults hold valuable knowledge, they may be generous with sharing that information. But restricting access with a bit of strategizing can go a long way. Elders may be unwilling to share the information, or at least not give it away for free—especially if scarce information is a valued commodity. For example, Simmons reported that older Navajo men could fetch a high price for sharing secret information about songs, legends, and medicine, apparently worth "more than a herd of thousands of animals."[13]

Another example highlights how elders control the spread of cherished knowledge. In a small atoll community on Etal Island in Micronesia, everyone fishes and farms, but some people also have specialized trades. They work as navigators, canoe builders, healers, and knot diviners, to name a few. Each trade requires dedicated training and knowledge. As knowledge is viewed as property of the clan, members cannot teach anyone, including other clan members, without approval from the clan elders. Such knowledge is a commodity that turns a profit. Healers, for example, receive gifts of land and other valuables from people they cure of major illnesses. Old people are critically important to their kin and the larger community because they hold (and transmit) the secrets to building much-needed skills.[14]

One common expression of elder knowledge is in the form of technology production. Older folks love to tinker. They make and fix things. Apart from getting the materials, tool and craft manufacture is compatible with being in camp. Even if elders aren't better at craftsmanship than others, their labor in this domain is universal. Elders were universally listed as having "expert craftsmanship, advice and supervision" in Simmons' seventy-one-society sample, and in most they were also deemed skilled makers of "toys, implements and utensils."

Older Chipewyan women from western Canada were skilled in sewing and beadwork. They also prepared animal hides to make leather. Older women were known to be better at these activities than younger women, and would partake in these until their eyesight was no longer good enough. Even then, though, older women might advise younger women and supervise their work.[15]

While living among the Ache, I collected information on all the hunting arrows and bows men had in their houses. After going from one hunter to the next, it quickly became clear that the best *wachi* arrows were said to be made by a couple of the older men. The same goes for the long black palm wood *rapa* bows. Contributing tools to others is more than a favor. Supplying key equipment needed for work can buy you entitlement to shares from kills. This is a general pattern among Ache and many other foragers. Among the Ju/'hoansi, anyone making a kill with arrows made by others is obliged to share the largesse. Surprise, it's the older men reported to make good arrows.[16]

According to anthropologist Charles W. M. Hart,[17] who lived in the early part of the twentieth century with Tiwi aborigines of Bathurst and Melville Islands in north Australia, men's "hunting days were over" by their mid-40s because of declining eyesight. Older men thus spent a lot of time in camp making things. They had the skill, but also the leisure to do this when others were busy doing all the other work (among the polygynous and gerontocratic Tiwi, it helps if you have six wives and fifteen children). Elder men wouldn't just make your run-of-the-mill digging stick or bow and arrow. They made canoes, ceremonial spears, and elaborate graveposts. These beautiful, decorative objects were a form of artistic expression. But importantly, they were usually gifted to other groups during funerals and other ceremonies. Can't show up to a party empty-handed. Old men spent a good deal of time in preparation or in attendance at funerals and other ceremonies. One of these, the *kolema* ceremony of life, was like Burning Man—a multiday and night celebration with body painting, dancing, and singing. Except more yam eating.[18]

Sing Me a Song, Tell Me a Story

Without screens and books to tell and retell stories, the oral tradition of storytelling and singing are mainstays of entertainment and learning. Sharing stories allows people to experience the vicarious thrills of other group members, and to safely avoid the folly of their bad choices. Storytelling has likely been with us since the early days of language. All hunter-gatherers tell stories—about people they've known, about themselves, about animals and plants, about tricksters and heroes, myths and legends. Stories instruct and act as moral guides. They're a form of "cognitive play" that includes plenty of drama and heated commentary, and when stories are told well, they make useful information stick.[19]

Among Ju/'hoansi, younger folks will protest when asked to tell stories, arguing that they "haven't grown old enough to have learned the things that old people know." Instead, every old person is able and motivated to tell stories, replete with ornate flourish and vigor. The anthropologist Megan Biesele has long studied storytelling and education among Ju/'hoansi (and gave us the quote from the elder Ju/'hoansi woman at the opening of this chapter). She explains that elders are good storytellers because they're "masters of dialogue" with their "verbal ability perfected over a long life with the details of early times usually known only to the old."[20] Like spoken word at the late-night coffee shop, the language of stories is "rhythmic, complex and symbolic," with deliberate repetition of phrases, gestures, and sound effects. It erupts in spontaneous song, bringing "characters right to the hearth and into the hearts of listeners," leaving listeners "stunned with suspense, nearly in tears, or rolling with laughter."

The anthropologist Polly Wiessner has also worked with the Ju/'hoansi. In one study, she meticulously recorded and coded the content of conversations throughout the day. While people spoke of practical matters during the day, at night they would gather around the crackling sparks of the fire and connect over stories that came to life. Stories make up the vast majority of these firelit conversations. It's in these illuminated settings that people focused on the "big picture" through engrossing stories that evoke the imagination, aided by singing and dancing.[21] Wiessner suggests that the regular use of fire by ~400,000 ya may have helped provide the space, time, and intimacy for storytelling and, by extension, trust and culture-building. (Even the most urban city slickers are easily mesmerized by a campfire, with embers dancing to the rhythm of conversation.)

A story from the Tsimane illustrates the ways in which this form of communication can both educate and entertain. In this case, the story reveals both the miracle of steel tools and the dangers of greed. Other than cut bamboo, there's little in the forest that can be used to chop and slash. Iron tools are so effective that many hunter-gatherers already had them, sometimes stealing them from logging or mining camps, well before any official contact or trade with other groups. In the story, a large tree called Chuchillotujsi, located above a large rock by the Maniqui River, produced not just fruit from its leaves but steel knife fruit. Knives of different sizes would fall into the water, where people could collect them and fit them with handles, all so that the Tsimane could learn to farm and prepare wild meat. A man named Tonoj would always show up late to the tree, and couldn't find a knife. So he cut the tree down, waiting to collect all the knives still left. But the tree sank and so the man got nothing. Some versions have one of the creator gods transforming the man into a crayfish.[22]

Another example that conveys how stories can educate, but also uphold social norms, comes from the Agta of the Philippines. Among the Agta, stories help promote cooperation and egalitarian social relations, especially between the sexes. By broadcasting norms, stories help coordinate group behavior.[23] The anthropologist Daniel Smith shares a beautiful example of a story emphasizing cooperation and equality: a monkey and his friends wanted to set up camp near a river, but each time they tried a terrible giant would get angry and attack them, shouting, "I will eat you all!" So the monkey and his friends huddled together and hatched a plan. The monkey lured the giant into a cave where there were many angry bees and ants. The giant perished from the onslaught of stings. The monkey was congratulated for his cleverness and bravery by his friends. But they were also quick to remind him not to get cocky. After all, even though he was the smartest animal in the forest, he could still be taken down by the monkey-eating eagle![24]

Smith and his colleagues did more than just document fun stories. They explored the implications of storytelling for group welfare. They found that camps with more storytellers were more cooperative. They measured cooperativeness in different camps using games where people can divide tokens (each worth a small amount of rice) between themselves and ten other camp members. People from camps containing more skilled storytellers gave a greater share of the rice-tokens to other group members.

And who are these storytellers? The skilled ones are older: roughly fourteen years older than ones rated as "unskilled."[25] The coup de grâce is that Agta preferred to live in camps that contained more storytellers. Given the choice to pick up to five people they'd like to live with, skilled storytellers were named twice as often as unskilled ones, even after taking into account that people likely name kin and sharing partners. More surprising, if you think food matters most, is that these verbal wizards were more desirable as social partners than skilled hunters, foragers, or fishers. Lest you think that storytellers are simply altruistic raconteurs, it turns out that the best Agta storytellers were more likely to be recipients of others' generous bounty, and, more importantly—they have the highest reproductive fitness (that is, more kids—the currency of natural selection). Again, correlation doesn't mean causation (if you have more kids, maybe it pays to keep them entertained through ornate verbal sagas), but the study's findings are consistent with others reporting that people (especially men) bearing status-enhancing traits exhibit higher reproductive success.

One of my favorite activities whenever I'm with the Tsimane is listening to the old songs, which are chanted when people gather together, usually drinking vats of *shocdye'*. Tsimane stories and songs cover their origin, animals and plants, avoiding hazards, unique weather events, and the usual fanfare of melancholy country songs: love lost and gained, travails of marriage and children, and of course murder. These songs (called *jimacdye*) veer from the Tsimane-language church songs with written lyrics, songs that were invented by evangelical missionaries. The traditional songs erupt from the heart and involve improvisation, so people are sometimes nervous to sing from the outset. But the elders, *anic chij jimaquij, jam noiyij mu'in* (really know how to sing, without fear).

Working with my former graduate student, Eric Schniter, we asked Tsimane about who knows the old songs and stories, who delivers them, and when.[26] We found that elders (over age 60) were most likely to know and tell stories and sing songs. Over half of older adults perform music compared to just 20% of young adults. Not only do elders know more of these stories and songs, but they're also more likely to broadcast them, and when much younger kin are around. Younger folks consistently report that the stories they've heard came from elders, most often their own grandparents. It's not that storytelling necessarily requires many decades to master, but that the oral tradition is a specialized niche for many Tsimane in late life, coming at a time when productivity in subsistence is declining.

Older Tsimane also perform other services in the oral tradition, like interpreting dreams. In fact, 70% of older adults interpret dreams for others, over three times more likely than younger adults. What's the value of interpreting dreams? Apart from fun, analyzing dreams unveils important omens that influence daily decisions, often involving dangers in the forest. If a hollow log transforms into a snake and hides under some brush, you're better off staying in bed.

Those Who Can, Do.
Those Who Can't . . . Teach or Babysit

In the absence of schools, education and socialization in hunter-gatherer communities happens largely through observation and direct experience. This form of "natural pedagogy" occurs in a variety of ways, including gesturing, advising, correcting, demonstrating, adjusting pace, and encouraging (and sometimes even old-fashioned direct instruction).[27]

A few studies among Aka foragers of the Congo basin show that children learn much from their parents and peers, but older relatives are also vital. Eric Schniter and I found that Tsimane identified over ninety skills as having largely been transmitted vertically—that is, three-fourths of all Tsimane named as having influence on skill transmission were older, usually same-sexed, individuals. While parents are most commonly named, aunts and uncles are also important. Grandparents account for about 8% of the people listed as key purveyors of information and skills, which is still a great deal. But it's not just the quantity—it's also what they're doing. Grandparents are more likely to teach skills deemed "difficult" to perform, especially involving music and oral traditions, as well as other skills that are disappearing today, such as making bark clothing and ceramic vessels.[28] Young people also reported that grandparents and other older Tsimane relatives were essential to shaping their understanding of the medicinal and other uses of hundreds of plants.[29]

Even when they are not directly involved in instruction or pedagogy, older adults hanging around camp can make a huge difference for the little ones. Most importantly, they free parents to accomplish tasks that are much more difficult with a toddler on the hip, back, or straggling behind. This is division of labor at its best, working to improve efficiency of group production. Older adults often keep watch and entertain, and they also ensure safety, especially in tropical forests where biting and stinging insects, poisonous snakes, and other dangerous critters are real hazards. Among the Hadza, G/wi, Greenlandic Inuit, Ojibwe, Omaha, Hopi, and many other groups, older adults are steady companions of children.

While Coast Salish parents of the US Pacific Northwest were primary food providers for their kids, it was the grandparents who raised them. Grandparents taught many of the practical day-to-day skills, but also the less common but still noteworthy, like preparing for spirit quests and giving speeches at funerals and potlatches. According to ethnographer Pamela Amoss, a successful person is someone who "listened to his [or her] grandparents' words."[30]

The more relaxed and playful relationship often observed between grandparents and their grandchildren reflects a privileged pattern between "alternate generations." First proposed by the British anthropologist Alfred Radcliffe-Brown, this "friendly equality" enjoyed between grandparents and grandchildren is observed throughout Australia, West Africa, and North America. The effect this relationship has on everyday divisions of labor is profound. First, parents clean up all the proverbial messes, while grandparents (and even great-grandparents) are the fun caretakers who are also responsible for passing on cultural worldviews and molding the little ones into model citizens.

Elderly Asmat women would mind their grandchildren while their parents were away foraging in the mornings. Among the Mbuti of the

Ituri rain forest in Congo, elders also stay in camp to look after young children while able-bodied adults go hunting and foraging. According to the anthropologist Colin Turnbull, "By playing with them, acting out great sagas of the hunting and gathering days of yore, or just by lying back under the trees and telling stories, old women and men filled the youngsters with their own love of the forest, their trust in it, and their respect for the forest values that made life so good."[31]

With Great Wisdom Comes Great Responsibility

Old age and wisdom are intertwined. The Iroquois believed that long life and wisdom were highly connected. The Hebrew word for elderly, זקן (zaken), is an acronym for *zeh shekanah hochmah*, or "one who has acquired wisdom." But as the French novelist Marcel Proust once said, "We do not receive wisdom; we must discover it for ourselves after a journey that no one can take for us or spare us." In other words, wisdom reflects knowledge gained through experience, not just age. And such wisdom is often harnessed to help lead. Our world leaders are chosen for their capabilities and experience, with age standing in as a handy surrogate. While the 2024 US presidential election witnessed for a time the oldest final candidates in its history (81 and 78 years old), the majority of the world similarly elects older leaders: the median age of world leaders across countries is still a venerable 62 years old (same for men and women leaders). The oldest is President Paul Biya of Cameroon at 91 years, and only five of 187 leaders are under age 40.[32]

All societies have leaders, even in the absence of formal hierarchy and institutions. Whether leading groups on raids, providing direction and purpose to the group, organizing feasts and rituals, or even just helping family members plan their daily schedule, leaders with extra influence on group affairs are ubiquitous (and often necessary to win a raid, steal cattle, move camp, or make sure that everyone gets along). Recent cross-cultural studies by Zach Garfield show that leaders across fifty-nine nonindustrial societies are viewed by members of their group as intelligent, knowledgeable, and experienced.[33] These traits are associated with older age, but moderate associations leave room for exceptions:

you could possess all three as a precocious 30-year-old, and possess none as a daft 74-year-old. Nonetheless, perhaps because elders are believed more often than not to carry these positive traits, old age itself was mentioned by group members as a contributing factor to strong leadership in over half of these societies.

Let's look at one extreme of the political spectrum to illustrate the pervasiveness of elder leadership. The Batek of Malaysia are an egalitarian group of hunter-gatherers with no formal leaders, and the group actively suppresses violence among its members. Sharing is widespread, and people vote with their feet when there are major disagreements in the group. But even here, some voices go further than others. The anthropologist Kirk Endicott says Batek "natural leaders" are often older men or women respected for their "intelligence, experience and good judgement." Younger folks like to be in close proximity to these older adults, and often rely on them for advice.[34]

Other egalitarian foragers show similar evidence of elders playing key leadership roles.[35] In the Andaman Islands, older men and women directed community affairs. When there'd be conflicts with younger adults about where to move camp, the elders usually had the upper hand. Sometimes these elders rewarded themselves with the best shares of food.[36] The missionary-anthropologist Paul Schebesta observed that, among Mbuti foragers of central Africa in the 1930s, an elder's authority "derives from his age, by virtue of which he represents the lineage ancestors. . . . He takes precedence in all matters; the best part of the game is his, whether or not he took part in the hunt; he is responsible for settling disputes with other bands or with villagers; he leads at the head of the dance."[37]

While elders are often leaders, leaders needn't be the oldest. Among Okiek foragers of southwest Kenya, the leader (called the *kiroukindet*) of the clan council was "not necessarily the oldest member of the clan, for such a man might be too senile to act properly; but someone fairly senior in age, with his faculties intact, who is liked and respected."[38]

Hunter-gatherer groups referred to as "complex" tend to have hierarchical social relations, permanent settlements, large populations, elaborate food storage, and more intricate notions of property ownership. Here's where we see formal chiefs, elder councils, and the like.[39] Coast Salish leaders in the US Pacific Northwest maintained broad alliances among villages, networked through marriages and ceremonial "potlatches," where many gifts would be lavishly displayed and distributed among invited allies. While war leaders among Coast Salish may be brave young men, the real political leaders were expected to be "wise, gentle, courteous, forbearing and old," according to expert Pamela Amoss. In order to carry influence they also needed to have large families, presumably to cultivate a trustworthy and loyal set of allies and source of labor, and to effectively maintain extended households. Amoss tells us that "an old person without grandchildren was in as pitiable state as a child without grandparents."[40]

In hierarchical societies, elder leadership can take radical forms, as in gerontocracies like the *gerousia* council of ancient Sparta, whose members must all be over the age of 60. In gerontocracies, wealth, power, and influence are concentrated among the old (though mostly old polygamist men who monopolize the women). While these elders no doubt lead, they do so often at the expense of the young.[41] Among the pastoralist Tiv of Nigeria, only the elderly could qualify to be chief, and power increased with age. Younger folks may have all the requisite traits to be good leaders, but they rarely stand a chance in gerontocracies. (There are exceptions—a rare 27-year-old chief on the island of Samoa complained to the anthropologist Margaret Mead about how awkward it was to be a young *matai* leader: "I must act as if I were old. . . . Thirty-one people live in my household. For them I must plan, I must find them food and clothing and settle their disputes, arrange their marriages, . . . and the old men shake their heads and agree that it is unseemly for one to be a chief so young."[42] The poor guy's hair had already turned gray after just four years as a chief.)

While elder women's authority and influence may appear weaker than men's in gerontocratic societies, older women and men alike can have strong impact in forager and farmer populations. Once women reach the age of menopause, their influence increases considerably. Many ethnographies describe how menopause is a marker of a transition to women having greater community-wide impact. Among the

Mundurucú horticulturalists of Brazil, for example, a woman passing through menopause becomes a "leader in her household and among the village women—a personality in her own right, whom not even her husband will challenge in haste."[43] I've noticed this among Tsimane as well. Postreproductive women are very visible, have a lot to say, and are effective at making themselves heard, especially in village-wide meetings.

Mediate, Alleviate, Try Not to Hate, Love Your Mate

Another way that elders lead is through bolstering community. They help mediate important conflicts that might otherwise split families and group members. In order to reap all those glorious gains of cooperation, people need to be able to weather the storms provoked by social tensions. When conflicts of interest arise between siblings, for example, people may weigh their own interests over those of their brothers and sisters. By being equally related to conflicting factions, elders may be less biased and better motivated to make sure that their adult children get along.

My own hunch is that elders provide the critical benefit of mediation. I first got this idea from observing many Tsimane elders mediating different types of conflicts within the community. From accusations of stinginess to issues of theft, cheating, drunkenness, and wild runaway gossip, elders are either approached by members of the group or their own family, or they step in to intervene. Tsimane love living with their kin, but close proximity presents challenges. One out of five Tsimane adults reports having had a serious conflict with someone in the prior three months, and four out of five report a serious conflict within the last year. Most of these conflicts are with kin, and don't go unnoticed. Tsimane report getting support from others roughly half of the time, and elders, including those over age 60, were more likely to help.[44]

Similarly, Ju/'hoansi elders were known to "best maintain community life" among families living in core areas around particular water holes. They didn't do this by force or fiat, but by working with younger folks to generate group consensus.[45] The "quiet voice of the elders" would also help "put disputes into perspective" for the Mbuti, by "show[ing] why things had gone wrong, . . . appealing to the supreme value of *ekimi* and citing precedent after precedent." Mbuti elders had free rein to intervene in disputes of all kinds without fear of retaliation. Semang elders of Malay Peninsula also settled quarrels and disagreements, though unlike the Mbuti, might sometimes scream and shout to get the job done.[46] There are many more examples like these reported in the old ethnographies of small-scale societies.[47]

Of Sorcerers, Shamans and Ancestors

While church and state are largely separated in many countries, there is no clear divide between religious and secular worlds in small-scale societies. Hunter-gatherers are animistic: they believe that other natural beings, whether animals, plants, or concepts like "forest" or "thunder," can be intentional agents that affect human lives. Beliefs in afterlife are common, as are shamans.[48] Shamans can be charismatic healers, leaders, mediators, and multipurpose problem solvers. They use their hard-earned powers to bridge the material and immaterial worlds. They attract much attention in the public imagination because their process usually involves visible performances and altered states of consciousness. Of the seventy-one societies studied by Leo Simmons, the sociologist who I featured earlier, old men and women were found to be shamans in almost all of them.

Providing food, making tools, and caring for children seem like legit ways of making a living, with tangible results. But nullifying evil curses, communicating with invisible spirits, and removing the bones from someone and replacing them with stones? Some think that trance states help connect disparate parts of the brain, giving the shaman valuable insights and abilities. Or maybe their actions improve the well-being of patients because of our old friend, Mr. Placebo: we get better because we believe going to the shaman will make us better. A more cynical view is that shamans help fill the demand for a market that they created. When misfortune surrounds and looms on the periphery, anything that helps you gain control and reduce uncertainty will be a good thing. Shamans fill that niche.[49]

Being able to prevent or reduce harm from malevolent spirits is a pretty big deal if there are enough believers to keep you employed. Among Akamba agropastoralists of Kenya, if a woman's young child died, her breasts had to be ceremonially purified by a qualified elder to prevent any harm to her future children. A terrible curse could befall people too if a hyena were to defecate in the middle of the night. Again, elders need to sacrifice a goat and enact purification rites.[50] As a native Inuit in Greenland remarked, "We believe our Angakut, our magicians, because we wish to live long, and because we do not wish to expose ourselves to the danger of famine and starvation. . . . If we did not follow their advice, we should fall ill and die."[51]

Being a shaman-priest was viewed as "one of the choice economic strategies for ensuring well-being in old age" among the Wikina-Warao foragers of Venezuela.[52] Among Wikina-Warao, experience with the supernatural increases with age, unlike strength and stamina. Elder men are shamans, but so are women, at least after menopause. Different religious practitioners were involved in different types of curing activities. The *wisatu*, for example, helped ensure the well-being of children, protecting them from the "poison that causes pain." But the *bahanarotu* specialized in sucking poisons out that come from evil magic arrows. Dieter Hienen, who worked among the Warao for years, was convinced that the "manipulation of religious power and witchcraft" was how people made their old age secure.

Shamans don't just cure: they are the ombudsmen of traditional society. Among Lengua hunter-gatherers of the Paraguayan Chaco, shamans could frighten ghosts from the village and could save the spellbound by sucking the bone splinters out of their wounds. Palaung elders of Burma didn't just treat sickness, but concocted love potions, helped women conceive, and made charms to promote success in various life domains. They could also predict the future and, most importantly, bewitch annoying people.[53]

Shamans could play offense as well, using magical powers to sorcerize others. Asmat sorcerers of the swampy mangroves of Papua New Guinea were dreaded. They'd combine someone's hair or nail clippings with animal feces and shoot the mixture using an invisible bow and arrow at the victim. Once the poor victim was told what had been done to him, he would purportedly die in a few days.[54]

Part of what enables older adults to take on these privileged roles is that they're free from the taboos that limit younger folks. Among Akamba, only senior men and women could handle the wood of a particular tree that had powerful properties. The young would get sick or die if they tried to handle the wood. Similarly, older Ju/'hoansi are free to do many things denied the young. Spiritual energy, called *n/um*, is thought to be too dangerous for young to handle. This enables elders to play key roles in many initiation rites.

Tsimane shamans are called *cocojsi*. In the old days (*porojma* they say), cocojsi would drink native beer shocdye' with sky spirits in a special hut called a *shipa*. Cocojsi were religious gurus, educators, advisers, and respected group leaders. In addition to communicating with ancestors and helping to heal others, only cocojsi could treat victims of sorcery, a constant threat in Tsimane life. Certain illnesses after returning from the forest were also believed to be of supernatural origin, and so only cocojsi could treat them. They helped ensure that game and fish remain available by communicating with animal and fish guardians and by enforcing respectful behavior by community members. Cocojsi also knew where to find desired amber charms that are believed to promote successful hunting. Some Tsimane blame declining numbers of wildlife on the loss of shamans.[55] Over the past century of Catholic and evangelical missionary exposure, shamanism practice has almost disappeared. One of the last cocojsi died in 1999 at the ripe age of 84. He once asked me for some of my own magic for headaches and fevers (acetaminophen).[56]

Traditions, Ritual, and Rites of Passage

Some hunter-gatherers believe that deities and ancestor-spirits are like elders exerting their power beyond the grave. While alive, elders maintain prestige and authority as a medium between the living and the dead. Among the Ainu, a population of complex hunter-gatherers in Japan, the aged were viewed as the most sacred and revered group members, close in status to the deities. As such, the elders not only led, but were the authorities on most traditions and all things religious.[57]

And as authorities on tradition, elders are known to uphold the status quo. Long before Boomers or Millennials were blamed for rocking society's foundations, even Socrates was complaining about the recklessness of the young. Being "conservative" means to literally conserve traditions and values, a role usually attributed to elders. As a Greenland Inuit man defended, "we observe our old customs in order to hold the world up, for the powers must not be offended." Were that not the case, he said, "snowstorms would lay us waste, the sea would rise in violent waves while we are out in our kayaks, or a flood would sweep our houses into the sea."[58] There are serious consequences for violating tradition.

Elder Yaghan foragers from Tierra del Fuego play a similar role, according to Austrian anthropologist-priest Martin Gusinde: "They demand strict observance of the traditional order and good ancient custom. They also admonish anyone who in any particular point becomes negligent toward universal custom. . . . Everything that was handed down by their forefathers is much too sacred to them to have it destroyed, even in small part, by thoughtless young people."[59]

It makes sense that if elders once made the rules, then perhaps older folks benefit more from traditions being followed than by changing them whenever the youth get rowdy. A comprehensive cultural system works because all of its components fit together well and are keenly adapted to the long-standing features of the physical and social environment. Being a guardian of the old way helps to support a functional system that works, if not for everyone, at least for the elders.

Part of protecting traditions is being actively present for major life transitions. This includes birth, puberty, marriage, and death. Older adults, especially women, have prominent roles bringing new lives into this world. As we saw in chapter 3, mortality among mothers and their infants is high in hunter-gatherers and other subsistence groups. Childbirth is a risky affair for both mother and baby. Older women play important roles as midwives in small-scale societies, including the Hadza, Navajo, Ojibwe, Hopi, Arunta, and many others. Thanks again to Leo Simmons's cross-cultural scoring of traits across societies, we can be more precise: elder women acted as midwives in 92% of the forty-seven societies that contained relevant information. When an elder woman acted as a midwife, she was usually a close relative of the woman giving birth. With absent or minimal medical services, such midwives are essential.[60]

Likewise, Ju/'hoansi elders help ritually transform children into adults (the verb used to describe this process is n≠um, to create).[61] Old men initiate boys in the *tshoma*, a ritual in which, over the course of a month, initiates face cold, hunger, thirst, and prolonged dancing. During this initiation, elder men share secrets about the animals with the boys who are present. Elder women help with rites of passage for girls. The old women "beat axes together for sound and bare their buttocks" when dancing to celebrate a girl's first menses. Soon after first menses, Ju/'hoansi women usually get married. During a period of seclusion, an old woman, often the grandmother, stays with the girl while she transitions into being a woman. Many ethnographies tell similar stories about how elders are involved with these important rituals. Even among groups like Sirionó hunter-gatherers of eastern Bolivia, who lacked shamans or priests, old men and women organized the initiation ceremonies and scarification rites for the young.

Coast Salish elders were caretakers for folks in these "liminal" states, be they girls at menarche, women in childbirth, or spirit questers. Often grandparents were ideal attendants at these rites of passage, because of their experience and wisdom and their special privileges. In the case of girls' first menses, elders were "impervious to the girl's sacred contagion."[62]

One 21-year-old Hopi initiate into "manhood" describes the consequences of elders not being involved in the ceremonies: "The old people . . . said that when this ceremony was not performed correctly, famines occurred. They also cautioned that people who perform their parts carelessly often either die soon after or lose a relative. I resolved never to neglect the ceremony or to fail in its proper performance."[63]

Rights, Privileges, and Property

As we've seen, gerontocratic societies concentrate privilege and power in their elders, especially men. Among herders and farmers, especially those living in large homesteads with extended family, specific people claim ownership over land and animals, and those owners are often older adults.

Even among hunter-gatherers with minimal formal ownership, elders often hold important rights. Charismatic Asmat leaders (called *tesmaypits*) were usually elder men who controlled rights to use certain rivers and to specific groves of sago palm, the main source of starch in the diet. Among Ju/'hoansi, the oldest man or woman represents the group to outsiders who may desire access to the water hole or other local resources.

Social relations are also a source of power. Ju/'hoansi maintain long-distance trading relationships based on delayed reciprocity, called *hxaro*. These become essential when things get rough in your neighborhood. Hxaro partners accumulate over time, and elders tend to have the most partners. That not only puts elders in a position where they can regulate resource access, but social partners are a form of wealth that they pass to their children. Even in egalitarian hunter-gatherer societies, wealth passes from one generation to the next.[64]

I have discussed how the ability to acquire food peaks in middle adulthood. But what about status and prestige—those slippery, immaterial accolades that people seem to value as much, if not more, than material resources? In relatively egalitarian societies, status is earned through effort and skills. The more difficult or in-demand the skill, the better. Though you can't eat status or prestige, the esteem others hold for you brims with potential. It can be converted into food, allies, mates, and deference in important decisions. Status confers privileged access to contested resources. It sometimes has a *dominance* aspect, shared with our primate relatives. This reflects the ability to inflict costs on others. But the *prestige* aspect of status is more uniquely human. Prestige reflects the ability to confer benefits on others. Given the many different commodities juggling in play in interdependent human societies, prestige carries a lot of weight.

While limited in supply (by definition, status is relative), status tends to increase throughout adulthood in line with economic productivity. But as productivity declines, status in subsistence populations usually remains elevated, for at least a decade. Status and prestige don't come for free no matter how many years you've lived. While prestige, respect, and even deference often come with age, it is rare that elders are valued simply for being old. Among Yakut herders of Siberia, for example, the important *sesen* were "white-haired, honored old . . . who know everything." Not every old person was considered sesen. Sesen need to earn recognition for their knowledge, experience, and wisdom.[65]

Status is an important facet of daily life, and of older lives, but how to measure something so intangible? Chris von Rueden, now a professor of leadership studies at the University of Richmond, meticulously interviewed Tsimane villagers about their opinions of others back when he was my graduate student. He asked them to do paired rankings over who is listened to more, who might win in a direct confrontation, and other aspects feeding into local understandings of "status." He found that men's political status, and the respect that men garner from others, peak in their 50s, and then remains at a decent level up through the seventh decade of life. Women's political status shows a similar trajectory, though it is much flatter with age compared to men's. To date, political leadership is not usually the domain in Tsimane culture where women prosper. Though more general, the overall respect women receive with age is more revealing of status. Tsimane respect older women more than younger women. Beyond menopause, women's respect only increases, and more so than men after the 60s. In the Tsimane case, skill in food production matters, but access to allies matters most—for both attaining and holding on to status.[66]

When Disaster Strikes

Older adults are also valuable members of their groups, as walking libraries, experienced decision-makers, mediators, ritual masters, and culture preservers—roles that really shine when disaster strikes. When researchers and community health workers consider contemporary disasters, the usual focus on older adults is about their unmet needs and the fact that they are often more vulnerable to illness and death. For example, the death rate associated with Hurricane Katrina and Typhoon

Haiyan was five times higher among adults over age 60 compared to those of younger ages.

Natural disasters may disproportionately harm older adults, especially when they live under poor conditions with little support or infrastructure. But there is another angle to factor in when we consider the lives of elderly people during a natural catastrophe. A lifetime of experience helps build resilience in the face of adversity. Maturity and perspective build with experience, but this growth can also occur over time, and without direct prior exposure to natural disasters. In one study of hurricanes in South Carolina, experience and age were both associated with greater levels of disaster preparedness, like stockpiling bottled water, flashlights, and canned food.[67] Older adults have been shown to be especially motivated to protect family and their community. In another study of perceptions about disaster threats in the rural US, older adults said their motivation to prepare for disasters was a desire to protect their children, grandchildren, and neighbors, rather than their own lives.[68] I list some of the ways elders have helped their community in contemporary natural disasters in table 6.1. Other studies of elders in the context of disasters show psychological resilience: better coping and less anxiety, depression, and post-traumatic stress disorder. Elders provide emotional and other forms of support in the aftermath of a disaster.[69]

Margaret Mead once pondered that women living long past menopause would "know things that no one else knew," and could use that knowledge and experience to enhance survival. Even in some nonhuman primates, old animals have been observed leading the troop to water outside the traditional home range during severe droughts.[70] When considering the value of elders, Jared Diamond writes about how an elder woman was the only one left on Rennell Island (in the southwest Pacific) who knew of the fruit "eaten only after the *hungi kengi*." Hungi kengi referred to the biggest cyclone to have hit the island in living memory (in 1910). As their gardens were destroyed, and people faced starvation, they had to rely on foods that they normally ignored.[71]

There are good hints that elders in subsistence populations come in handy. As masters of their territory, Ju/'hoansi elders were relied on to

TABLE 6.1. Contributions by older adults during natural disasters

Emergency	Contributions of older adults
Hurricanes, tsunamis:	
Aceh tsunami (2004)	• Helped families in evacuation and cared for children during recovery • Told stories to children and cared for them in camps • Reached out to women and children to offer support and aid
Cuba hurricanes (over 155 yrs)	• Participated in all aspects of community emergency planning, response, and recovery (e.g., information and education on evacuation and home safety measures; weather watches and dissemination of local emergency directives; identification of local risks and safe, secure areas; cleanup, reconstruction, moral support to others)
Floods:	
Manitoba flood (1997)	• Served as volunteers (cooking, baking, donating money and clothing, fundraising, hauling sandbags, helping in shelters, socializing with evacuees)
Mozambique flood (2000)	• Provided traditional knowledge in predicting weather • Participated in community-based rehabilitation projects (e.g. home visiting vulnerable persons, organizing reconstruction efforts, planning and managing seed distribution in the community)
Earthquakes:	
Kobe earthquake (1995)	• Acted as models of resilience and resourcefulness • Acted as witnesses to relate the disaster and provide lessons for the future • Set up mutual aid and support projects in temporary housing • Offered ongoing outreach and peer support to other older people still affected by the earthquake
Kashmir earthquake (2005)	• Provided wisdom and coping skills learned from previous hardships • Cared for children and those who were ill and took in orphans • Used traditional position of honor and respect to keep families and communities intact and functional (e.g., taking responsibility for admission of camp children to the public school outside the camp) • Older imams provided counselling and teaching • Established a tented mosque for community worship
Other disasters:	
Bophirima drought (2002–2005)	• Supported families economically with their government pensions • Deprived themselves of food to feed children and grandchildren • Cared for grandchildren when adult children went to work in cities • Shared traditional knowledge and farming skills to cope with drought
Chernobyl power plant accident (1986)	• Served as historical witnesses of the event and as examples of taking control over personal destiny (by returning to home area) • Facilitated social and economic revitalization of previously evacuated area • Shared knowledge on how to minimize exposure to radiation in the soil
Lebanon armed conflict (2006)	• Provided care for others, including other older persons, children, and grandchildren, during and after the conflict

Source: Adapted from (WHO 2008).

know where to go to find food during rough patches. Among Warao of the Orinoco Delta in Venezuela, ethnographer Heinz Dieter Heinen describes how elders helped with seasonal shortages of their palm starch staple. During the multi-day *nahanamu* festival, starch from the moriche palm, called the "tree of life," is processed into a flour called *yuruma*. The great thing about flour is that it can be stored. Which it is, in a temple, where wise elders "guide [their] people through periods of scarcity." The festival begins only after the elder shaman-priest has the right kind of dream, and it's through dreams that the elder knows when to distribute the stored flour. The big organization and coordination involved in mass harvesting and processing showcases elder influence, and an elder's authority derives from direct contact with the supreme spirit called *kanobo*.[72]

You might be thinking this all sounds nice, but where is the hard evidence that elders affect the survival of other group members? One possibility comes from the work I have done with colleagues through the Tsimane Health and Life History Project. In early 2014, many Tsimane villages experienced massive flooding during one of their wettest rainy seasons in decades. Across the expanse of the tropical Beni region, hundreds of thousands of livestock died, and thousands of hectares of crops were ruined. As water levels rose, many Tsimane families abandoned their houses and fled to higher ground, to other villages, or to temporary encampments in the outskirts of the nearby town, San Borja. The Tsimane Health and Life History Project was fortunately well positioned to help with some evacuations and brought much-needed supplies to people affected by the floods. During the course of relief efforts, we discovered that over a third of adults had prior experience with flooding—and the people who had been affected by previous floods were less likely to relocate and set up fields in new locations.

More remarkable was that communities with more Tsimane elders fared better. People living in villages with more older adults (age 80+) lost fewer crops, and their children were less likely to be sick. These findings didn't change even when taking into account the size of villages and other factors.[73]

While 2014 was a terrible flood year, the Maniqui River floods some regions more regularly than not. And with the many hazards and risks of the Tsimane environment, I wondered whether being in the presence of older adults might improve well-being more generally. Does having more elders (say, age 60+) living in your village lead to a lower death rate among those under age 60? Yes, it does. An extra life is saved about every five years, when comparing living in the villages with the fewest (bottom 10%) to the most (top 10%) elders in the community. As above, this result takes into account other factors that might also contribute to the death rate.[74] Based on this (crude) correlational analysis, having more elders around a village of average size, say 150 people, means six more people are saved in just a single generation!

Utility Is in the Eye of the Beholder

Even when elders aren't ample contributors to the communal coffers, there are cultural expectations that they still make a difference. According to Incan law, elderly unfit for manual labor should, if nothing else, work as scarecrows to ward off critters from agricultural fields. According to Heinen, elderly Warao women would "try to be useful by incessantly weaving hammocks," whereas elderly Warao men would gather juicy moriche grubs, or fish from shallow pools.[75] Simmons remarked that old Chippewa (or Ojibwe, of the Great Lakes area across southern Canada and northern United States) women tanned hides, wove fish nets, winnowed rice, and peered over the shoulders of young girls while they worked; old men kept watch at night while others slept "with their feet toward the coals," and by day, carved ladles and pipe stems.[76] Simmons also reported that, among the Hopi, "retirement is impossible at any age" and that older adults "expressed the desire to keep on working until they died."[77] In no traditional society are able-bodied elders full-time vacationers. As Simmons once remarked, "any participation [is] preferable to complete idleness and indifference."[78] And participate elders do, in many different ways that add up. According to the ethnographer Henry Sharp, when a Chipewyan man can no longer hunt or trap animals, "his days as a complete adult male are ended." Sharp claimed that the Chipewyan didn't value stories, myths, handicrafts, or other activities that might otherwise provide status and respect to the no-longer-hunting elder man.

But you have to look carefully. Hunter-gatherers value much more than hunting. Among Tsimane, I've seen elders tell stories, sing songs, kill poisonous caterpillars and spiders crawling near infants, adopt a granddaughter when her mother died of tuberculosis, help grand-daughters learn to weave beautiful handbags, make medicinal brews from a dizzying variety of plants, help settle conflicts with (and be-tween) their children and grandchildren, scold kids for not sharing, be vocal advocates for a new well in community meetings, show their grandkids how to make bark clothing, and much more. When I for-mally studied several of these types of contributions (along with Tsi-mane Health and Life History Project colleagues), my impressions were confirmed: middle-aged adults and elders are most likely to re-solve conflicts and aid kin when their crops fail or when they fall ill. But they contribute in these ways only up through their mid-70s.

These activities that benefit others are not just filling time, but are the reason there's time to fill in the first place. Certainly elders may overpromote their prestige and glorify old age in folklore. Even Sim-mons thought that elders "embellish their tales with qualities and in-ferences which built up their own prestige." Probably everyone does that, not just elders. But elders behave strategically to help others. They often act in ways to maximize their impact. If elders prefer to benefit their youngest kin (who may need the most support), then you'd expect elders to live in camps where they could have the greatest impact. To my knowledge, this has only really been explored among the Hadza. And this is precisely what the Hadza do. Hadza move camp every two to four weeks. In tracking who lived with whom, older adults were found to be living with close kin having small children—exactly the descendants who might benefit most from having grandparents around.[79]

I've laid out an admittedly cherry-picked smorgasbord of the many ways that older adults make valuable contributions in traditional societies—what any elder does will vary by region, subsistence, social structure, health, and other factors. But some contributions do seem universal, like craftsmanship, giving good advice, providing information and entertainment, and babysitting.

What matters for evolution is the sum total of all the contributions aged people make. For the Tsimane, with whom I'm most familiar, that includes old-fashioned food production, making and fixing things, babysitting, teaching, engaging with song and stories, mediating conflicts, and applying knowledge when catastrophes strike. As suggested by figure 6.1, these contributions are likely to be impressive—at least through the seventh decade of life.

Among the Ju/'hoansi at Dobe, dozens of active older adults were in camp back when Richard Lee and his colleagues studied them a half century ago. They were still contributing food and acting as camp leaders. They were collective owners of prized waterholes and beacons of ritual and healing know-how. And they were still fed and helped by younger people once their productive years were behind them.

The Only Constant Is (Culture) Change

Every generation thinks the previous one was the golden age for elders. From Aristotle to Millennial/Gen-Z-hating Boomers, the popular view is that younger generations don't respect the old ways, or they brashly ignore what elders have to say. You'd think that in a rapidly changing world maybe the contributions I've mentioned in this chapter would become obsolete. Does modernization and culture change elder contributions, and their value?

No culture remains unchanged, and so what matters is how older adults contribute in a changing world. When groups like the Ju/'hoansi adopted farming and small-scale livestock herding in a more sedentary setting, elders were able to be *more* productive than they would have been under a more physically demanding, mobile hunting-based livelihood. Add to that the sales of crafts to occasional tourists, and settled life may carry advantages for elders. After declines in the 1980s, foraging saw a resurgence in the Dobe region. During that time, elder ecological knowledge was highly prized. Contrast this with an unfortunate situation: In 1990, many !Xū and Khwe (both are groups of !Kung) were relocated from Namibia to South Africa by the South African Defense Force. Being so far away, and with totally unrecognizable environs, they were unable to practice their more traditional livelihood. Remarking on their situation, !Xũ elders were described by one observer as "bored, dependent, marginalized and perceived themselves to be largely worthless."[80]

Elders may be losing out in other ways too.[81] While knowledge, wisdom, and rights helped support leadership and status of elders, leaders in many transitioning populations today are younger folks. These leaders speak nonnative languages and have experience interacting with neighbors, government officials, and other outside interests. While elder Tsimane, especially shamans, used to be local leaders, today elected leaders (called *corregidores*, or "correctors") are often men in their 30s or 40s. These younger men speak Spanish, have some schooling and some experience working with loggers or ranchers, and know their way around town. While elders still wield authority in other ways, they're less privy to the occasional wheeling and dealing that comes with being a corregidor in villages with resources others want (for example, mahogany, cedar wood, and *jatata* palm).

As some types of information become valuable, others grow obsolete and irrelevant. With missionization and greater regional integration, the Asmat of PNG no longer raid or head-hunt, which, combined with schooling and a partial cash economy, has resulted in a "ritual void" previously occupied by old men, the high-status tesmaypits. New rituals involving drumming and chanting are performed by women, who now play a larger role in contemporary ritual life.[82]

While many traditional rituals have indeed faded in recent decades, cultural revitalization of Indigenous identity in many parts of the world has placed renewed enthusiasm and attention on elders and their irreplaceable knowledge about the old ways. This is especially the case once traditions and language start disappearing. Elders maintain important roles in many Native American groups, like the Coast Salish. Coast Salish elders maintain prestige and high status by controlling information about old rituals, and by being imbued with hard-to-get spiritual power.[83]

Even when losing some usefulness, elders may gain access to newer resources of value to others. Among Tsimane, Ju/'hoansi, and other groups, elders now receive cash pensions from the state. In Bolivia, the BONOSOL social security program started in 1997 to provide extra income to adults age 60+ with no other state compensation. Now called Renta Dignidad (Income with dignity), elders can receive up to 4,550 Bolivianos per year (about $661 US). That's about two month's wage labor. Given that wage labor is still hard to come by, this is a hefty amount that can make a big difference for household purchases (and is probably why ages on government-issued ID cards are usually inflated). Among Ju/'hoansi, these pensions are even more lucrative, especially for those living in Namibia compared to Botswana. When adult children fight over who gets to be your caretaker, maybe it's because of your great sense of humor, but it might have something to do with that pension.

As I'll reveal in chapter 7, feeling useful and needed, and being seen as useful and needed by others, are key for how elders are treated.

7

Help Others and You Will Help Yourself

For old age is honored only on condition that it defends itself, maintains its rights, is subservient to no one, and to the last breath rules over its own domain

—CICERO, *CATO MAIOR DE SENECTUTE*, 38

A CLASSIC Brothers Grimm fairy tale tells of a poor frail grandfather whose "eyes had become dim, his ears dull of hearing, his knees trembled." When eating, he'd make a mess. His son's family was disgusted by him and made him eat paltry scraps out of a wooden trough behind the stove, out of sight. One day, the 4-year-old grandson was observed collecting wood. When asked what he was doing, the child said, "I am making a little trough for father and mother to eat out of when I am big." The boy's parents were mortified—and Grandpa was quickly invited back to eat at the table.[1] In a different version of the fairy tale, the father weaves a basket to throw the elder into the river. His son asks for the basket so that it can be used for the father one day.

As these stories illustrate, what goes around comes around. Indeed, the moral fiber of a society is sometimes measured by how the elderly are treated, especially when they're perceived as a burden. But insightful stories and concerns about our moral failings do not solve the problem of how to best treat our elders. Instead they just remind us of how hard it is to grapple with life-and-death decisions related to elderly members of society.

Many of us have been, or will one day be, forced to make decisions on behalf of our beloved elders. What factors matter most in the cold calculus that determines whether to preserve an elder's life? In chapter 6, I covered the many ways that older adults may be helpful in subsistence societies. Now we're in a good position to address some key follow-up points that will help us face the problem of late-life caregiving: How do elders in nonindustrial societies balance their need for support with encumbering their families amid declining health and function at later ages? How do these patterns compare to end-of-life decisions in the rest of the world? Our first tack is to see how older adults are treated within forager, farmer, and herder societies, in terms of cultural expectations and in daily experience.

Before we start waxing nostalgic for the way things used to be, historians will be quick to bark that there never was a golden age of aging. Wherever elders are held in high regard, with prestige and respect bestowed on them, we also find younger members of society treating them with irritation, hostility, and outright antagonism. In summarizing this apparent contradiction, demographer Barbara Logue remarks that the old have been "respected and loathed, loved and feared, supported and abandoned."[2]

In this chapter, I unravel the ostensibly cruel but practical cost-benefit logic that explains how older adults are treated in subsistence societies. We'll revisit the amorphous concepts of "prestige," "status," and "respect" to explore how these privileges affect the welfare of elders later in their lives. We'll see that the ways elders are treated begin to twist as they transition from asset to burden. As you probably guessed, that transition generally occurs in the 70s. But as people, situations, and cultures vary, this transition isn't always obvious. Given this fuzziness, we'll explore myriad ways that elders negotiate their own treatment— how they best support and are supported by the interdependence of family and community. When death doesn't come quickly, others step in to help expedite the passing. We'll explore the spectrum of these types of morbid "death-hastening" behaviors. This includes the sometimes sensationalized, and oft-misunderstood, killing of elders, or geronticide. Though such death-hastening seems grim, my exploration informs current debates about how to navigate our lives when healthspan can't keep pace with lifespan.

The Good Life?

A meaningful older age arguably entails "influence and fulfillment," which Leo Simmons took to mean the desire to live a longer life, to be free of physical hardship and excess exertion, and to maintain social rights and sustained participation in community affairs, all ending with a "timely and honorable closure."[3] This notion of the good life is certainly not unique to small-scale societies.

The good life, as told by elder Ju/'hoansi to the anthropologist Pat Draper (who once taught me how to make the click sounds in their language when I was a student at Penn State), is simple: "If you have a child, you have a life." Elders and their kin are bound together, especially parents, children, and grandchildren. A bad life for elders is when there are no adult children around, or when they do not support the older generation.[4]

This core idea about the value of children has been documented in many societies. Among the Yakut, having as many kids as possible was believed to help ensure the best social security in old age.[5] Among many groups, like the Herero, adult children were the primary caretakers of the elderly. The Chipewyan preferred to live with a daughter in their old age to receive the best care (believed to be a better situation than being stuck with a daughter-in-law).

Family is most important in traditional societies, but this doesn't mean that there's no strife. Conflicts of interest between parents and kids can be frustrating and painful. In studies from the 1960s, middle-aged Warao complained to anthropologist Johannes Wilbert that the worst misfortune to befall them would be to grow old without daughters and sons-in-law around. Yet even when aged Warao lived with one of their children, the elder was "never treated very well by any of them," and they were "often beaten when their nagging, complaining and begging become unbearable."[6]

According to ethnographer Henry Sharp, when no adult children were around, Chipewyan elders in the 1960s would sometimes adopt someone else's child, like a grandchild, to help lighten their load.[7] Many groups, like the Herero and Blackfoot, would also adopt children when no one else was available. The special relationships between grandparents and their grandchildren that I mentioned in chapter 6 are win-win opportunities for guiding and instructing kids while those kids help their elders. Among Gros Ventre (agrarian maize cultivators of Montana) this relationship was more formal: elder grandmothers would help train their granddaughters from around age 7 until marriage, in exchange for elder care and assistance.[8] Even without such formal arrangements, Tsimane elders say they like to live near their younger adult children, who will themselves have young children to help their grandparents and keep them company.

Who Are You Calling Old?

In chapter 1, I explored how the timing of old age onset is subjective. But most societies distinguish between elders who are able-bodied, healthy, fully functional—what we might call "successful aging" today—from those experiencing a downward spiral of a loss of vitality. Gerontologists refer to these two periods of elderhood as the "third age" (marked by achievement and fulfillment) and the "fourth age" (filled with dependence, falling apart, and ultimately death). More bluntly, gerontologists contrast the mature "intact" elder versus the frail "decrepit" elderly, with these two terms linked to either being an "asset" or a "burden." In subsistence societies, the decrepit period of grim disability is relatively short, compared to what's possible with amenities of modern health care and technical support. As hinted by the mortality patterns of chapter 3, and as we'll see in chapter 9, functionality declines rapidly for the lucky few late-life survivors who reach beyond seven decades. Chronological age is certainly relevant in these transitions. But despite our numbers-based fixation, age is rarely used to determine who is too old. What matters are changes in capacities and social roles. In a cross-cultural study, at least a quarter of societies made explicit the distinction between intact old and decrepit old. And this distinction matters: the intact are largely supported, while the decrepit are not.[9]

Golden Age That Never Was?

Among highly egalitarian hunter-gatherers like the Ju/'hoansi, Mbuti, and Batek, growing old may sound idyllic. Elders are cared for and attended to by family members, as well as by many other group members. There's no obvious decline in their status as they age into elderhood. Among Ju/'hoansi, elders aren't ridiculed or feared. As one woman explained, "What's there to think about? You see an old person, . . . she can't walk, she can't do it for herself, so you do it."[10] According to ethnographer Colin Turnbull who lived among Mbuti in the late 1950s and early 1960s, Mbuti elders didn't lack food or shelter, and any person who disrespected a Mbuti elder or ignored his or her request would be criticized or abused.[11] Among the Batek in the 1970s, even the blind old people were "helped by everyone."[12] The Okiek of Kenya treated their aged parents with "much kindness and consideration. . . . The more helpless the parent, the greater the trouble taken to make them comfortable."[13] Such golden treatment is reassuring, assuming seasoned ethnographer impressions accurately capture what goes on. So let's take these cases as an ideal of what is possible, at least when there are only a few very old people in the community.

On the flip side of these warm and welcoming geriatric havens are cases of intolerance, scorn, and rejection. As I mentioned in chapter 6, hunting was men's main gig among the Chipewyan of western Canada. As observed by the early explorer Samuel Hearne in the eighteenth century, "when he is past labour, he is neglected and treated with great disrespect, even by his own children." Without recourse, the elder man "submitted patiently to their lot, even without a murmur, knowing it to be the common misfortune attendant on old age."[14] Most groups tend to fall somewhere in between the extremes of all-is-great and all-is-terrible. And within any one group, elders may be treated in a variety of ways, just like we observe in any society.

From Asset to Burden Revisited

Intact to decrepit. Asset to burden. Young-old to old-old. The status and treatment of elders in traditional societies is based on a practical logic: you are better off being useful, influential, productive, and needed. As Simmons lucidly stated, "Security and survival in senescence are not a boon of nature, nor a gift of the gods; they depend upon the contributions which old people can make or the rights which they can command."[15] As we saw in chapters 5 and 6, the nature of those contributions can vary, even over adulthood. Among many groups, like the Ju/'hoansi and Asmat (of Papua New Guinea), aging hunters often shift to become political leaders or healers. Typically, women are able to keep up with foraging activities more easily than men can with hunting, so as long as a woman can see, hear, and walk, her efficiency may shift but she can still carry out most tasks.

Are elders appreciated for the many things they do, or are some activities not valued enough to be recognized? Babysitting, for example, is a vital responsibility—but unfortunately doesn't garner much prestige, even in modern urban societies, where reliable CPR-certified babysitters can earn decent wages. While high prestige warrants fair treatment from a broad range of group members, making yourself useful doing appreciated (but low-prestige) tasks can still motivate others in your social networks to look out for you.

Who Gets No Respect?

Is respect greater where elder activities are valued? An activity can be considered valued when it's deemed integral and necessary, like being part of a council of elders in the more gerontocratic societies like the Tiv of Nigeria. But this also includes household food production and other activities deemed locally valuable. A cross-cultural study of 135 small-scale societies tallied up all indications of "high respect."[16] These include when there are cultural obligations to aid elderly, when there are religious or legal sanctions against neglecting the aged, and when elders are fed and sheltered even if physically helpless or senile. "High respect" also includes when old age is viewed as a sign of good fortune, and elder knowledge and experience are held in high regard. On the other hand, "low respect" refers to when elders are neglected as their abilities decline, when they have to beg for food and shelter, and when caring for them is viewed more as a burden than as a duty.

About two-thirds of societies bestow high respect on their elders, and in a similar percentage of societies, elders participate in valued activities. As you might expect, these two occurrences are correlated: elders receive high respect in societies where they're reported to engage in valued activities. The probability that elders earn high respect is much higher in those societies where their activities are deemed more valuable.[17]

The general picture sketched from this type of comparison is that elders largely remain active and relevant, and garner prestige. Their active participation in important and necessary tasks makes earning prestige more probable, consistent with Simmons's conjecture in the 1940s.[18] That older adults remain engaged in important and diverse cultural activities well past the age of 65 flies in the face of antiquated and erroneous ideas that continue to permeate public imagination (and fuel public fears) about the impacts of a global elderhood that "disengages" or "withdraws" from community life.[19]

What's the Value of Respect?
Or, Better to Be Feared Than Loved?

There is a vague sense of honor or respect for elders everywhere, but how does that reflect actual treatment? Can respect and prestige fill your belly or shelter you from the rain?

James VanStone, an archaeologist working among the Chipewyan in the early 1960s, believed many of the ethnographic reports by informants about taking care of the elderly was cheap talk, meant to appease naive anthropologists. He thought the old were often disrespected, but tolerated for their government pension payments.[20]

By the 1970s, researchers were engaging with these ideas, bent on gauging how deference or respect actually played out in daily life. The sociologists Philip Silverman and Robert Maxwell made the best attempt to document what daily life might be like by identifying the many forms of favorable treatment. They focused on the ethnographic literature and discovered a host of examples. When Mbuti refer to elders as mother and father regardless of kinship (what is called *linguistic deference*), they are addressing their elders with respect. When Samoan elders are the first to be served *kava* drink at public gatherings, and Turkana elders are given the cherished liver of a roasted goat, the members of the community are engaging in the preferential (edible) treatment of elders, which is referred to as *victual deference*. When members of the society cook for an elder, maintain the household of an elder, and do other chores on their behalf, this is called *service deference*. One example: cooking and sewing torn clothes for elder Pawnee of the Midwest Plains. *Spatial deference* occurs when elders get the best seats at the concert or the comfy seat by the fire, and *presentational deference* is a hodgepodge mix of preferential treatment with respect to clothing, grooming, and other facets of physical appearance. Lastly, there's old-fashioned gifts and favors (*material deference*).

Table 7.1 reports the frequency of these varied forms of deference in older men and women across thirty-four societies.[21] There's a decent amount of preferential behavior in small-scale societies. It also appears that deference is greater among older men compared to older women, though it is difficult to tell if that is due to bias from (mostly) male ethnographers. Even if there was a bias, it's not consistent across deference categories. Victual and linguistic deference are the only types more commonly reported among men than women.

Sometimes elders might be treated well, but not out of deep respect or cultural ideology of elder appreciation. Rather, they are treated well because elders are believed to have a direct line of communication with spirits or supreme beings. Not a bad gig if you can convince others of your spirit-realm connections.

TABLE 7.1. Deference toward elderly men and women

Type of deference	Men (%)	Women (%)	Significantly different?
Victual	76	38	Yes
Linguistic	71	29	Yes
Spatial	56	35	No
Presentational	53	18	No
Service	56	59	No
Material	44	26	No

Source: Adapted from Silverman and Maxwell 1978.

Notes: Tabulation based on 34 societies. "Significantly different" refers to statistical differences in reports of deference between men and women.

Because of elders' privileged access to the spiritual world, children among Gros Ventre were taught to be kind to their elders, and to put elder interests above their own (as one adult explained to the anthropologist Loretta Fowler, "Old people—right or wrong, they are right").[22] Elder command over ritual, religion, and all things supernatural may help shift the interests of the community toward the well-being of the older people in the group. As I discussed in chapter 6, elders can be protective about sharing esoteric knowledge. A cynical interpretation is that elders limit supply to help maintain demand and, in effect, hold the reins of influence a while longer.

Magical powers and secret knowledge providing esteem and social security are examples of the human knack of separating dominance from prestige. If everything rode on your ability to fight schoolyard fisticuffs, status and treatment would peak in early adulthood and decline thereafter with the waning of muscular strength. In such a world, elders would be at a huge disadvantage. But the privilege of prestige rests on not needing to fight for every morsel, nor needing to extract favors through coercion. Why is prestige rewarded so freely?

An extraordinary feature of human social systems is that prestige rests on the skills, knowledge, and abilities that others find valuable. These include many of the things mentioned in chapter 6 that older adults are known to excel at, like healing, ritualistic services, leadership, information control, and conflict mediation. But if there are rewards for prestige, it's not just payback from past feats of glory. While it's nice to be appreciated for what you've done in the past, the world doesn't really work that way. Good treatment comes out of expectations about what you can still do for me now, and in the future.

But there's another dimension at play here that goes beyond a repertoire of valued skills and activities. As we saw in chapter 5, the multigenerational and cooperative nature of human groups requires long training periods throughout early life, lots of practice, and lots of learning. While kids often learn basic skills from their peers, more difficult ones, and those that are less common, require attention from proficient adults and older experts. By virtue of their success (based on past reputation and being alive), there may be much to learn from elders.[23] Co-residence, close proximity, and deference to elders are important factors that enable a young person to observe and learn skills from elders.

The same reason that some elders earn respect might explain why other elders get little to none—they are the unfortunates without notable traits, abilities, or connections. Members of the community (and even their own families) usually no longer prioritize the well-being of elders—no matter the past glory—once they have transitioned from intact asset to decrepit burden.

(Elder) Membership Has Its Privileges

One of the most commonly observed benefits is the ability to eat foods that are otherwise deemed off-limits. Older Ju/'hoansi can eat steenbok meat and ostrich eggs, tasty foods that are believed to make younger people crazy if eaten. Elder Onge of the elusive Andaman Islands get the most prized parts of wild pigs—the kidneys and juicy surrounding fat.[24] Among the Bolivian Sirionó, elders can eat anything they want. Some foods are reserved exclusively for the old, like howler and owl monkey, anteater, and the ever-grizzly harpy eagle. Maybe not the highest priority foods, but, beats leftovers. Chuckchi reindeer herders of Siberia claim that drinking reindeer milk can cause impotence in young men and flabby breasts in young women. Young adults avoid drinking reindeer milk, while the aged are actively encouraged.[25]

Older adults are often released from other prohibitions that might constrain younger folks. Elder Ju/'hoansi can better wrestle with spiritual energies (*n/um*) thought to be too dangerous for the young. Not only does that help them eat the foods just mentioned, but it gives them the ability to initiate the young into different life stages. Coast Salish elders also guide people through all the transitional or "liminal" stages of life, from girls at menarche to women in childbirth, and dreamy spirit questers. In these and other groups, grandparents are held in high regard as ideal attendants, not just because of their collective wisdom but also because of their unique ability to avoid contagion.[26]

Women of reproductive age in many hunter-gatherer groups are viewed as potentially dangerous, whereas postmenopausal women are considered safe, which affords them certain privileges denied others. For example, pregnant and menstruating women are believed to pollute things or people they come into contact with, especially hunters and hunting gear. Among Tsimane, they're believed to carry an odor that offends forest guardians and scares game away. This is not the case for elder Tsimane women, whose status as *chañej'* (dried up) affords them more liberties. Among Hiwi, a menstruating woman touching a man's bow can render him an impotent hunter. Menstruating Tsimane women are also not supposed to make manioc beer, *shocdye'*, and are limited in how they can prepare food and who they can be in contact with.

While living with Tsimane as a doctoral student in the 1990s, British social anthropologist Rebecca Ellis learned this lesson the hard away. A large group of Tsimane were eating food that she had proudly prepared while on trek, when she casually slipped into conversation something about her menstrual status. The family spit out the food, dumped the rest, and were furious with her, fearing that illness was imminent. Were she to have made shocdye' while menstruating, the situation would've been dire. Shocdye' is viewed as an extension of the woman who makes it. As a Tsimane woman explained to Rebecca, "*Jäsi ra mi jun'si chäshi mi - chaija' ra tsun', a'chis shupqui*" (If you make beer when you are bleeding, we will all vomit, it will turn out bad and ugly").[27] Perhaps for this reason, postmenopausal women are recognized for making huge delicious vats of (safe) shocdye' that attract thirsty enthusiasts far and wide.

Coast Salish took this type of privilege a step further. Pregnant and menstruating women had to also avoid fishing sites, clam beds, and even shamans. But once postmenopausal, none of these taboos apply.[28] After menopause, women sometimes do things normally viewed as men's activities; among the Pomo of northern California, they can smoke and go to sweat lodges. Or postmenopausal women exhibit new abilities, like elder Omaha woman who could make game animals appear during periods of famine.[29]

Deliberate Death-Hastening

Leo Simmons glibly referred to frail elderly as "decrepit," but also more prosaically as "living liabilities," "over-aged," or in the "sleeping period" or the "useless stage," even "already dead."[30] And it is during this phase of extreme frailty and diminished capacity that death-hastening behaviors come into play. On the mild side, these behaviors include insulting elders, serving them scraps at the dinner table, and neglect. On the more extreme side, temporary or even permanent abandonment, and, of course, geronticide.

These behaviors have been documented in a wide range of societies. Though geronticide has been reported in many societies throughout history, the most reliable reports are based on insider and eyewitness accounts. Reports about sensitive subjects by explorers, and sometimes careful scientists, may not always be reliable, especially if based on hearsay or quick impressions. Biologist and geologist Charles Darwin may have been right about a lot of things, but his impressions of native Tierra del Fuegans during his travels on the *Beagle* were way off. Among other things, he thought the old women were killed once they could no longer work. On the other hand, Austrian priest-anthropologist Martin Gusinde, who spent almost two years there in the early twentieth century, proclaimed "this much is certain, no one is abandoned because of his infirmities."[31] About the Selk'nam, he wrote, "The entire neighborhood is concerned about them [old people]. One keeps them company, and it soon becomes a rule that the neighbors gather in the hut of an old couple for evening conversation."[32]

TABLE 7.2. Cross-cultural survey on treatment of the aged

Treatment type	# mentions	%
Supportive	49	35
Nonsupportive	32	23
Own dwellings	3	2
Nonspecific	1	1
Insulting	3	2
Property	10	7
Witches	15	11
Death-hastening	59	42
Forsaken	17	12
Abandoned	16	11
Killed	26	19

Source: Adapted from Glascock and Feinman 1980.
Notes: n = 60 societies. Supportive behavior is deliberate action to help elders. Nonsupportive behavior includes elders living apart in their own dwelling, insults or ridicule directed at the old, elders forced to give up their property, being regarded as witches, or "nonspecific" mild maltreatment. Death-hastening behaviors are more severe actions.

Table 7.2 assembles information organized from a systematic sample of sixty societies by anthropologists Anthony Glascock and Susan Feinman, showing that supportive community treatment of elders is evident in over a third of these societies. However, nonsupportive community behavior, like insulting elders, regarding elders as witches, or forcing elders to live apart from the family or to lose property, was evident in about a quarter. The more severe death-hastening behaviors were present in 42% of societies. One in ten showed evidence of abandonment, and one in five included evidence for geronticide.[33]

The main reason that elders are held in contempt (or negative deference) across societies is physical infirmity, which renders elders unable to carry out their share of the burden of work and unable to care for themselves. When elder contempt is evaluated by summing up the full range of negativity in a society, from feeding on food scraps to geronticide (as shown in table 7.2), physical and mental frailty and loss of skill together explain much of the contempt documented across nonindustrial societies. Another important factor for holding elders in contempt is when their appearance deteriorates, not just with wrinkles and sag, but an overbearing "ugliness." But the biggest single predictor of contempt level was the absence or loss of children.[34] Elders are more likely to be defended, protected, and cared for if they have close family living nearby.

Another Side of Geronticide

An old Khoekhoe woman abandoned by her family was found by a missionary in the mid-nineteenth century. Khoekhoen were nomadic pastoralists from Botswana. She told him, "Yes, my own children . . . have left me to die. . . . I am very old and am not able to serve them. When they kill game, I am too feeble to help in carrying home the flesh. I am not able to gather wood and make a fire, and I cannot carry their children on my back as I used to."[35]

Mobile, nomadic life in the jungle was also rough, moving camp every few days. Not being able to keep up with the band had dire consequences, especially for elders. Usually it meant being abandoned. One old Ache man was abandoned in his sickness. The man was too weak to move and protect himself from the vultures that were circling overhead in anticipation of his death. Luckily, the man recovered and returned to the band, though didn't come out of it unscathed: he was given the nickname "vulture droppings" because he returned to camp covered in bird feces. Other elders experiencing similar circumstances were sometimes buried alive, which was deemed preferable (by others in the community at least) to being scavenged by hungry vultures.[36]

Whereas feeble Ache men and women were sometimes abandoned if they couldn't keep up, only old women were said to have been killed outright. The usual way was a clubbing on the head with an ax while the woman was distracted. Among Chukchi, the elder was usually strangled by two men pulling on either side of a rope around the neck, while the victim was wrapped in a shawl with their head resting on the knees of their spouse. Or the victim helps another push a spear or knife into their heart.[37]

In many of these cases, the elder voiced a desire to die, and others discussed and tried to talk them out of it. Among Siberian Yupik, it was custom for an elder to ask three times for someone to assist, with the third request being the one that can't be refused.[38]

And usually a relative, like a son, would make the fatal blow. One sorrowful tale makes my Bar Mitzvah sound like a gentle and comforting rite of passage:

> A [Chukchi] hunter living on the Diomede Islands related to the writer how he killed father, at the latter's request. The old Eskimo was failing, he could no longer contribute what he thought should be his share as a member of the group, so he asked his son, then a lad about twelve years old, to sharpen the big hunting knife. Then he indicated the vulnerable spot over his heart, where his son should stab him. The boy plunged the knife deep, but the stroke failed to take effect. The old father suggested with dignity and resignation, "Try it a little higher, my son." The second stab was effective.[39]

When the mortal blow was delivered as depicted by his son, the whole event was deemed less painful to the elder Chukchi man.

In other cases, no one person overtly deals the fatal blow, and so no one is really to blame. And, if sufficiently able-bodied, the frail elder might not perish after all. For example, among the Tiwi of Australia,

> The method was to dig a hole in the ground in some lonely place, put the old woman in the hole and fill it with earth until only her head was showing. Everybody went away for a day or two and then went back to the hole to discover, to their great surprise, that the old woman was dead, having been too feeble to raise her arms from the earth. Nobody had "killed" her, her death in Tiwi eyes was a natural one. She had been alive when her relatives last saw her.[40]

Where geronticide occurs, elders are usually respected and treated well, at least up until their demise. This is not a contradiction. The killing is often done by a close family member, not by ruthless ageist murderers. And it's done by consensus among an intimate group of family and familiars, including the elder, and often in a ceremonial way.

For example, among Labrador Inuit, geronticide reflected wishes of the elder in question. Succumbing was believed to be "proof of devotion" that would provide "heavenly compensation for dying voluntar[il]y."[41] The elder's spirit could reach the Aurora Borealis, where in the afterlife they'd be reunited with "courageous souls" and, even better, be able to "play football with a walrus head."

These decisions were not taken lightly, as is the case in the contemporary context of euthanasia or assisted suicide. Where distinctions between intact and decrepit exist, it is the decrepit who are vulnerable to geronticide or abandonment. Prolonging life at all costs has never been a goal in most societies. In the early twentieth century, Samik, a Netsilik man from northern Canada, explained to the polar explorer-anthropologist Knud Rasmussen why other group members' seemingly harsh treatment of a frail elderly woman named Kigtaq made sense:

> And if he [a hunter] has to be at the breathing holes next morning at the proper time to secure food, he cannot travel backwards and forwards between the old and the new camp to salvage an old woman. He has the choice between helping one who is at death's door anyhow, and allowing his wife and children to starve. This is how it is, and we see no wickedness in it. Perhaps it is more remarkable that old Kigtaq, now that she is no longer able to fend for herself, still hangs on as a burden to her children and grandchildren. For our custom up here is that all old people who can do no more, and whom death will not take, help death to take them. And they do this not merely to be rid of a life that is no longer a pleasure, but also to relieve their nearest relations of the trouble they give them.[42]

Geronticide may seem like an indicator of moral bankruptcy, and the ultimate disrespect to elders. But it conveys a sense of how a culture views aging and death. Throughout history, societal fears and anxieties about aging manifest through works of art and fiction. There were no decrepit old people living in the fictional utopias of the sixteenth century.[43] In the 1976 dystopian movie *Logan's Run*, about life in the twenty-third century, anyone reaching age 30 was met with a quick demise to help keep the population in check. That movie strikes a chord with its audience because 30 is so young—an age that is certainly not when most of us think we've reached our prime. The more recent movie

Plan 75 has a related premise: in the near future, Japan institutionalizes the euthanizing of elders once they have reached the age of 75. Again, this screams "horror" because it arbitrarily targets people based on a number, and one that is well below Japan's life expectancy.

The 2019 folk horror film *Midsommar* introduces a cult in the rural Swedish countryside that considers it an honor to throw themselves (or be pushed) off a high cliff once they reach the age of 72. If an elder miraculously survived the fall, community members would shriek in empathetic pain, before crushing the elder's head with a hammer. Though far-fetched, this practice is based on a Nordic legend of the Ättestupa ritual, whereby elders would be killed in a similar fashion once they were no longer deemed useful.

But geronticide, as it occurs in subsistence societies, isn't based on reaching an arbitrary milestone age. Nor is it systematized extermination.

When Does It Pay to Delay (Death)?

Though abandonment and killing have been common throughout history, their patterning reveals a careful cost-benefit logic. To paraphrase Simmons, such behavior reflects the hardness of life, not the hardness of hearts. Family members face an underlying tension between competing sentiments: the heavy investment and care needed to keep an elder alive, compassion toward loved ones, and mercy to end suffering.

So we'd expect geronticide to be most targeted toward the feeble "decrepit" elders, like we saw in the expression of contempt—and that it would be more common in societies or contexts where harsh living conditions make the support of elders beyond their warranty period an overtaxing burden. As you might imagine, nomadic groups living under precarious conditions with erratic food supply are the groups that are most ill-equipped to take care of their elders.

Anecdotes like those I have described are both shocking and revealing, but cross-cultural studies are our best shot if we wish to understand the conditions leading to geronticide. I compiled some useful relationships between different factors and death-hastening behaviors from several studies by a team of sociologists (including Robert Maxwell and

Philip Silverman who we met earlier when discussing types of elder deference) that included a hundred or so societies (see Appendix table A7.1). These studies confirm that elders were about four times more likely to be abandoned and killed when there were complaints about their physical weakness (26% versus 7%).

As expected, geronticide is reported twice as much in more nomadic groups than in those living a more sedentary existence. Remember above, with the highly mobile Ache, and also with another tropical forest group in nearby eastern Bolivia, the Sirionó (where ethnographer Allan Holmberg believed the very old were considered "excess baggage, . . . having outlived their usefulness they are relegated to a position of obscurity. . . . They move at a snail's pace and hinder the mobility of the group"[44]). Similarly, mobile Inuit living in the Canadian Arctic sometimes abandoned the elderly, whereas the more sedentary Inuit living on St. Lawrence Island did not. Yet both groups held similar attitudes toward elders; both bestowed honor and prestige on the old.[45]

Farmers are more sedentary than foragers and herders, and usually have a more predictable food supply. So it makes sense that geronticide was three times more commonly reported in foragers and herders than among farmers. And among farmers, complaints about the physical deterioration of the elderly was only half as common. (Remember from chapter 5 that elders in farming societies maintain productivity until later ages than among foragers.)

Other societal features seem to matter, too, when it comes to predicting gerontocide. Societies with greater social hierarchy or rigid social structure are less likely to have reported cases of geronticide, probably because these societies are more sedentary, do more farming, and may be more likely to have a gerontocratic political slant. Similarly, societies with bilateral descent (who trace property and inheritance rights to both maternal and paternal sides of the family) show more geronticide. One possibility here is that bilateral descent, now common in much of the Western world, was once favored as a form of risk buffering in harsh environments. Another possibility is that, in the case of bilateral descent, networks are more diffuse and so any one member may be more expendable.

Two results seem inconsistent with the idea that more severe forms of death-hastening, like geronticide, will occur under less forgiving circumstances. The first one shows no significant relationship between environmental harshness and either gerontocide or complaints about the physical capacities of the elderly. A simple, direct measure of "environmental harshness," though, is tough to capture. It was assessed in a crude way in the holocultural studies of the 1970s and 1980s. The authors of the studies (compiled in the Appendix) had simply grouped together populations living in the Arctic, tropical regions, desert, and nontemperate forest as inhabiting "harsh" regions, and anywhere else as "moderate or mild." The other puzzling finding was that those living with a more regular food supply bore witness to more geronticide than those living with unpredictability. This is the opposite of what is found based on Simmons's sample, but who knows how Simmons came up with his measures.[46] Methodological issues aside, maybe elder knowledge is more useful when resources are less predictable, as suggested in chapter 6.

Memento Mori: Slowly Dying or Quickly Living?

So much seems to turn on the rapid decline in physical condition and function. Is it so obvious when a person shifts from intact to decrepit, useful to useless, asset to burden? Witoto hunter-horticulturalists of the Colombian Amazon abandoned their seniors unless they still "possess[ed] great wisdom and experience."[47] Among the Ju/'hoansi, elders were well cared for, and folks would often say to multiple seasoned researchers that geronticide simply didn't occur. Further targeted questioning eventually revealed that elders were left behind, but only under very desperate circumstances.[48] The burden sometimes was very real, literally carrying elders on one's back when moving camp. During a drought, famine, or other period of hardship, care of individuals once deemed manageable would quickly become unbearable.

In many societies, a distinction is made between being sick with the hope of recovery, and dying. People who are dying may experience a social death prior to physical death. In some groups like the Tsimane and Ache, extremely frail elders are sometimes referred to as already dead, beyond personhood, beyond help. This stage between living and dead is captured best by the word *mate* that appears in many Melanesian languages. It's usually translated as "dead," but it also means "very sick" and can be used to describe someone who is, according to the anthropologist W.H.R. Rivers, "healthy but so old that, . . . if he's not dead, he ought to be."[49]

Bring Out Your Dead . . . I'm Not Dead Yet, I Feel Fine!

Based on his experiences among the Sirionó, Allan Holmberg claimed that the "aged and infirm are weeded out shortly after their decrepitude begins to appear." Other ethnographers make similar claims about other groups, suggesting that severe frailty is a sorry state of no return. But when is that period exactly? It varies within and among groups depending on living conditions, food supply, and other amenities, but there may be some common features. For example, in the case of the nomadic Blackfoot hunter-gatherers of the US Great Plains, this was when an elder was physically weak, and could no longer walk long distances or see well. That's when "a canopy was erected wherein they sat and gradually died of starvation."

The struggle to still feel relevant, involved, and necessary is real. Old age may be honored, but like the Cicero quote at the beginning of the chapter, good treatment requires some effort. My longtime colleagues at UCSB, John Tooby and Leda Cosmides, called the struggle to get others to take notice the "banker's paradox."[50] The desperate person in need of a loan is exactly who a banker might least desire to lend money. Meanwhile, those who are rich are the safest bets for the bank. Thus, even if elders hold prestige for what they've done in the past, they may be "credit risks." To ensure others are deeply invested in your well-being, you've got to make yourself needed, to cultivate those desired skills, abilities, or activities in high demand but short supply. Brian Wood, an anthropologist with many years of experience with the Hadza, told me about a very old Hadza man whose back pain, among other physical hardships, reduced his foraging to a minimum. But he was well liked and respected, fed and cared for by everyone without complaint. More to the point, he was a charming talker and entertainer, and known to be a whiz at resolving disputes. Wood's example nicely illustrates that costly care is not rewarded freely. The cost may even seem negligible when the elder is viewed as essential.

When inheritable property is at stake, there's additional tension, and opportunities for leverage. Give up the property too soon, and your eager-to-inherit children may show less enthusiasm in tending to your needs. Hold on to it too long, and you may drain your dwindling energies and build runaway resentments. The Etal Islanders of Micronesia, with their chieftain system, resolved this matter of what to do with valuable property by holding on to just enough land and animals to ensure sustained care by their kids who stand to inherit.[51] Among Tallensi farmer-herders of northern Ghana, such tug-of-war over familial resources held by the elder patriarch was resented by the eldest sons eager to inherit. These impatient sons were said to hate their fathers, unlike the latter-borns. According to the ethnographer Meyer Fortes, eldest sons wished their aging fathers would die, thereby freeing up their resources. But guilt over such feelings was channeled into a mystical fear of the ancestors. If a parent dies angry with his children, you can imagine how that might instigate a fear of the ancestors.

Living far too long may raise suspicions, as if it were only possible by having supernatural or magical powers. (Keith Richards I suspect falls into this category.) The fear and distrust don't seem to have earned older folks any bonus points, but may have at least kept the greedy relatives at bay. And maybe it serves to offer up some care and attention.[52]

As Fortes writes, "one can live without the help of a living parent, if necessary, but not without the mystical protection of one's dead parents."[53] What better way to improve how your family treats you than to threaten to haunt them from beyond the grave?

Complain, Complain, Complain . . .

Elders do not usually suffer in silence and passively wait to be noticed. When asking Ache or Tsimane about who helped them get by during nasty bouts of illness or debilitating injury, I'd often be met with the same answer: "Nobody helped me." That jarring response seems to shatter the image of widespread cooperation and support in hunter-gatherers. If true, that would imply that people didn't eat or drink anything for days. When they are prodded, however, it's clear that various folks do help, including spouses, siblings, other relatives, and neighbors. Sometimes the help may be viewed as inadequate, but it's apparently so obvious that they would be supported if something bad were to happen that it's not even mentioned. Just like there's no traditional Ache or Tsimane words for "thank you"; why thank someone for behaving like you're supposed to?[54]

As explained by Koka, an 80-year-old Ju/'hoansi woman, "Old people have long complained; it is an old thing. Even if the child did everything for them, they would complain."[55] Voicing complaints is common, but doesn't necessarily convey true frustration and hardship. A common response to "Who looks after you?" is "No one! Can't you see that I am starving and dressed in rags?" Such "complaint/neglect discourse," as it's called, guards against hoarding, and keeps goods and services in circulation. Indeed, Ju/'hoansi elders have honed their complaining to levels of entertaining performance art. In a telltale example, the elder Kasupe complained emphatically in glorious detail how no one took care of him when his foot got caught in a hunting trap at night. But this exact event was prominently featured in Richard Lee's famous 1979 ethnography on the Ju/'hoansi. That book even includes a photo of Kasupe surrounded by a large group of family and healers frantically helping him. When the anthropologists pointed all this out to Kasupe, and showed him the photo evidence, he and his audience burst into raucous laughter (figure 7.1).[56] There was no shame having been caught lying, as it was an enthralling story that captivated his audience. But as Ju/'hoansi expert Harriet Rosenberg explains, the story itself illustrates what it might feel like to actually be neglected and abandoned.

Rosenberg and others who have spent a long time with the Ju/'hoansi say that the Ju/'hoansi complain all the time. Complaining seems to be a way for older adults to signal "I'm still here," still active and embedded in mutual care networks. As my mother used to say, "I'll stop complaining when I'm dead."

FIGURE 7.1. The elder Ju/'hoansi, Kasupe (standing), laughing after caught lying about not receiving any help after a tragic accident that had occurred about two decades earlier. His friend /Twi (seated) is clearly amused. *Photo credit:* Rosenberg 2020.

Eat, Drink, and Remarry

In the short poem "Risk," Anaïs Nin observed that the risk involved with blossoming can outweigh the pain of remaining a "tight bud." One way that elders without support can blossom is by remarrying after the passing of a spouse. This is an opportunistic strategy often reported in the ethnographic record.

FIGURE 7.2. **Tsimane marriage.** The man, age 68, and woman, age 58, lived in a relatively isolated community deep in the forest back in 2003. Over a four-decade long marriage, they had six children survive past childhood, four of whom as adults lived in the same community. The woman died at age 68 in 2013, and the man was still alive (aged 86!) in 2021. His four children still live with him in the community. *Photo credit:* Michael Gurven.

On average, husbands tend to be some years older than their wives, and so an older widower may find it easier to remarry than a widow. Being married is known to extend lifespan (especially among men) in contemporary urban populations, while being widowed without remarrying is associated with shorter lifespan.[57] Among Tsimane, your spouse is expected to be your main partner—working together in all things, despite marked divisions of labor by sex. If your spouse dies, it's usually desirable to find another partner if you can. The man in figure 7.2 may have been too old to remarry when his wife passed (he was 78 at the time). Fortunately, he is being looked after by his four adult children and their families. Two of his daughters, co-wives of the same man, never had any children, and enjoy the company with their jovial father.

But women remarry as well. Among the gerontocratic and polygynous Tiwi of Australia, it was believed that females of all ages should be married, including girls even before they're born. These yet-to-be-borns were betrothed by their fathers to an adult man. The marriage itself would typically occur once the girl was about fourteen, when her husband was then about forty. The large age gap meant that by the time the woman was fully adult (say ~25), her husband was close to geriatric. He might get some support from his wife, but when her husband passed, her son or brother would remarry her to a younger adult man.[58] Not unlike Lonely Island's classic hit song "Motherlover," brothers might even swap mothers in widow remarriage. Though men's accumulation of wives, in a polygynous system where all females should be married, may sound morally questionable to Western sensibilities, it at least helped ensure security for postmenopausal women.[59]

A Day without Me

As a last resort when there are no other options, elders sometimes take matters into their own hands. Among Inughuit (Greenlandic Inuit), this might happen when one's life "became heavier than death."[60] Suicide, and assisted suicide (sharing a blurry boundary with geronticide), have been most documented among Inuit groups from the circumpolar north (for example, northern Canada, Greenland, Alaska, Siberia). As expressed by Samik, the Netsilik man in the opening quote, a common theme to many cases is the view that one is no longer useful.[61]

I heard several stories about elder Tsimane men who poisoned themselves with charcoal or other toxic substances soon after their wives had died. In two cases, those elders had no living adult children who might otherwise provide care. I estimate a high suicide death rate of about 1,900 per 100,000 among Tsimane aged 60+. In the anthropological demographic land of small numbers, this rate reflects only four deaths, but that's still over ninety times higher than US suicide rates. I guesstimated those Tsimane adults to have been roughly 68 to 82 years old when they died. Suicide among older Tsimane often followed spousal loss (that is, a widowhood effect), along with feelings of being a burden due to disability and poor health.

Suicide among elderly people may also have occurred among Hiwi but was unheard of among Ache and Hadza.[62] Information on sensitive topics like suicide are hard to come by, but in his magnum opus, Simmons reported evidence of suicide among elderly in eleven of the seventeen tribes with available information.

This situation may not look terribly different from that in modern nation-states. An important predictor of suicidal ideation is the nagging perception of being a burden to others, especially loved ones. In many studies, this perception of being a burden predicts elevated suicidal ideation and suicide attempts even when taking into account other common risk factors like depression and loneliness.[63] Psychologists have pointed out that suicide in this case may be a distorted extension of the self-sacrificial behavior often displayed in other highly social and cooperative species.[64]

While the impression given here is that elders may willingly end their own lives to help relieve the burden on kin, sometimes suicide attempts among elders may reflect the ultimate protest against real or perceived maltreatment by relatives. Among the Etal, old men and women might "cast themselves adrift in a small canoe" as a public protest against perceived neglect. If resulting in a death, there would be a lot of shame on the accused relatives. The mere threat of suicide acts like a negotiation tactic. If shame alone isn't enough, the transfer of property of an Etalese who died by suicide is quite contentious.[65]

Interdependence and multigenerational cooperation set the stage in human foraging groups for how adults were treated once past their prime. Throughout most of human history, there were no 401(k) accounts, no retirement savings. Social security depended on others. Elders belonged and were needed. Sans libraries and internet, many defer to elders, given all the different ways that elders contribute to their extended families and communities. Once their main gig is up, elders switch things around, changing their role from, say, gatherer to mediator to storyteller (that is, if the stories are still worth listening to). When care is not forthcoming, elders use their wits and wiles, if necessary. If you own property and can control territory and other resources, as among many herders and farmers, even better. If you are living a more sedentary existence, you can probably catch a break. But once physical frailty overwhelms, elders find themselves increasingly dependent on others, and this dependence increases when the elder lives beyond seven decades. About this time, and certainly by eight decades and beyond, if not privy to an accident, illness, or battle, there is a good chance that a merciful release awaits, a pragmatic and final expression of familial love.

Part III

8

Saving the One-Hoss Shay

Just the hour of the earthquake shock . . .
The poor old chaise in a heap or mound . . .
You see of course, if you're not a dunce,
How it went to pieces all at once.
All at once, and nothing first.
Just as bubbles do when they burst.

—OLIVER WENDELL HOLMES, *THE DEACON'S MASTERPIECE*
OR, THE WONDERFUL "ONE-HOSS SHAY": A LOGICAL STORY

Though I look old, yet I am strong and lusty; for in my youth I never
did apply hot and rebellious liquors in my blood; and did not, with
unbashful forehead, woo the means of weakness and debility: therefore
my age is as a lusty winter, frosty but kindly.

—WILLIAM SHAKESPEARE, *AS YOU LIKE IT*, ACT 2, SCENE 3

WE AGE and die. But death isn't programmed, and aging is neither adaptive nor an adaptation. As I described in chapter 2, physiological aging occurs as a by-product of "decisions" that affect how our body performs in its productive (and reproductive) years. Natural selection favors youth and prime adulthood, but the good works of elders may boost the force of selection against some aging in later adulthood (at least until those assets become burdens, like we saw in chapter 7). That transition

279

affects how we're treated, and our perceived quality of life. But what shapes the pattern of physical decline during this critical transition? None of the existing evolutionary theories can shed much light on the nature of aging throughout the body. We know our reproductive machinery senesces more rapidly than our soma, but what of our lungs versus our liver? Our skin tone versus our kidneys?

Intuition would suggest that a well-designed machine should see its components fall apart at the same time. An apocryphal story about the moving assembly line inventor Henry Ford captures this spirit perfectly. As the story goes, he'd survey scrapyards to inspect the parts of his defunct Model T car. Inspectors reported that many parts would fail, except kingpins—a critical part of the steering mechanism. Kingpins apparently had years of life left. This angered Ford, who considered it a waste that kingpins would remain faultless amid a wreck of broken parts. He ordered the kingpins to be made more cheaply, so that everything would fall apart at the same time. This simultaneous collapse is captured perfectly by Oliver Wendell Holmes in his epic poem about the one-hoss shay (that is, horse chaise) described in the opening quote. The shay was built to run without fail, and to fall apart all at once after exactly a century of good life. It was designed this way by the brilliant and logical deacon who simply fixed up all the parts that might otherwise fail.

Since natural selection often acts like an effective but blind designer (without any blueprints), many biologists once believed that biological organisms would follow the same rule as the one-hoss shay. Economical design, after all, shouldn't favor waste, and therefore our physiological systems should be synchronized in their decline. Throughout this book, I've argued that the human body was built to last seven decades under traditional conditions. But in much of the developed world today, it's more like eight or nine decades. *What has changed that now shifts our warranty period by over a decade?* Does everything still wear out at the same rate, but we just survive longer now? Or does everything grind down more slowly? We're finally getting some answers to these questions.

In this chapter, I present a framework to help us think about the timing of aging in all its diverse manifestations. While there are major gaps in our understanding of why different organs and systems senesce the way they do, especially in subsistence populations, the framework helps us think through possible reasons for observed patterns. We'll first take a whirlwind tour through some of the more noticeable changes that occur with aging. Then I'll outline general design principles borrowed from engineering to help us think about rates of declining function in different parts of the human machine. Given the competing demands on our body, and the costs and benefits of performance in relation to fitness over the life course, we will engage in some speculation about the mosaic character of physical aging knowing the body is built to last at least seven decades. That being said, we'll bounce between what mosaic aging might look like with a seven-decade modal lifespan versus the eight- or nine-decade lifespan we see in high-income countries in the Global North.

The Aging Body

From afar, all humans, despite living under vastly different conditions, show similar patterns of biological aging. But if we zoom in on the fine-grained details, we can see that some folks age faster than others, especially when considering different parts of the body. There are thousands of potential biological variables ranging from the micro (like the molecular "hallmarks of aging" such as telomere attrition and loss of protein homeostasis), to the macro (specifically, the functional aspects of how organs, and the organism as a whole, performs and survives). The "father of gerontology," Nathan Shock, who was fanatical about age changes in hundreds of arcane physiological parameters, helped launch the longitudinal study of aging at the National Institutes of Health (NIH) long before NIH formalized a separate subdivision that today funds a large percentage of US research on aging (the National Institute on Aging, NIA).[1] What we now take for granted, Shock helped establish: measuring healthy, "normal" people over time rather than studying aging strictly by evaluating donated cadavers. His experience, leadership, and vision led to the creation of the longest-running aging study to date, the Baltimore Longitudinal Study of Aging (BLSA). Its ongoing mission since 1958 is to "observe and document the physical, mental, and emotional effects of the aging process in healthy, active people." So far, more than three thousand adults of all ages from the Baltimore area have been studied, many sampled repeatedly every one to four years.

One takeaway from the study's thousands of findings is that certain changes once believed to epitomize normal aging may be absent in healthy people. For example, among healthy Baltimoreans with no heart disease, many hemodynamic measures of heart function don't change with age. But when biomarkers from heart-healthy folks are combined and analyzed jointly with the biomarkers from people with heart disease (as is often done in representative population surveys), average stroke volume, cardiac output, and other factors all appear to decrease with age. When healthy and nonhealthy people are analyzed separately, only systolic blood pressure increases with age in both groups. Cardiac output reflects how well the heart meets all the oxygen demands of the body. While healthy folks showed no age differences while at rest, maximum workload doing exercise on a stationary bicycle did show a decrease with age. This is just a simple glimpse into the stickiness of studying aging, and with attempts to identify what is "normal." Intertwined with the biological changes that occur with age are the many factors that modulate those changes. These sometimes get classified as disease, socioeconomic status, and "lifestyle" behaviors.

Defining and characterizing aging can sound like a philosophical debate: Where does aging end and disease begin? Lots of observable changes in preclinical biomarkers can predict later disease onset. Even by age 38, there are differences in health trajectories, and these can reliably be predicted by biomarkers in your 20s—well before any sign of clinical disease (at least in Dunedin, New Zealand, where this unique study of aging in healthy young adults occurred). Young folks who exhibited more rapid biological aging in eighteen physiological measures showed more cognitive decline and brain aging and more mobility limitations by the time they reached their late 30s. They even looked older, and felt older.[2]

Some elements of our later fates may be set even earlier. Low birth weight from undernutrition and other stressors while in the womb and in early infancy have been linked to obesity, insulin resistance, and ischemic heart disease in adulthood. Deficits in early life predict chronic disease risks in adulthood above and beyond the circumstances of adult life already known to be risk factors. From this "fetal programming" we know that aging depends on resource allocations made throughout development. Growth is about building reserve and maintenance—both highly relevant to aging. In this case, a predisposition to chronic diseases in adulthood is a consequence of adaptive shortcuts made early in life to help cope with deprivations in early life.[3] As my friend and longtime collaborator Tuck Finch once chortled, "Aging begins in your grandmother's ovaries!"

Singing the Body Electric

When we talk about normal aging, we talk about oodles of changes small and large. These are well catalogued in any human physiology textbook. Aging affects our ability to see, hear, taste, remember, balance, eat, digest, and heal.

At the cellular level, some cells self-sacrifice through an orchestrated process called apoptosis; others are arrested in suspended animation, lingering in the body in a senescent state. With cellular aging comes aging in the organs these cells call home. Most organs lose some function with age. Kidneys filter the blood less efficiently. The brain and spinal cord lose nerve cells and mass. The lungs lose air sacs and capillaries. Even the spleen may not maintain its splenetic youth. Your spleen doesn't just store pent-up anger—it performs a variety of quality-control tasks for your red blood cells, stores reserve blood in case of emergencies, and helps prepare immune soldiers called lymphocytes. Other changes throughout the immune system diminish our ability to defend against new invaders. Vaccines often become less protective in older adults with compromised immune systems.

Changes also occur throughout our skeleton and muscles. The skeleton is so much more than just a support structure. It's an active organ. We lose bone density with age, and some bones become weaker than others (for example, at the wrist, in parts of the spine, and at the hip). Bone marrow decreases, affecting the production of new blood cells. Cartilage between joints can thin, and ligaments lose elasticity. These translate to arthritic pain and loss of flexibility. We lose muscle mass and brute strength. The lens of our eye stiffens, making it harder to see things up close. And the lens becomes denser, making it harder to see in the dark. Our hearing declines, even if we've never been front row at a Megadeth concert. Then there are the more obvious surface changes we notice in the mirror. Hair thins on the head but starts appearing in other places. Skin wrinkles.

Join the Reserves

While changes to our body may be universal, the rates at which they occur vary with exposure and luck. Yes, sunbathing everyday will increase wrinkling, and hardcore couch potato-ing will atrophy our muscles. But some changes affect us more than others. Consider the spleen again. Some people are born without one, or have it removed if it is damaged or diseased. We can live without a spleen because the liver takes over some of the spleen's functions. Related, many of us live healthy lives after our gallbladder is removed. We lose a kidney to cancer, but we can still function with its leftover twin. The pancreas, necessary for blood sugar regulation and as a digestion aid, has much built-in reserve. Its function is only slightly compromised even when much of it is removed. One reason pancreatic cancer is so deadly is because we often don't notice it until most of the pancreas is destroyed and the cancer has already spread throughout the body.

Having an extra organ, or cells, or other component parts in reserve makes sense from a design standpoint. Having more brain cells means you can get by if you accidentally fry some in a bender. Another way to ensure reserve is to maintain the integrity of the cells you do have.

In the world of machines, reserve is captured by the concept of "safety factors." Consider the following safety sign in an elevator: "Maximum weight capacity: 800 kgs, 10 people." What happens if the eleventh person to enter the elevator pushes the load past the 800 kg mark? Luckily, the cables supporting elevators have safety factors of ~12. A safety factor reflects the ratio of capacity to load, which means that elevator cables have the strength to carry twelve times the certified load before breaking. Three angry gorillas could get in the elevator and you'd still be fine (at least safe from falling). A freight elevator, on the other hand, has a smaller safety factor, of about seven. Losing freight isn't as awful as losing lives, so less is invested to ensure that freight elevators don't fail.

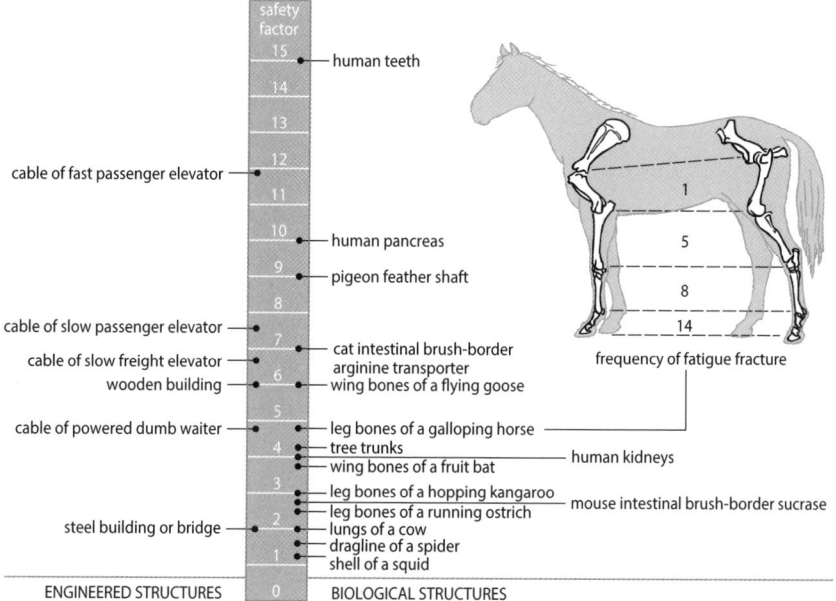

FIGURE 8.1. **Safety factors.** Larger numbers mean the structure can withstand greater load, but likely at higher cost to build and maintain. Shown for a range of biological structures and engineered structures. The hooves of a horse have a similar safety factor as human teeth. From Piersma and van Gils 2011, © 2011 Oxford University Press.

It turns out that our organs are like these engineered parts. They have safety factors as well. Bones, for example, may have safety factors ranging from 1.3 to 5. The resilient pancreas? A safety factor of ten. The winner goes to our teeth, with a safety factor of fifteen! If you think about it, this makes sense. Sans dentures, we can't regrow new teeth. Figure 8.1 shows safety factors for a variety of parts of both organic bodies and inorganic machines.[4]

The Safety Dance

If a safety factor is too low, an organ can fail, or a structure could break. If it's too high, that could be a wasteful expense to make things indestructible. Optimal safety factors seem to make sense. But should every organ have the same safety factor? We'd hope for a design that will satisfy but not exceed what's needed to get 'er done. The old adage that a chain is only as strong as its weakest link suggests that having organs with different safety factors would be wasteful or inefficient. The maximal uptake of oxygen to your body's cells depends on the rate at which your circulatory system can move oxygen through the blood, and the rate at which muscles can use oxygen. Everything is connected. But it turns out there are good reasons for safety factors varying among your body's building blocks.

The main reason is intuitive: safety factors vary when anticipated loads and capacities on organs differ. Load is the strain on the system, and capacity is the ability of the system to function. If a weak elevator cable has to lift a gaggle of skinny third graders, no sweat. But for a team of sumo wrestlers, disaster strikes. The greater the variation in load or capacity, the more likely we'll encounter deadly combinations, like large loads coupled with low capacities. Under these conditions, when the cost of failure can be fatal, we expect a larger safety factor. Higher safety factors are also needed when capacity decays over time. This is why wooden beams used for house construction need higher safety factors than similar components of brick or steel.

All else being equal, higher safety factors are a good thing. But it costs money, energy, or time to make things more robust from the outset, and to maintain integrity (let alone in the human body, where all the organs have to squish together to fit). If you thought Santa Barbara's housing costs were high (I assure you they are!), rents on the limited space in our body are outrageous. In our zero-sum world inside our skin, more space for one organ means less for another.

Aging Process or Processes?

Do different safety factors for different organs or somatic systems help ensure that they decay at the same rate? It certainly appears like changes are synchronized in our late years. Many elders might feel that "when it rains, it pours" applies to their bodies and their physical ailments. The famous evolutionary biologist John Maynard Smith argued that natural selection should lead to the simultaneous deterioration of all body parts. The evolutionary giant George Williams agreed: "Senescence

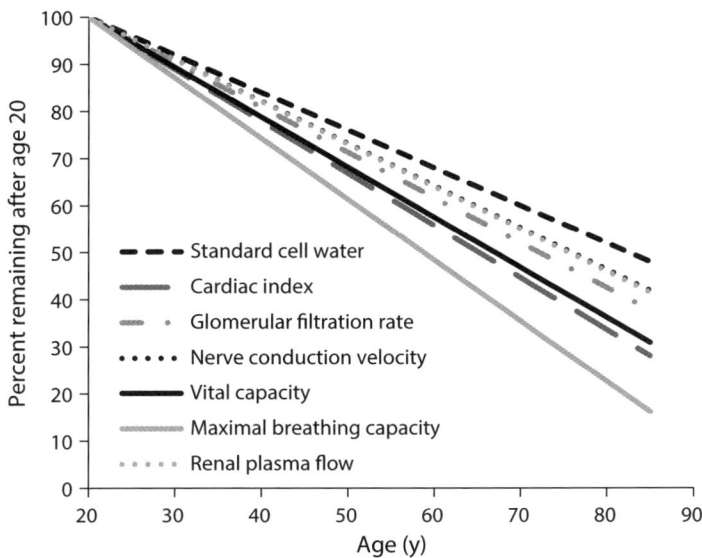

FIGURE 8.2. **Percentage of physiological function remaining after age 20.** Adapted from Strehler and Mildvan 1960, using Baltimore Longitudinal Study of Aging (BLSA) data from Nathan Shock.

should always be a generalized deterioration, and never due largely to changes in a single system." But if this were true, there'd be no point in donating your organs after death, as they'd all have failed simultaneously and be of no use to anyone. And we'd expect a lot more dying from multiple-organ failure. But less than 5% of hospital emergency department visits are due to organ failure. Brain failure carries the worst prognosis, and liver failure the "best" prognosis. The chances that three or more organs fail at the same time is rare. Usually this type of systemic failure occurs following sepsis, or is a secondary consequence of major trauma or inflammatory cascade.[5]

As I mentioned earlier, Nathan Shock's claim to fame was to document the age trajectories of many different biomarkers that describe organ function and bodily processes (see figure 8.2). Some traits, like muscular strength and the maximum amount of oxygen your body can use during intense activity (what your Fitbit calls VO_2max) reach their peak in early adulthood (say, 20s or 30s) and decline thereafter. Others have been shown to change very little throughout adulthood, like the mean acidity in the stomach. Shock showed that the most general pattern is to be at the top of your game by about age 30, with a linear reduction continuing over the rest of the life course. But the steepness of these declines—showing how rapidly things wane—varies a lot. Figure 8.2 shows that maximal breathing capacity declines more steeply than nerve conduction velocity or kidney filtration rate.

Physiological Aging in the Wild

Maybe you're still thinking that variable aging rates for different body parts is due to the weirdness of industrial life. From two decades of studying Tsimane health, we've learned a great deal about age-related changes throughout the body under nonindustrial conditions. For body size and fat composition. Immune function, kidney clearance rates, short-term memory, and so much more. And also in higher-order performance tasks, like normal walking speed, balance, and strength. Aging rates among Tsimane appear slower than Americans and Europeans for some traits, faster for others, and for still others no difference.

But despite variability, some patterns are clear. Bone mineral density on the spine and VO_2max decline with age in men and women, despite Tsimane being more physically active than urbanites throughout their lives. Activity is usually protective against bone mineral loss, but not enough to ward off effects of infection and having many kids in relatively rapid succession.[6] Physical strength, as measured by gripping as hard as you can on a handheld dynamometer, peaks in mid-20s to early 30s then declines steadily thereafter, again despite Tsimane actively using their bodies for work and mobility every day. The amount of air a person can exhale in a forced breath is a useful measure of lung capacity and can be compromised with respiratory ailments. This is the same breathing capacity line we saw among the Baltimoreans in figure 8.2. At their peak, your lungs can hold three large soda bottles worth of air. Given frequent respiratory infections and bronchitis among Tsimane, declines with age may not be surprising. Though your lungs can age even without asthma or disease.

Morning blood sugar levels remain low and flat with age, as to be expected with so little insulin resistance and diabetes (see chapter 9). Though obesity is rare among the Tsimane, body mass index (BMI: weight divided by height-squared) still creeps up throughout adulthood. By their 60s or so, BMI crawls back down, just like it does in the US and other places where obesity is common. Cortisol, a steroid hormone that affects inflammation, metabolism, blood sugar, and sleep-wake cycles (often known at parties as the stress hormone), shows no clear trend with age. That's probably a good thing. Another flatliner is a measure of oxidative stress, which used to be touted as one of the holy grails of aging. Oxygen is essential for our bodies but leaves grenade-like by-products in its metabolic wake. Persistent oxidative stress was once thought to be a primary cause of cellular aging (revolutionarily called the "free radical theory of aging"). Accumulated oxidative damage can certainly lead to aging in our cells and tissues, (including dysfunction of our sacred mitochondria—the powerhouse of our cells), but it no longer seems to be the golden ticket. Indeed, this marker of biological aging does not show any increase with age among Tsimane. In a study of wild chimpanzees, markers of oxidative damage didn't increase with age either, except in the lead-up to death.[7] The growing consensus about the role of oxidative stress on organismal aging is that there is no real consensus. As scientists love to say, more research is needed—to better understand how oxidative stress is connected to other molecular and cellular mechanisms of aging.[8]

Less Is More, Don't You Think?

Different organs and systems seem to senesce at different rates, but all those parts are caged together in the same organic ship that must sink or float. We tend to think of an organ's optimal function in isolation, but recognizing that organs work together in sync suggests coordination and compromise. We move based on actions of muscles and joints on a skeleton of bones, directed by our nervous system, fueled by energy we've consumed, digested, and processed aided by our oxygenated breaths. In other words, many interactions and potential compensations govern how our body performs. How to best capture a holistic view of physiological aging in light of this complex dynamic?

One approach is to consider multiple snapshots (that is, biomarkers) simultaneously. These types of metrics usually show linear changes with age, and they reliably predict all-cause mortality. So they're meaningful in a statistical sense, and act as a summary measure of how many things might be off-kilter. A different approach aims to capture the concept of dysregulation. Imagine that all the vital metrics of a healthy person hang together in a coordinated fashion, like a form of homeostasis. As you get older, your biomarkers might shift, but if they do so in harmonized ways, health can still be maintained. Being able to bounce back to a healthy norm once there's a disturbance in the system is the essence of resilience. This process occurs at cellular, tissue, and whole-organism levels. If aging reflects diminishing resilience, then aged individuals should find it harder for blood pressure to lower after a stressor, for a bone to heal after it breaks, and to survive after an assault on their health.

In practice, we can measure this sense of aging taking advantage of the many biomarkers available across different systems. If the values of biomarkers are like stars positioned in the sky, then those stars should move relative to your position throughout the day. But if Pegasus starts galloping across the sky on a whim, and its head travels north while its legs move south, then we have a problem.

The positioning of all those biomarker stars can be summarized in a single statistical measure, called D_M. Larger values of D_M mean more dysregulation from a healthy baseline. It tells us how strange or abnormal an individual's whole biomarker profile is—the straight-line distance from healthy in multidimensional biomarker space. As biogerontologist Alan Cohen explains it, this measure embodies the Anna Karenina principle: "All happy families are happy in the same way, but each unhappy family is unhappy in its own way." In other words, there are many ways things can go wrong. D_M represents the inability of a physiological system to return back to its healthy baseline state. Wherever D_M has been measured, it increases with age and predicts disability, chronic disease risk, and death.[9] Unlike the "deficit index" and other similar approaches that attempt to aggregate multiple biomarkers into a single index, D_M doesn't assume that, say, high blood pressure is problematic but low blood pressure is not; it doesn't prejudge bio-marker levels by relying on arbitrary thresholds to mean poor health (for example, that BMI $\geq$ 30 = obesity).

A cool feature of this approach is that beyond fourteen or so biomarkers, very little information is gained about global dysregulation. And the composition of biomarkers doesn't seem to matter much. Having enough snapshots inside the body tells you enough about whether there's systemic falling-apart dysregulation.

Tsimane are probably the only subsistence population with enough measures collected, and many among the same individuals over time, for which we can assess D_M. Figure 8.3 shows that, like everywhere else, D_M increases with age. It's not just the population average that increases, but each individual's trajectory shows greater physiological dysregulation with age. And here's the exciting part: the relative amount of dysregulation that occurs with age from a healthy baseline is similar, if not even a bit faster, than what's observed in industrialized populations, like the United States and Italy. Despite the big differences in mortality and causes of death experienced by the Tsimane, Americans, and Italians, fundamental aging at the physiological level seems to be roughly similar across vastly different environments.[10] Any health gains from being fit while eating tasty omega-3-rich river fish and breathing fresh air are off-set by the otherwise harsh conditions of rural subsistence tropical living experienced over decades.

Because Tsimane D_M is measured relative to a healthy Tsimane base-line, and US D_M is measured relative to a healthy US baseline, we can't yet directly compare whether Tsimane age faster or slower than Americans using the same yardstick. Remember that the epigenetic aging measures described in chapter 2 suggested that Tsimane age more slowly than Americans (except for their immune systems, which age faster). The meaning of epigenetic aging, especially in comparison with other metrics, is still unclear. This field is advancing rapidly, with new measures being made available each year. Studies comparing these metrics against each other have shown that they are sometimes only weakly correlated with each other. This suggests that they at best capture dif-ferent aspects of the aging process(es).[11]

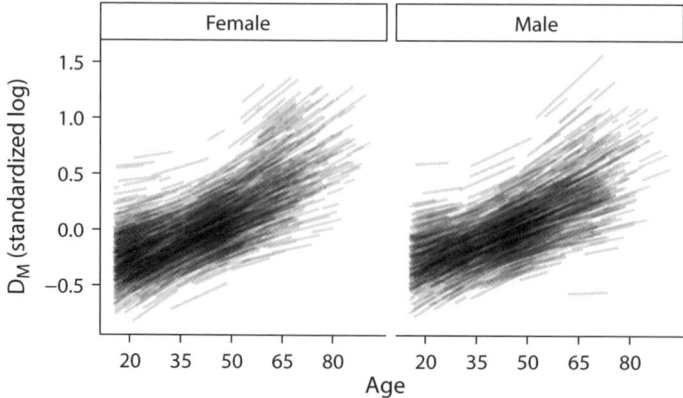

FIGURE 8.3. **Physiological dysregulation among Tsimane women and men.** Each line models the statistical composite of physiological dysregulation, D_M, for an individual sampled at least three times. D_M reflects dysregulation, or the degree of departure from a healthy state, based on information from multiple biomarkers. Individual D_M trajectories show accelerated dysregulation at older ages. See Kraft et al. 2020 for details.

But Really, Just Tell Us Who Ages Faster

I've watched 70-year-old Tsimane chop wood and clear fields with machetes for hours. Yes, I've also seen elders rest for much of the day, occasionally taking walks with the aid of a sturdy gnarled stick. On the whole, however, Tsimane are committed hard workers, and often insist on laboring in their fields, even when they are sick. Their persistence and endurance is impressive. The main reason Tsimane often abandon their four- to six-month tuberculosis treatment regimens at a health clinic located just outside town (standard protocol is that all drug administration must be directly observed by health officials), or cut short their recovery from surgeries performed with great organizational effort in distant towns, is the pressure they feel to return to their village and harvest rice or attend to other work needs.

Even if they are ailing, elders are still the life of the party. When I was visiting the community Tontumsi' in 2008, I met 81-year-old Caramelo Tayo. One evening, Caramelo started singing, and when he sings, there's no end in sight until he either passes out or everyone eventually leaves (when the pot of *shocdye'* is empty). Anecdotes in other subsistence groups suggest a longer healthspan relative to lifespan than experienced by the rest of the world saddled with the chronic diseases of aging. Among the G/wi, seasoned observers noted that men and women who lived to late ages were pretty fit up until the end. Hearing and eyesight deteriorate, but "less so than would be normal for a European of the same age." When things do start to fall apart, it can happen quickly. As the veteran observer of the G/wi and other San groups of the Kalahari, George Silberbauer, wrote in the mid-twentieth century, "The development of the senile state is rapid, within six months to a year the old man or woman is physically and mentally debilitated, with little physical strength, poor memory and short temper."[12]

Beyond the daily toil of subsistence, an easy way to gauge the bare minimum self-sufficiency is whether you can take care of your basic needs—to bathe, dress, and feed yourself, go to the bathroom, and move about. In the gerontological world, these "activities of daily living," or ADLs, are used to assess whether a person needs assistance or is no longer capable of independent living. As shown in figure 8.4a, most older Tsimane have no problems with ADLs. By age 70, they start to have some challenges, mostly with mobility. The ability to accomplish ADLs is worse for women than men, and gets worse for everyone by the time they reach their 80s. But even as some difficulty is common, the overall level of disability is low, with an average ADL score of 3 for women, and 2 for men (out of 24).

Going further, we can gauge how people see, hear, and walk. Sight and hearing are critical for navigating the Tsimane (or any) environment. By their 70s, almost half of Tsimane have problems with their hearing, and the proportion only increases as they reach their 80s. We've used fancy audiometry equipment to measure people's hearing, and it shows that living free of Iron Maiden concerts and subway rattling may indeed preserve your hearing for longer, compared to people living in a noisy city. We asked people to compare their current hearing and vision against when they were younger. Despite not wearing glasses, and no history of watching TV and reading books under dim

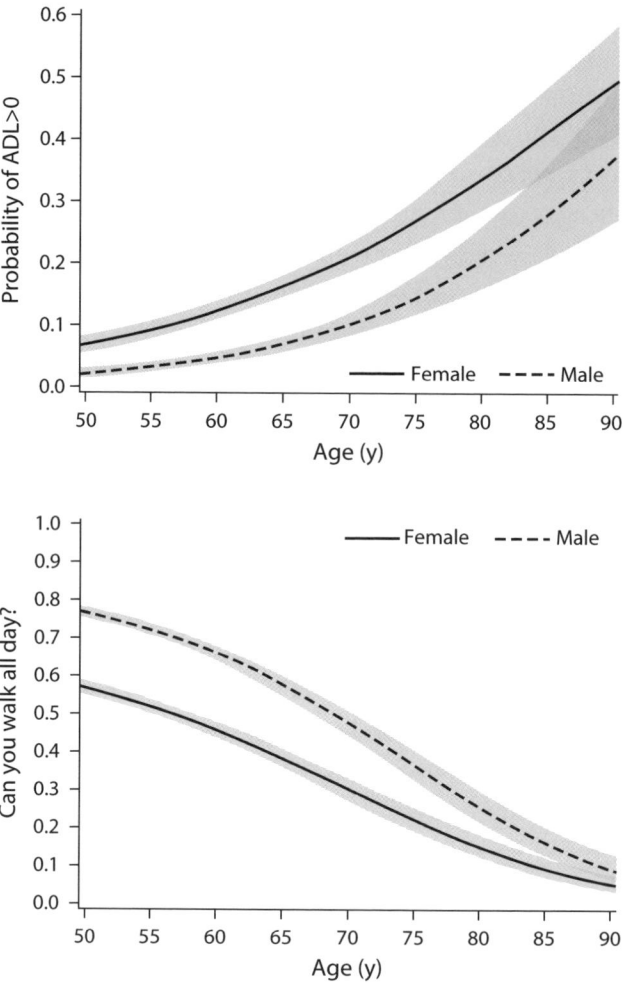

FIGURE 8.4. **Changes in activities of daily living, and walking ability.**
(*top*) Probability of any kind of problem with at least one activity of daily living
(ADL); self-reported difficulty in (*bottom*) ability to walk all day (at least ~8 hrs).
Based on over 4,000 self-reports among Tsimane ages 50 and up, as part of the
Tsimane Health and Life Project medical rounds.

light, Tsimane report that their close-up vision declines with age. By age
70, half of Tsimane adults say they have at least some trouble seeing up
close to, say, thread a needle. By their 80s, few can see well up close.

As shown in figure 8.4b, older Tsimane adults also have a hard time
walking long distances. It's not uncommon for Tsimane to walk much

of the day when visiting other villages using a combination of trails, dirt roads, and along riverbanks, or when on trek foraging, fishing, or hunting. By age 80, Tsimane are able to walk about two hours per day. Similarly, 80-year-old Tsimane walk about 5,000 steps per day according to accelerometers worn on their waists, and an impressive 14,000 steps when the accelerometer is worn on the wrist (which registers non-walking movement as steps).

Improvements in survival among Tsimane over the past couple decades are likely an indication of more robust health in older age (remember the "prospective ages" I discussed at the opening of the book). Given recent increases in Tsimane survival rates, we would conjecture that older Tsimane are now about eight years biologically "younger."

Mosaics, Shays, and Fords

So far, we've seen how aging can occur at different rates across the body, and how physiological decline can be described systemically via measures like D_M. Experts haven't come to a universally accepted conclusion regarding whether aging is caused by an underlying single biological process. As John Maynard Smith argued, the uncharged battery in your car will lead to sputtering decline in the ignition, lights, and starter motor, not necessarily at the same time. But all these problems stem from a single underlying cause. Similarly, aging could manifest across the body at different speeds but be maneuvered behind the scenes by regulatory processes that act like faceless puppeteers. An entertaining alternative is that there is no unitary biological phenomenon of aging.[13] Aging these days is all about systems biology. When we don't understand something, we say it's complex, and talk about hierarchical networks of nested parts, feedback loops, and such. We draw elaborate flow charts where everything is connected. And we can come up with composite measures that incorporate eclectic groups of biomarkers and call that aging.

Cynicism aside, the truth of the matter is that aging may be best considered a complex biological process.[14] Not just because there are lots of parts and at different scales (molecules, organelles, cells, organs, and so on). Complex systems have a hierarchical structure consisting of

modular networks serving important functions, like glucose metabolism and stress response. Redundancy and flexibility are built-in features of the system, which helps make the organism resilient in the face
of stressors. The modules both serve important functions *and* communicate information across the system, which results in regulatory feed-
backs and feedforwards. No single biomarker exerts top-down authority
when control is diffuse.

In complex systems, the whole is bigger than the sum of the parts,
resulting in what are called emergent properties. Dysregulation is an
emergent physiological process reflecting the breakdown of these dynamic regulatory systems that maintain law and order. It may be hard
to grok but it can at least be quantified crudely with measures like D_M.
Decline in function often appears abruptly—it is not quite a gradual
process. As concrete examples, proposed emergent aging-related phe-
nomena span levels of biological hierarchy: from immuno-senescence,
frailty, and dementia to falls. These developments can't be easily explained
by adding one building block atop another.

There are other surprising design principles about aging that can help
illuminate how and why it manifests. For example, age-related decline
of a physiological system can emerge even when its component parts
don't age. When non-aging parts are connected to each other like in a
parallel circuit, then breakdown of one or two parts might have only
a small impact on system function. Only when enough parts break down
does system aging become noticeable and performance decline. Multiple
parts connected in parallel thus increase organ reserve and enable the
system to tolerate some damage. Much of the gradual deterioration that
comes once that reserve dwindles (i.e., what we call aging) owes itself
to the natural redundancy of biological organs and systems.

Reliability theory from engineering builds on these ideas and shows
that the famous Gompertz exponential increase in mortality (which
fits humans and so many other organisms) emerges when (1) cells, organs, and other parts have built-in redundancy, and (the Biblical book
of Genesis notwithstanding) (2) organisms aren't born perfect and
flawless; they come with some damage. Hamlet called such defects the
"thousand shocks that flesh is heir to," but the influential evolutionary

biologist Theodosius Dobzhansky declared, "the shocks are innumerable." Unlike machines, which are built from well-inspected parts by an external agent, organisms are self-assembled organically from an embryonic soup (echoes again of the importance of early life for setting the pace of adulthood). In just one study of mice, for example, the health of parents (measured by resting metabolic rate) was a better predictor of a mouse's longevity than that mouse's own health.[15]

The dysregulation that I mentioned earlier is the breakdown of these coordinated systems. The robust nature of the system declines with the loss of resilience. The manifestation of it all before our eyes is what we usually think of as aging. Separate functional systems with their evolved "safety factors" decline at different rates. But if our bodies are organic machines designed to maximize their fitness (rather than just live forever), these varied rates are still part of a sweet package deal (or better at least than alternative packages).

What does system-by-system dysregulation look like among subsistence societies? To address this question, I worked with my former postdoc Tom Kraft (now a professor at the University of Utah) and colleagues to group biomarkers collected among Tsimane into broad categories, as we wanted to capture potential dysregulation in three distinct systems: musculoskeletal (including grip strength and bone mineral density measures), cardiometabolic (including markers like glucose, cholesterol, BMI), and immune (including white blood cells, antibodies, and other immune cell types). Biomarkers within a system statistically hang together better than a random grouping of variables across systems, suggesting that these systems are reasonably independent. And dysregulation in any one system (again using the D_M metric) was not associated with dysregulation in the others, which also suggests some degree of independence.

Dysregulation occurred most rapidly with age in the musculoskeletal system, followed by the cardiometabolic system. What was interesting and initially surprising was that the immune system showed very little increase in dysregulation with age. We know immune senescence occurs when looking at numbers of certain circulating immune cells, like naive T-cells that help combat new infections the body isn't yet familiar with.

And with infections being such an important cause of death throughout the lifespan (see chapter 3). It made sense that immune burnout might follow a lifetime of pathogenic blitzkriegs. Rather, immune defenses are so important that the immune system, as a coherent, coordinated whole, doesn't fall apart.

There are many compensations that can still make most immune responses still sufficient. For example, consider the important component of our early-line innate immune defenses called natural killer cells—a type of patrolling white blood cell that destroys harmful cells affected by viruses or cancer, and without any prior exposure to specific pathogens. Natural killer cells increase with age and their numbers are much higher in Tsimane than observed in other populations.[16] Another surprise we found is that cardiometabolic dysregulation is notable with age, yet cardiovascular disease and diabetes are rare among Tsimane (see chapter 9). It's possible that the Tsimane baseline is healthy enough that some dysregulation with age doesn't result in salient health problems (yet). Dysregulation among industrialized populations with a less healthy baseline is more likely to veer into clinical disease territory.[17]

When organs or biological processes are connected in a series (instead of parallel), the weakest link in the organic chain should be the limiting factor. And with a weak link, shouldn't the other organs or processes dumb it down, lest more inefficient wasting? Not so fast (or slow). Back to the guiding insight shaping an organ or process's safety factors. What matters is using limited resources wisely, and thinking about the whole person's survival, not just isolated organs. Organs vary in how much damage they may receive, and likely in how expensive it is to make repairs. Extra capacity in an organ can be favored if a "weak" element in the chain is cheaper to produce or more variable in its performance, especially when damages are unpredictable. Taken together, engineering principles suggest that organs that break down are not necessarily "underbuilt," nor are perfectly working kingpins "overbuilt" (despite Henry Ford's protests).[18] Variable failure rates make sense from this engineering perspective. Everything failing all at once, like what happened to the one-hoss shay, is a special case and does not describe how most living organisms age.

In fact, there's good reason to suspect that damage and repair costs are not uniform across organs. Most of our cells are far younger than we are. Cells in our brain and teeth may come close, but cells more directly exposed to the elements, like in our outer skin and gut lining, turn over very rapidly (like every week or two). Red blood cells turn over every few months, and your fat cells maybe every ten years. (Some lizards have tails or legs that easily snap off. How could that ever be favored by natural selection? Predators! Better for a predator to steal your tail than your whole body.)

These and other design principles are useful heuristics for helping us think about patterned ways our body erodes. In addition to the examples I have already mentioned, regarding the ways in which body part organization relates to functional goals, remember that many organs have multiple functions. It's hard to optimize across different goals at the same time. Lungs circulate oxygen to the blood, but also help remove by-products of metabolism (for example, carbon dioxide) from the blood. Most success in making sense of the details of organismic design by combining these heuristics have focused on the respiration chain (air flow → pulmonary diffusion → cardiac output → muscle diffusion → ATP generation) and food-to-fuel chain (eating → digesting → absorbing). Gaining a holistic sense of whole-body aging from this engineering perspective is new territory.[19]

Use It or Lose It

Hang out in zero-gravity space and your muscles atrophy, bones lose mass, your heart shrinks, and blood pressure drops. That can also happen if you're bedridden for weeks. Plenty of exercise, as well as pregnancy, are both linked to an increase in the size of your heart (especially the left ventricle) to accommodate the need for greater cardiac output. Climb Mount Everest, and your VO_2max goes down, along with your maximum heart rate, in response to lower oxygen concentration in the air. You'll no doubt still be tired.

Despite gazillions of calories available to us, we don't grow to be 12 feet tall. All that extra energy can't be channeled to double our brain size (even when we're trying to write a book) or increase the rate at which nutrients are absorbed through our gut, and the like. Our total energy expenditure hits a metabolic ceiling. Our muscles alone consume about a fifth of our daily energy budget, our brain another fifth. We make do on our constrained energy budgets, like we saw in chapter 2 when I introduced evolved life history "decisions" shaping the life course. Instead, our bodies are designed to store extra calories, like a fanny pack of flesh around our middle. Our high-quality diet and cooking (i.e., extra-somatic processing) enable us to get by with smaller guts. Can we cheat the system by joining the "calorie restriction" diet bandwagon? Severely restricting our caloric intake has been touted by the hopeful as making us live longer. Maybe, but even if that were true, extended survival via a maintenance boost doesn't come for free. You're likely to also experience brain fog, become sterile, and lose the ability to regulate your body temperature. Even when there are successes—like selectively breeding dairy cows to more than triple their milk production—there's always a catch: their fertility and lifespan are cut short.[20]

A Crude Framework

Armed with these ideas, we can venture some guesses about how physical changes in our body affect its function. Reproductive fitness is the ultimate goal from the viewpoint of natural selection, but what gets us there? It's hard to envision what the marginal fitness impact of 10% more liver function would be, especially in exchange for more oxygen diffusion. One high-level goal that's a good starting point is energy production, or work. And leading up to work in subsistence societies are basic elements of your eighth-grade physical fitness test: speed, endurance, strength, balance, and flexibility. Substantial command of your senses is important for successful production as well. I've known a couple deaf Tsimane fishers, but never a deaf hunter. And blind Tsimane (rare) aren't doing any food acquisition. Evading predators, competing with competitors, and the like are also important fitness-serving functions, but you have to start somewhere. I like food.

Figure 8.5 maps different somatic systems to these more functional outputs of the body. The multi-system regulation described earlier helps

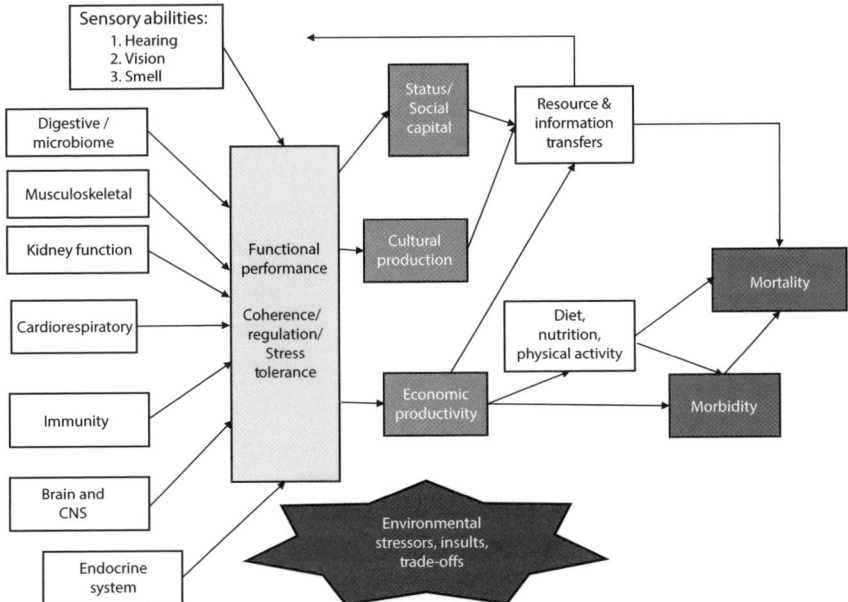

FIGURE 8.5. **A simplified conceptual model of aging in nonindustrialized societies.** From left to right: physiological systems are organized in ways that help serve functional performance, by ensuring robustness and resilience to common environmental stressors. Abilities and capacities lead to all the forms of production that affect biological fitness: economic productivity, knowledge, social status, rituals, and other expressions of cultural capital. Food, coalitionary support, mate access, and the like jointly impact fitness via morbidity and mortality and reproduction. Transfers of food, knowledge, and other resources benefit others, and also help provide protection against physical and social stressors.

connect separate systems together to the extent that they serve midlevel functional goals related to physical fitness (which itself serves the purpose of finding, procuring, and transferring food to others). Beyond food, I also include knowledge, morals, and other aspects of cultural production that can affect fitness, as we saw in chapters 6 and 7.

To my knowledge, a model hasn't been framed like this before, much less tested with hard-earned data. Better models and detailed tests await. But for now, we can use this guide to sketch what happens with age along several key dimensions. Again, I present Tsimane as a case study. Figure 8.6 shows age curves—all derived from real data collected by the Tsimane Health and Life History Project—that inform different

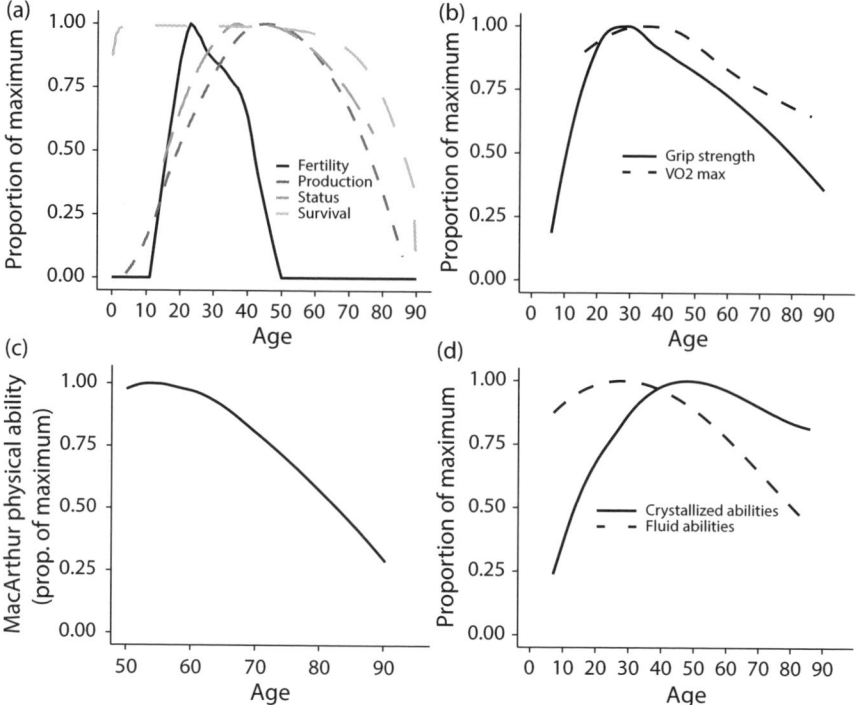

FIGURE 8.6. **Drivers of physical aging among Tsimane.** (a) Fertility drops to zero before age 50, and survival probability drops rapidly by the seventh decade. Food production and social status peak in the 40s and decline in advance of survival. (b) Muscular strength (grip strength) and cardiorespiratory fitness (VO_2max) peak in early adulthood, then decline thereafter. (c) Physical function—a composite measure based on tasks borrowed from the MacArthur Study of Aging, assessing balance, flexibility, and walking speed—declines by the seventh decade. (d) Cognitive scores on a battery of standardized tasks. Effortful "fluid"-like abilities peak in early adulthood and decline thereafter, while "crystallized" abilities increase throughout adulthood with only slight declines in old age. *Source:* Tsimane Health and Life History Project.

elements of the flowchart in figure 8.5. First, survival declines rapidly in the 60s and 70s. As I outlined in chapter 5, Tsimane food production and social status peak in their 40s, but then eventually decline in sync with survival (figure 8.6a). Physical strength peaks in early adulthood, while endurance (as gauged by VO_2max) peaks in the 40s, and both decline thereafter (figure 8.6b). By age 70, Tsimane are still at about two-thirds of their peak strength and three-fourths of their endurance. By 80, both drop to around half their maximum. Note that production and status still increase with age despite the loss of physical vigor in the 30s and 40s, due to the large role played by cumulative skill and knowledge. Even as productivity then declines, Tsimane maintain basic physical capacities well into their 60s and 70s (figure 8.6c). The ability to maintain productivity, and especially elder roles as teacher, adviser, leader, and ritual specialist depend critically on our smarts. Tsimane brains lose volume with age, as is the case for adults everywhere. But their brains do not lose a lot of volume. Consistent with the finding that Tsimane experience a lower rate of dementia than other populations, Tsimane also lose brain volume more slowly than Americans and Europeans.[21]

(Slowly) Walking Library Revisited

Shifting from brain volume and zeroing in on cognition using a standard battery of eight tests (like short-term recall of words, visual attention, timed naming of animals and fish) reveal an interesting pattern. The kinds of effortful abilities clustering as "fluid" are those that allow us to plan, focus, process quickly, and multitask. They help us sustain goal-directed behaviors. These mental abilities decline with age, just like our strength and endurance. But "crystallized" abilities don't. These reflect cumulative knowledge from prior experience and learning. These skills *increase* throughout adulthood, with only modest declines at late ages. These patterns have been widely observed in areas where formal schooling is the main way that people learn. But what's telltale is that these same patterns are found among Tsimane who never had any exposure to schools (see figure 8.6d).[22]

Given the skills-intensive nature of human livelihoods, it may be no surprise that parts of the brain have a high "safety factor" and that cognitive function is prioritized. The only cases of dementia we encountered among the Tsimane and the closely related Moseten (six people out of 435 adults over age 60) were all over the age of 80. Including those cases, Tsimane and Moseten have one of the lowest rates of dementia ever reported (see chapter 9). But globally, dementia tends to strike later than any other chronic disease, usually after age 80.

In all human groups, elders maintain roles as providers and investors in the "embodied capital" of their younger kin. Chapter 6 showcased elder instruction, advising, conflict mediation, storytelling, ritual leadership, healing, and information sharing, especially during the period of life when economic productivity may be plummeting. Natural selection may have shaped our brains to shift from sustained learning earlier in life to the crystallized application of that knowledge—for economic and cultural production and sharing. The decline in effortful fluid domains (involved with attention, processing, and learning), but preservation of crystallized domains well into late age, may reflect their changing relative contributions to fitness with age. If the broader roles of elders mattered for fitness, cognitive skills needed for knowledge retention, memory retrieval, and information transmission should have been selected to stay online for longer. This includes the category fluency just mentioned (also called semantic memory) and other crystallized abilities that enhance one's ability to generalize and abstract things from the contexts in which they were first learned.

Maybe the cognitive shift makes it harder to learn new things in late adulthood. You may never appreciate Dua Lipa if you matured on Lady Gaga, or on Madonna—but these are the sacrifices we pay to effectively wield accumulated knowledge like a battle axe. This shift may also contribute to that elusive you-know-it-when-you-see-it quality we call wisdom. Elkhonon Goldberg, neuroscientist and author of *The Wisdom Paradox*, succinctly summarized the positive effects of cognitive aging by reflecting on the changes in his own thinking: "What I have lost with age in my capacity for hard mental work, I seem to have gained in my capacity for instantaneous, almost unfairly easy insight."[23]

From Gonorrhea to Grandparents

Once an individual reaches mid-to-late adulthood, evolutionary engineering logic would posit that further maintaining fast, effortful cognitive processing may not be worth it. The expected gains to productivity and/or fitness are outweighed by the costs of keeping all that machinery in tip-top shape. And while fitness gained at older ages is snubbed by natural selection relative to earlier gains, remember from chapters 2 and 3 that the classical assumption, postreproductive life = zero fitness, doesn't apply (for us lucky humans at least). Some evidence supports this idea about slower cognitive aging relative to other systems, and conserved (nay, enhanced!) crystallized abilities in particular (figure 8.6d).

One idea supporting the cognitive shift away from learning rests on changes in network connectivity across different brain regions with age. Changes in the brain's functional network architecture include greater buy-in from prefrontal cortical regions related to cognitive control, and less suppression of the so-called default network that taps into prior knowledge about how we see ourselves in the world.[24]

From a gene's-eye view, protective alleles that help maintain cognitive function up through late ages are unique to the human lineage and are believed to have spread rapidly in our species. Rapid spread like that usually occurs under positive selection (that is, when there's a big benefit that outweighs any costs). One of these genes codes for a different form of the CD33 receptor expressed in immune cells, including those in the brain, called microglia. Microglia are the immune scavengers that survey and protect the brain from harmful infections and trauma.

A mutant form of the CD33 allele is unique to humans, not being present in the great apes or among other hominin cousins like Neanderthals and Denisovans. This human form is especially good at protecting against inflammation in the brain. It helps microglia do their clean-up job in the brain. The CD33 receptor gene is also associated with lower risks of Alzheimer's disease. The advantage leading to the spread of this gene, though, probably had little to do with preventing Alzheimer's among prehistoric 80-year-olds. The best guess is that it fostered protection against the fertility-crumbling effects of gonorrhea. As if scrotal pain and cruel discharges weren't bad enough, untreated gonorrhea can lead to infertility in both women and men. Gonorrhea bacteria dress up in a false disguise to evade our immune system, but the human CD33 gene recognizes the trick and positions the immune system to attack.[25]

CD33 isn't the only gene to protect against cognitive decline. Our genus *Homo* evolved a friendlier version of the apolipoprotein-E gene that lowers the risk of heart disease and Alzheimer's disease, part of a complex of genes oriented toward more meat-eating in our hominin diet.[26] Other genes have variants unique to humans that also seem protective against dementia. They have cute names like *SPON1*, *BIN1*, *PICALM*, and *ARID5*.

What If Tithonus Asked for Just Two Extra Decades?

In the eponymous Tennyson poem based on a Greek myth, Tithonus mistakenly asks his lover, Aurora, goddess of the dawn, for immortality. Like a bad *Twilight Zone* plot twist, he ends up with eternal life, but realizes he should've asked for eternal youth. He remains forever old, a "white-hair'd shadow roaming like a dream." Whereas Aurora is renewed with youth each morning. He laments ever begging to live beyond the "goal of ordinance" (surely, seven decades)! In reference to this myth, the evolutionary biologist George Williams calls our obsessive focus with lifespan the "Tithonus error." He reminds us that death is not a programmed event, nor is aging. Aging and death are consequences of programming for life (and reproduction). In this view, and the one I have taken throughout the book, aging isn't just what happens to your body at age 67. It's the result of our lifetime of inputs, investments, allocations, and decisions—and the lifetimes of our ancestors over past generations.

A central theme of this book is how the mountain of adult lifespans peaked at about seven decades for much of human history. As I explored in chapter 3, it now peaks closer to eight or nine decades throughout the industrialized world. What does that mean for the timing of the demise of different organs and systems? We know the "reasons" have to do with fewer infections, more food, and medical interventions—that is, reducing exogenous mortality. Their impacts extend well beyond helping us live from age 65 to age 75. Effects titrate throughout the life course. Improved nutrition and fewer pathogens means a bigger energy account to spend more on growth and immunity. And maybe more reserve capacity, with higher organ/system safety factors. If so, we'd expect a slower rate of deterioration throughout life. That's different than coasting at the same rate but patching up the occasional tire that bursts. Such patches may indeed outweigh or at least stave off the damage left in the wake of sedentism, obesity, air pollution, and other residues of modern living.

What is different about populations shifting from a seven-decade to a nine-decade modal lifespan? As a crude start, I make use of a simple model created by the Swedish statistician Anders Ledberg. Rather than consider a complex, hierarchical structure of organs and systems, each with varying rates of decay, he squints his eyes and considers a holistic biological system to age whenever damages accumulate throughout. Damages are repaired, but not perfectly, and the ability to repair declines with age. This is a cartoonishly simple setup but nonetheless leads to the beautiful exponential (Gompertz) increase in adult mortality. How damage accumulates doesn't change the fact that mortality will increase exponentially with age when there's more damage through the system. Using this model, I compare two cohorts—say Swedes born in 1910 with those born in 1770—and determine the extent to which differences in their adult mortality profile are due to differences in the rate of damage accumulation versus the ability to tolerate such damage. The 1770-ers had modal adult lifespans around 70, while the 1910-ers had a steeper peak at age 83. So this comparison is great for thinking about differences between our evolved species-typical prehistoric modal lifespan and the postindustrial reality of now.

Using Ledberg's framework, I find almost no difference in the rate at which damage accumulates in the two cohorts. Instead, differences in their mortality rates are explained entirely by differences in the "threshold" beyond which death occurs. A greater threshold means greater system reserve. The threshold among 1910-ers is 37% higher than that of the 1770-ers.[27]

Better nutrition, less infection, and medical intervention all help us live longer lives, but the model gives us a sense of how. This modern trifecta helps our bodies better endure the "damage" we experience. There's no need to invoke a better initial repair rate right out of the womb, repair mechanisms that remain intact for longer, or a lower amount of total damage experienced. Making the initial repair capacity greater or ensuring that repair mechanisms keep working only serves to increase our damage tolerance even further, rather than decrease the total amount of damage that accumulates with age.

More realistic models get more complicated, but still show a similar dynamic.[28] They apply damage and repair rates to specific nodes (organs, and biological processes) that are connected to each other in a network. Some nodes are heavily connected in the network and so damage there can lead to death, while others are moderately connected in the network. Most nodes, though, are only weakly connected to each other.

This pattern of most nodes being weakly connected and few being highly connected obeys a mathematical power law called a "scale-free network." Damage accumulates across nodes spread across this network. What's fascinating is that, in these models, the familiar Gompertz-like exponential increase in mortality emerges, but without assuming that repair ability breaks down with age. Damage occurs haphazardly from stressors in the environment, and damaged nodes affect nearby nodes in the neighborhood. This complex network approach also explains the increase of cumulative "deficits" with age. Deficits include health problems, wild biomarker values, and other ways of quantifying things going awry in the body—similar in spirit to some of the frailty and aging indices mentioned earlier. Everywhere they've been studied, and regardless of the combination of measures, these holistic measures of "how many things are off?" increase in adulthood, and accelerate in the 60s and 70s.

Contemporary Horsemen of the Apocalypse

Our ability to tolerate damage has improved, non-aging-related causes of death are lower, and more of us now survive up to and past seven decades. Aging is a mosaic, as we've seen, but how does all this manifest in terms of the diseases we get, and what eventually kills us?

Figure 8.7 shows death rates, grouped by biological systems, for Americans in 2018. Even with Death Mountain centered around almost

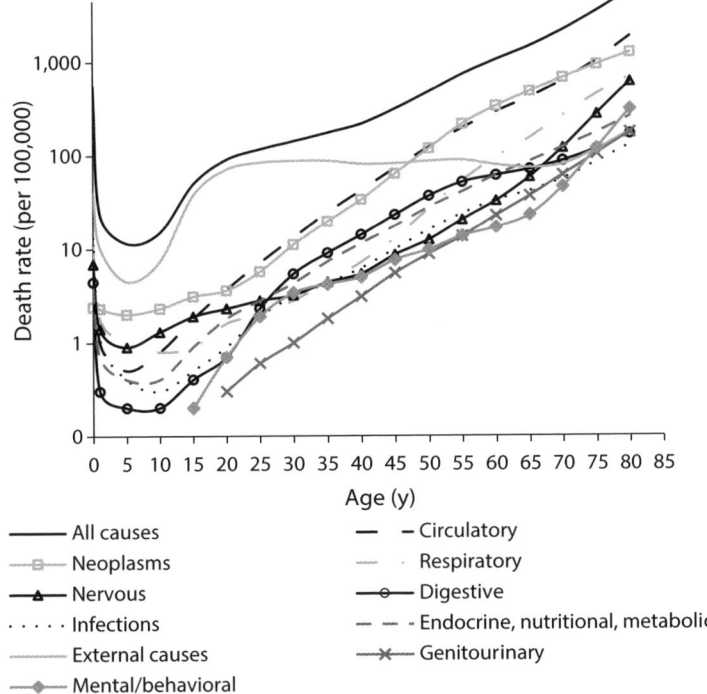

FIGURE 8.7. **Death rates by age and cause, United States in 2018.** Causes are grouped into large categories defined by International Classification of Diseases. Death rate (per 100,000 people) is logged. *Data source:* CDC Wonder.

nine decades (recall figure 3.16), the rate of dying from any cause accelerates in the 50s and 60s, and surges by the 70s and 80s. On the log scale, the increase is like a straight line from ages 40 (if not even 20) onward. That means exponential growth. In fact, all causes seem to grow exponentially and at roughly similar rates, consistent with a pattern of mosaic aging that nonetheless leads to similar changes in mortality across systems. These parallel curves only differ in their intercepts, meaning the baseline levels of risk. The proportional changes you see in mortality between ages 20 and 30 are the same you see between ages 70 and 80.

The two most important causes of death—ischemic heart disease and cancer—are reflected in the top two lines of figure 8.7. Circulatory system deaths also include hypertension, pulmonary heart disease, and strokes. Next in line comes the respiratory system (chronic lower respiratory problems, the flu, and pneumonia), endocrine/metabolic (largely diabetes, obesity, and metabolic-related disorders), nervous system (mainly degenerative diseases, systemic atrophy of the central nervous system, and movement disorders), and then the genitourinary system (mostly kidney failure).

However, three categories have age profiles of mortality that look remarkably different—"external causes," "infections," and "mental." External causes are a catchall that includes accidents, homicides, self-harm, and car crashes. As a whole, these affect adults more than children, with death rates being relatively constant over adulthood—consistent with their "exogenous" origin having less to do with intrinsic physical aging. Prior to the COVID-19 pandemic, infections mainly include "intestinal infectious diseases," "other bacterial diseases," HIV, and viral hepatitis. Rate of dying from infection increases throughout life, but slows down in middle adulthood, then speeds up again by the seventh decade. "Mental" includes mental and behavioral disorders due to psychoactive drug use among younger adults, along with Alzheimer's and other dementias in later adulthood. Consistent with the argument about slower brain aging above, death from dementia accelerates only by the seventh decade.

Death Disco with Cancer

Cancer vies with heart disease for top billing as the leading cause of death. Unlike other causes, cancer stems from multiple changes in our genes that affect how our cells function. In the US, 40% will develop cancer over their lifetime, and 20% will die from it. Cancer manifests throughout the body in a patterned way. With rare exceptions, most cancers occur after age 45. The median age of diagnosis in the US is about age 66. The reason for a late onset can be threefold. First, a series of mutations have to occur for tumors to form and spread, and that can take decades. Second, an immune system weakened by age can let cancer cells escape detection, and form tumors. A poorly functioning immune system might also let certain cancer-causing viruses, like Epstein-Barr and human papillomavirus, wreak havoc on our bodies. Lastly, exposures to known carcinogens like the sun's ultraviolet radiation, tobacco, processed meat, alcohol, asbestos, and your basement's radon all build with age. Smokers carry higher risks of getting cancer, but smokers and nonsmokers alike show similar mortality-by-age curves for lung cancer. Issues of diagnostic technology aside, median age of diagnosis for pancreatic, bladder, and lung cancer is in the early 70s.

Focusing only on death rates doesn't tell us when cancer is likely to first strike. New cases of a disease are reflected in epidemiological measures of incidence. Figure 8.8a shows cancer incidence rate by age in the US from 2014 to 2018. The first thing to notice is that the three most common cancers (breast, prostate, lung) are linked to modern urban environments. Others are linked to infections like human papilloma virus (HPV) (cervical cancer), the bacteria *Helicobacter pylori* (stomach cancer), and viral hepatitis B or C (liver cancer). In order of appearance (when passing, say, 25 cases per 100,000), first comes cancer of the breast, then thyroid, prostate, colon/rectum, kidney, skin, lung, brain, pancreas, liver, stomach, and esophagus. Thankfully, cancer of the bones and joints, and of the gallbladder, never cross this threshold.

As absolute rates vary among types, a different way of looking at things is to consider when half of the total number of cases occur for each type. This is between ages 65 and 75. One curious pattern is that incidence peaks in the 70s and 80s, then declines, rather than burst off the chart at the oldest ages.

Figure 8.8b shows the death rate from the same types of cancer. Despite remaining a top scourge, cancer mortality has been declining since the 1990s. Overall, breast and lung cancer are the biggest killers, followed by colorectal and pancreatic. The gap between diagnosis and death varies by site and by types of accessible treatment. Overall, two-thirds of Americans diagnosed with cancer are still alive after five years. Unlike what we saw in figure 8.7 for most other causes of death, death rates for many specific cancers don't seem to increase exponentially with age; the rate of increase after ages 50–60 tends to slow down. That being said, the chances of dying from *any* type of cancer increases in a more exponential

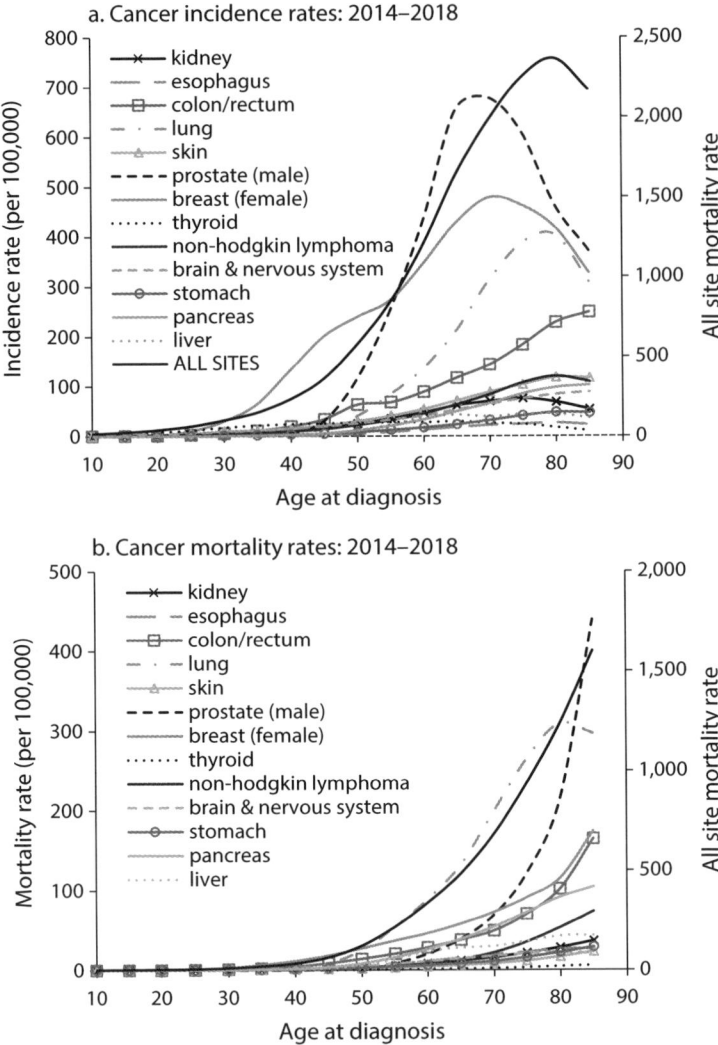

FIGURE 8.8. **Cancer incidence and mortality in the United States, 2014–2018.** *Data source:* National Cancer Institute SEER Explorer. Incidence and mortality for all sites combined are shown in black bold using right y-axis.

fashion. Curiously, few centenarians die of cancer. Many men die with prostate cancer but not from it. Some of the differences in mortality might lie in how quickly cancer can spread from its origin to other vital organs. Once it destroys organ function, death is imminent.

Environmental exposures clearly matter, especially for organs in direct contact with our surroundings, like skin (and sun), lungs (and air), and the gut epithelia (and remnants of our food). The heart, arguably the most vital of all organs, sees almost no cancer. And the brain, relatively little. For reproduction, what's more important than the uterus, testes, and ovaries? Also relatively low lifetime cancer risk there. Consistent with the logic developed earlier in the chapter, selection should lead to stronger anticancer defenses in organs more vital for fitness, and those that can't function with tumorous growth (e.g., small organs, not those coming in pairs). This generally bears out.

But another factor seems to affect where in the body cancer manifests: bad luck. Organs with more lifetime divisions of stem cells are more likely to see more mutations by chance, and more cancer. While most organ cells are specialized and relatively short-lived, the sacred stem cells are like superheroes: they can renew themselves, generate new cells with specific functions, and act as the master internal repair system. There are orders of magnitude more stem cell divisions in the colon and rectum than in the testes, thyroid, and gallbladder. Ironically, more stem cell divisions in an organ might be viewed as helping to maintain its function, and possibly increase its safety factor against other failures. But that comes with a nasty cost in later life. The relationship between stem cell divisions and organ-specific cancer incidence is so tight that it accounts for up to two-thirds of the differences in cancer rates across tissue types. While debates continue to rage over how much cancer might be avoidable versus a result of bad luck, cancer results as a compromise from the parliamentary proceedings among our cells housed in different tissues and organs all serving the same master.[29]

Cancer susceptibility is often viewed separately from the rest of the physiological workup of aging. Like an awkward Facebook relationship status, it's complicated. On one hand, the immune system doesn't just protect against a wide range of bodily assaults, but also helps identify and eliminate malignant cells in our body. Aging of the immune system leaves us vulnerable to nefarious pathogens *and* our own rogue cells. On the other hand, inflammation can also be harmful when it comes to cancer. It helps foster a favorable environment for tumor growth, where tumors co-opt molecules of the innate immune system. In this regard, tumors make themselves act like wounds that don't heal.

A good example of how defense against cancer may shape other aspects of aging takes us back to telomeres. Remember, these caps to our chromosomes erode with each cell division. Faster erosion due to a variety of stressors is believed to accelerate aging, and to push cells closer to a "senescent state." While taking worn cells out of commission sounds like a bummer, the failure to erode chromosomes can lead to cancer.[30] When Carol Greider and Liz Blackburn won the Nobel Prize in Medicine in 2009 for their discovery that the enzyme telomerase can elongate telomeres, many people wondered if this news meant we were close to slowing cellular aging. Unfortunately, while flooding cells with telomerase lengthens telomeres, it also increases mortality due to cancer. A senescent state may be a way of silencing cells against harmful mutations and potential cancerous growth, even though too many senescent cells also contribute to aging and compromised survival.

Parliament of Organs, One-Hoss Shays, and Other Mixed Metaphors

By now, I hope it's clear that aging and lifespan are not determined by optimal design for a long, healthy life. We must work with what we've inherited from our ancestors. The body may seem like a well-oiled machine, at least when it's working well. But the analogy falls apart upon inspection. All of life is a fight against entropy. How bodies do this has been shaped by natural selection, placing privilege on reproduction, and survival through the reproductive years. Aging is as much about what happens in early life as it is about what occurs after age 60. Like Kirkwood showed us in chapter 2, a long-lived soma is no guarantee of evolutionary victory. It's "disposable" in the service of reproductive success.

My contention throughout this book is that (at least two decades of) postreproductive life has also been subject to active selection, rather than an evolutionary afterthought. Natural selection privileges reproductive ages, but postreproductive ages in social animals like humans also affect fitness. Somatic maintenance is tuned to keep us functional past age 50 in nonindustrial environments. That means there may be physiological decline, but not enough to make us inept or hopeless.

On the whole, our postindustrialized reality is a mixed bag. We live longer, and functionality is reasonably good up through our 60s. Well into our seventh decade, we feel aging's wretched pull, but can manage if we're lucky. If needed, we wear glasses, insert hearing aids, get our backs serviced, our knees replaced, our polyps removed. Into our 80s and beyond, we're likely in better shape than our ancestors, and when compared to contemporary subsistence folks. These gains aside, we haven't slowed aging enough to stave off the stormy tempests of physical adversity. Once we first notice a slowing down, in our 40s and 50s, we may strive to diversify our professional portfolio, just like hunter-gatherers. The broader roles of elders—as educators, advisers, leaders, and the like—may themselves be adaptive shifts to our mosaic aging, where our minds are preserved and our physical prowess is not. By our 60s, multiple body parts might start ailing in sequence. Sooner or later, maybe a decade, maybe two—disability grinds us down. If we're lucky, we can still adapt to our changing circumstances enough to get by.

Like the demographer and longevity skeptic Stuart Jay Olshansky quipped, fighting aging is like "epidemiological whack-a-mole: when one disease is knocked down, two or more rise to take its place."[31] The reaper eventually finds us, hopefully well beyond our warranty period. No, our bodily shay didn't evolve to fall apart all at once. But we can still have a lengthy functional healthspan even if we age. Our "function" changes up through elderhood.

———

In *Natural Causes*, the late author and crusader Barbara Ehrenreich, an immunologist by training, wrote about how she was shocked to learn that the same heroic immune defenses that protect and save us could cruelly and traitorously promote the spread of cancerous tumors and help clog our arteries.[32] Our bodies are *not* designed, but are kluges built and modified by evolution. Organs and systems must take on multiple roles and are connected and networked in systems of systems. Amusingly, Ehrenreich also rages against systems science for its mystic-sounding holism and hints of crystal-cuddling positivity.

Not just a fad, in our case, a systems approach is unavoidable when trying to understand aging. Aging is neither a single global process nor a deregulated mess of wildly varying independent processes. It's something in between—system-specific processes linked together. An annoying implication of such a view is that there may be no prime mover or cause, no magic bullet that will alter everything, no miracle molecule to tweak. On the plus side, interventions don't need to "slow aging" in order to have big health-promoting effects over many years. No one would argue that a hip replacement surgery slows down aging. But if it enables more years of physical activity and comfortable engagement with friends and family, then call it a win. In fact, as we'll see in chapter 9, physical activity is probably the best intervention for a good reason. It affects many of our interconnected systems. It improves our muscles, bone density, heart, and brain, it reduces inflammation, makes us feel better, and so much more.[33] It's also the common feature of every subsistence population you've read about in this book.

9

A Mismatch Made in Heaven

A man is only as old as his arteries.

—THOMAS SYDENHAM, ENGLISH PHYSICIAN (1624–1689)

"WHY BOTHER studying something that none of us have?" a trusted Tsimane friend and research assistant asked me one day. It was an astute question, one I've asked myself many times over the years. The simple answer: studying diseases among people who don't appear to have them can reveal protective factors that could help the rest of us. Since we understand human aging within the context of chronic diseases like atherosclerosis, diabetes, and Alzheimer's disease, we don't have a good sense of what it might look like to be without these diseases.

When we think of diseases of old age, what springs to mind are what were once labeled "diseases of civilization" or "diseases of affluence": heart disease, cancer, diabetes, and dementia. But these are no longer restricted to high-income or developed countries. They're global killers. Then there are other aches and pains like arthritis, cataracts, hearing loss, and muscle atrophy. We tend to think of healthy aging as the absence of all these torments. But are these challenges the inevitable results of life well lived? Have these always been part of "time's doting chronicles"? In this chapter, we explore what we know about chronic diseases in hunter-gatherers and other subsistence societies. We'll cover the big killers, like heart disease, diabetes, cancer, Alzheimer's disease, and more. We'll see that there's no visible line that separates healthy from pathological aging in different epidemiological worlds. After experiencing changes in their environments and their habits, most populations will see their health change, but changes associated with industrialization and urban living (that is, "modernization") do not always lead to worse health. Modernization may help solve some health problems while creating new ones. As should be expected, shifting lifestyles affect health and aging in complex ways. But we'll wade through the morass.

If what we eat, how we move our bodies, which microbes nestle inside us, and other aspects of our lives are so different now compared to our deep past as hunter-gatherers, our health may be compromised by this *mismatch* between our genes and our current environment. According to this view, our genes evolve too slowly to keep up with the rapid pace of cultural, technological, and environmental change that we've experienced throughout human history, especially in the last 150 years. In other words, there's an evolutionary mismatch between our evolved genetic makeup (attuned to life as a hunter-gatherer) and the current environment we inhabit today (figure 9.1). To borrow the title of one of the first prominent commentaries by S. Boyd Eaton, Mel Konner, and Marjorie Shostak to promote this view, we are *Stone Agers in the fast lane*!

We'll explore how this concept of evolutionary mismatch sheds new light on the complex landscape of human aging. By the end of this chapter, you'll have gained a long view of human aging and disease that extends beyond the familiar terrain of urbanized lives. (You might also become skeptical about adopting the latest paleo diet fads, or accepting an invitation to a RePOOPulation party.[1])

When I'm trying to convey to my students how a foraging lifestyle might shape physical fitness and health, I like to show clips of *Naked and Afraid*. It's a reality show that sets well-fit survivalist-types in remote locations to fend for themselves for twenty-one days. The show is absurd but gives a feel for what it might be like for ordinary adventurers, with minimal tools, to suddenly find themselves immersed in a foraging world. Without fail, contestants lose weight. A lot of weight. They're more physically active than in their normal life, but for the most part they suck at finding food.

FIGURE 9.1. **Evolutionary mismatch.** Our hunter-gatherer genome may be ill-fitted to our contemporary urban lifestyle. Credit: Anders Nilsen.

The show is reminiscent of a small experiment that was done in Australia in the 1980s. Nutritionist and public health expert Kerin O'Dea enlisted fourteen middle-aged urban-living Aboriginal Australians with type 2 diabetes, most of them overweight, for a seven-week program wherein they returned to a traditional foraging lifestyle in the bush. Unlike virtually all of the contestants on *Naked and Afraid* (which once featured a student who did his PhD with the Ache, and who made it successfully to the end!), the participants in this study already knew how to live as foragers.[2] They spent over a week traveling from place to place, a couple weeks on the coast, and the rest inland along a river. They ate reptiles, kangaroo, yams, and freshwater fish, amounting overall to a lean 1,200 kcals/person/day, with over half of the energy consumed as protein and 13% as fat (a third of which was saturated fat).

As you might expect, the changes in their cardiometabolic health profile were profound. All participants lost weight (8 kgs, or 18 lbs, on average), reducing their body mass index by about 10%. Metabolic abnormalities usually associated with diabetes improved substantially. Their fasting glucose levels dropped. The insulin response to glucose in their blood also improved—all consistent with lower diabetes risk. Their blood triglycerides declined, as did their diastolic blood pressure.

Overall, there was a remarkable improvement that was consistent with the impression that, prior to contact, Aboriginal Australians probably didn't suffer from heart disease or diabetes. Yet with a Western diet and lifestyle, they've been hit hard by these diseases. Today, Aborigines have much higher rates of stroke, heart disease, and diabetes than non-Indigenous Australians, and are diagnosed with these ailments at earlier ages. According to O'Dea, by the end of the study, the participants felt renewed confidence, proud of the knowledge and skill they had mostly neglected while living in town.

Perhaps inspired by O'Dea's rare experiment, several groups introduced a few interventions in contemporary Aboriginal communities whose residents were experiencing poor nutrition and a high prevalence of cardiometabolic disease. The Minjilang Study on Croker Island off the coast of Australia's Northern Territory began by request of Senior Aboriginal Health Worker Daisy Yamirr, following the untimely deaths of two men. Both had succumbed to heart attacks from unmanaged heart disease. All aspects of the project involved full collaboration with the community. Interventions (reinforced by community elders) aimed at altering adult diets toward less refined sugar and processed foods, trimming fat off of meat, and eating more fruits and vegetables. All these changes were administered through a single community store. After a year, blood pressure, total cholesterol, and fasting blood sugar levels had declined, and these changes could be linked to the changes in diet.[3]

The Looma Healthy Lifestyle program was another Indigenous community-led and community-directed intervention designed in the 1990s to improve access to healthy food and nutritional knowledge in the remote Kimberley region of Western Australia. After four years, people ate more fruits and vegetables, and showed marked improvement in several risk factors for cardiovascular disease. The prevalence of high cholesterol, for example, declined in half (from 31% to 15%). Carotenoids also increased in the blood. These are the components in plants that give them color, help us make vitamin A, and, lucky for us, have antioxidant effects. Despite these benefits, there was unfortunately little change in rates of diabetes or obesity after four years.[4] Either more drastic changes were needed, or perhaps they needed to be started at earlier ages. After a longer period of follow-up (sixteen years later), the intervention finally showed some sustained effect: diabetes prevalence remained level, whereas it increased in surrounding communities that didn't receive the intervention. In any case, success of interventions targeting only diet may be limited without introducing changes in activity and other aspects of a traditional lifestyle.

The urgency to do something to improve health couldn't be clearer. What we do with our bodies affects our vulnerability to chronic disease. As recent as 2020, incidence of heart, stroke, and vascular disease remains at least 50% higher among Indigenous Australians compared to age-matched non-Indigenous Australians, and the average age of onset is almost eight years earlier.[5]

Ice Mummy Dearest

We know that some of our ancestors likely lived seven decades. What can be said about the condition of their bodies toward the end of their lives?

Long before *Top Gun*'s cocky Tom "Iceman" Kazansky, or the Fantastic Four's Johnny Storm, was the world's original Iceman, Ötzi, who lived in what is now the Italian Alps (figure 9.2). His ancestors likely left the Middle East for Europe, settling in regions of the Tyrrhenian Sea, like Sardinia and Corsica. Recent studies of his 5,300-year-old mummified remains, recovered from an Alpine glacier in the early 1990s, have revealed new stories about our past, and the condition of our bodies as we age. Not only is Ötzi the earliest person to be found with Lyme disease, we now know that he was lactose intolerant, was blood type O, and had brown eyes. He was a lean, physically active hunter-gatherer with sixty-one tattoos. His last meal was wild goat, some deer, and maybe some grains.

Yet, here's the rub. In spite of this diet, and the fact that he didn't smoke or drink, his arteries showed the same type of calcification that we commonly observe today in chest CT scans of older adults. This includes calcified plaques in his carotid artery and in the small arteries at the base of the skull. We know that these are risk factors for stroke, as these routes supply blood to the brain.

FIGURE 9.2. **Coronary artery disease in two mummies.** (*above*) Best-guess reconstruction of what Ötzi the Tyrolean Iceman may have once looked like, akin to Stanford neurobiologist Robert Sapolsky. *Photo Source:* South Tyrol Museum of Archaeology. (*facing page*) Examples of calcification from CT images of an Unangan woman aged ~49 y who lived in late nineteenth century, and of an Egyptian princess aged ~42 y who lived ~1565 BC. RCA = right coronary artery. Reprinted from *The Lancet* 381 (9873), Thompson et al., "Atherosclerosis across 4000 years of human history: the Horus study of four ancient populations," pp. 1211–22, Copyright 2013, with permission from Elsevier.

FIGURE 9.2. (*continued*)

In other words, he had atherosclerosis, the progressive thickening of artery walls due to sticky plaque buildup that can decrease blood flow and lead to blood clots, heart attacks, and ischemic stroke. Atherosclerosis accounts for much of the heart disease we see today, for strokes, and for other conditions resulting from blocked arterial roadways. Before it gets severe enough to afflict your heart and brain, it can reduce blood flow to your legs or arms, making walking painful and tiring.

Not only did Ötzi show evidence of atherosclerosis, but he also had a genetic predisposition to it. Ötzi shared at least six genes recognized to cause heart disease in modern populations. But he didn't die of a heart attack or a stroke. He died in his 40s of an arrowhead lodged in his shoulder.[6]

How did such an active, hunter-gatherer-y guy like Ötzi end up with atherosclerosis? Maybe he was just unlucky, an isolated case. As early as the mid-nineteenth century, however, pioneers doing autopsies and tissue microscopy of ancient cadavers showed evidence of calcified lesions in the arteries of Egyptian mummies. One evaluation of the Pharoah Meneph-tah, a descendant of Ramses the Great (who lived 3,200 years ago) noted "severe atheromatous disease [in the aorta], large calcified patches distinctly visible."[7] But surely if you're important enough to get mummified, you probably led a privileged, well-fed, and sub-mobile life (and afterlife in the Field of Reeds). If that were indeed the case, atherosclerosis could've occurred long ago for the same reasons that it is so pervasive today.

If only things were that simple. Luckily, a team of cardiologists and radiologists answered the question about whether atherosclerosis appears only in elite fat cats. Calling themselves HORUS, after the Egyptian falcon god associated with healing and protection, they used fancier, non-invasive technology—X-ray computed tomography (CT scans)—not just on Egyptian mummies (ranging from 3100 BC to AD 364) but also on the preserved remains of the Puebloan peoples of the American Southwest (dating from 1500 BC to AD 1500), of Indigenous inhabitants of highland Peru (from AD 200 to AD 1500), and of Unangan hunter-gatherers from the Aleutian Islands (from AD 1800 to AD 1930). In total they studied 137 mummies spanning four thousand years of history. They found evidence of probable or definitive atherosclerosis in a third of these mummies, and across all four geographical regions.[8] Since that landmark study in 2013, the HORUS group has found even more evidence of atherosclerosis in the ancient world. They've studied seventeen cultures and have found that atherosclerosis existed in all of them.

While atherosclerosis was observed everywhere, it isn't observed throughout the body. Calcified lesions were more commonly found in the aorta, the iliac, and femoral arteries, and only rarely in the coronary arteries. This profile would be more consistent with what is usually labeled "preclinical" atherosclerosis. In other words, the level of atherosclerosis observed wouldn't have resulted in any debilitating symptoms or hard events like heart attacks. Evidence of hard events unfortunately doesn't preserve easily in mummies.

In high-income countries, it is well known that atherosclerosis begins early in life, even in childhood, although clinical manifestations don't typically occur until middle age. Among young US soldiers killed in the Korean war (average age 22 y), over three-fourths showed evidence of coronary atherosclerosis in autopsy, though none had a clinical diagnosis of this disease. What was found varied from "fibrous thickening to large atheromatous plaques causing complete occlusion of one or more of the major vessels."[9] Fatty streaks—the earliest visible lipid-laden lesions that can form plaques—are observed in children and teenagers, even infants, in the aorta especially, but also in other arterial beds.[10]

Given their findings, the HORUS group argued that atherosclerosis is "an inherent component of human ageing and not characteristic of any specific diet or lifestyle." While the evidence clearly supports that claim, I don't agree with their conclusion that atherosclerosis is a "serial killer that has been stalking mankind for thousands of years."[11] Unless the stalking is just a potential predator that rarely catches its prey.

Heart Disease

How do we tell whether atherosclerosis (and heart disease in particular) would've been relevant in the day-to-day of our ancestors? Hopefully it's not a broken record to say it again—let's turn to contemporary subsistence populations!

Remember from chapter 3 that infectious disease, accidents, and violence were responsible for the majority of deaths in hunter-gatherers and other subsistence populations. If people experienced chronic degenerative diseases, they died of other things first. But many chronic diseases are hard to diagnose directly without fancy tools. Easiest is to look at common risk factors known to predict those diseases, at least assuming that like predicts like in different environments. That may not always be true, but it's not a bad place to start.

Among past and contemporary hunter-gatherers, risk factors common to both cardiovascular disease and type 2 diabetes, like obesity, hypertension, elevated blood cholesterol, and insulin resistance, are rare. Epidemiological surveys from the mid-twentieth century among Ju/'hoansi, Mbuti, Hadza, Aboriginal Australians, South African Bantu, Pacific Islanders, Turkana pastoralists, and other rural, subsistence-level populations with minimal exposure to markets all support the notion that these risk factors are rare.

Table A9.1 in the Appendix compares measures of these risk factors in hunter-gatherers, horticulturalists, and pastoralists against values in the United States. There's little doubt that subsistence populations now and throughout history had healthier levels of blood pressure, cholesterol, and blood sugar than we now see throughout much of the urban world.

Beyond Risk Factors

Having more risk factors increases your likelihood of heart disease based on prior studies in industrialized populations. But it's unclear whether risk factors lead to the hard clinical events like heart attacks and strokes in very different environments, which is a departure from the uniformitarian hypothesis I introduced in the first section of the book. Applied here, this hypothesis would suggest that the same exposures and biological processes should affect health similarly no matter where or when. Luckily, a few studies go beyond risk factors. Clinical studies of Maasai pastoralists of East Africa in the 1960s showed little clinical evidence of hypertension, obesity, high cholesterol levels, or ischemic heart disease. This was surprising because the Maasai eat a diet rich in milk and meat, foods usually associated with elevated heart disease. They do, however, show evidence of rheumatic heart disease, which occurs when untreated strep infection leads to rheumatic fever, which in turn can damage the heart's valves. Damaged valves can lead to heart failure. Though rheumatic heart disease has declined worldwide, it's still a cause of preventable deaths in poor regions where acute rheumatic fever still lingers. But it's unrelated to atherosclerosis.

Ju/'hoansi also showed no clinical evidence of coronary heart disease, and no history of symptoms like chest pain (angina) or silent infarcts detectable on an electrocardiogram (ECG). But, like the Maasai, some Ju/'hoansi showed some rheumatic heart disease.

A follow-up autopsy study by physician George Mann of the hearts and aortas of fifty Maasai men "dying suddenly" helped clarify things further. As part of the International Atherosclerosis Project designed to survey atherosclerosis globally, local medical personnel in eight health stations speckled throughout Maasailand in northern Tanzania and southwestern Kenya monitored any male deaths that had occurred "within a week of an accident or acute illness." The team reported extensive fibrosis and fatty streaks in the aorta, but only sporadic indication of complicated lesions or other signs of further developed atherosclerosis. Coronary blood vessels became thicker with age, but the opening (or lumen) was also bigger, acting perhaps as a form of compensation. In other words, the Maasai had atherosclerosis, but no compromised blood flow in their arteries.[12]

You're Only as Old as Your Arteries

From my years of studying health and aging among Tsimane, it didn't seem like they suffered from heart disease either. Looking at the same risk factors, it was clear that Tsimane didn't have many. For example, their LDL ("bad") cholesterol was low and obesity was minimal. But not all was favorable against risk. Tsimane showed high levels of systemic inflammation, primarily due to combating infection. While some of the elevation reflects acute bouts of sickness, many people show higher inflammation on follow-up visits, due to repeated short-term infections or from chronic infections dirty dancing with the immune system. Ordinarily, chronic inflammation is linked to all stages of atherosclerosis, and to progression of many chronic diseases of aging, and so the mix of factors makes Tsimane an interesting test case for pitting the role of inflammation against other risk factors. Levels of Tsimane HDL, or "good" cholesterol, are also low, with half of adults falling below the usual recommendations.

Here was an opportunity to improve on past studies. Older studies had small samples. It'd be easy to miss people too sick or tired to participate. In these cases, it is difficult to know whether samples are representative of the broader population. A more serious problem is that past studies are snapshots—they look at only a single point in time. If no one with heart disease is seen during a brief study, maybe that's because they already died. Without anti-hypertensive meds, statins, and life-saving defibrillators, people with heart disease might not live very long with it before succumbing. In addition, knowing risk factors are minimal is not the same as direct evidence of absence. Even in the US, people with few or no typical risk factors have heart attacks and strokes. By one estimate, one out of seven first-time heart attacks occur among those without any risk factors.[13] With biomedical surveillance done year after year, the Tsimane Health and Life History Project (THLHP) team has seen everyone in most Tsimane villages. If we miss someone one year, we likely saw them in another, and, armed with censuses, we can be sure we didn't miss anyone. And if someone misses a scheduled visit, we visit their house to make sure they're okay. From extensive interviews, it appears very few of the deaths that occurred over the past seven decades resembled heart attacks or strokes. Same goes for the deaths that have occurred since we started our study in 2002.

We then confirmed that blood pressure barely increases with age, unlike what's seen everywhere else in the industrialized world. We showed this looking at the same folks over time, rather than what all previous studies in subsistence populations had done: comparing people of different ages. Elevated blood pressure at one point in time was only persistently elevated for half of the cases originally believed to have hypertension. Being nervous, or any isolated stressful events, might suddenly elevate your blood pressure. In other words, hypertension is even lower than we initially suspected (but where it does occur is disproportionately among older adults over age 60).

Next we looked at blood pressure in the arm and across the tibial arteries in the ankle. Their ratio is used as an indicator of peripheral arterial disease (PAD). Lower ankle pressure suggests reduced blood flow due to the narrowing or blockage of arteries in the legs. PAD can cause pain in your limbs when you exercise, but more importantly, having PAD means you may be more likely to witness severe atherosclerosis in your future. Our results couldn't be clearer (or easier to calculate). The prevalence of PAD was a big fat zero—not one case among more than 250 Tsimane over age 40. Yet it affects up to 40% of adults age 70 in other countries. All indications would suggest no heart disease. Of course, "suggest" is still not hard evidence.

Be Proud of a Tiny CAC

The next stage, starting in 2013, was a fun, ongoing collaborative adventure in which we teamed up with the HORUS group of mummy-scanning clinician-researchers who argued that atherosclerosis is everywhere and everywhen. As you might imagine, it was a massive logistical undertaking to bring folks from remote villages to the distant Beni capital city, Trinidad, located several days' journey away by canoe and truck. There, our friends and colleagues at the German Busch Hospital helped us perform CT imaging of the hearts of elder Tsimane. This was also a nice opportunity to get older adults extra medical attention, as specialists are hard to come by in the resource-strapped towns nearest the Tsimane territory.

It took a few years and many hair-graying CT campaigns, but we eventually examined more than seven hundred adults over age 40. From CT imaging of the heart, you can measure the amount of calcification in the coronary arteries, probably the best noninvasive direct measure of atherosclerosis. Under most conditions, more coronary arterial calcification, or CAC as it's called, means more active plaques in the arteries, which all means a higher risk of heart disease and heart attack.

We found that 85% of older Tsimane had no CAC whatsoever. And only 3% had a CAC score greater than 100, the cutoff for moderate atherosclerosis. Even by age 75, two-thirds of Tsimane still have no CAC. For perspective, only one-fifth of Americans are CAC-free by this age, and half pass the CAC > 100 threshold. Only 8% of Tsimane by this age pass this threshold. What this amounts to is that Tsimane elders have an arterial age that is two and a half decades *younger* than that of Americans!

So how low does the CAC go? Tsimane don't just have a lower level of coronary artery disease (CAD) than freedom-fries-loving Americans, but have the lowest levels ever recorded anywhere.[14] Of course, that's also because the Hadza, Ache, and other groups haven't been studied in a similar way. I'm sure we'd find the same thing among other groups, and conclude that, despite varied diets and lifestyles, subsistence populations have minimal coronary artery disease. And when they do have it, it's not fatal or burdensome. As our studies were of older adults, and we track people over time, the lack of fatal heart disease is *not* because folks don't live long enough.

Tsimane hearts are strong and healthy in other ways. Like other subsistence populations, they show higher aerobic capacity, as gauged by VO_2max (that is, the maximum volume of oxygen the body can use per minute). While Tsimane have higher levels than age-matched Westerners, even their high rates show evidence of decline with age (see chapter 8, figure 8.6). Tsimane adults have low resting heart rates, much like ath-letes, often under sixty beats per minute. After physical activity, their heart rates are quick to return to baseline, even at later ages.

More direct measures of the heart's pumping efficiency show consis-tently elevated performance throughout adulthood. Your heart's ejec-tion fraction assesses how effective the heart is in squeezing out blood from its chambers with each lub-dub. Usually < 45% is an indication of impaired function, but Tsimane average about 65%–70% even in their 80s. We also find very little visceral fat surrounding the heart, something that when abundant can become inflammatory and further promote atherosclerosis. And if all that isn't convincing, after evaluating over a thousand ECGs, we found evidence of just one silent infarct—that is, a heart attack with mild or no symptoms—and only one case of atrial fibrillation, the main type of heart arrhythmia that can increase your risk of a stroke or heart failure.

Diabetes

The diabetes referred to as type 2 is usually first observed in adulthood, hence why it's also known as adult-onset diabetes. It shares some of the same risk factors as heart disease and is itself a risk factor for heart dis-ease. Diabetes refers to a loss in your body's ability to regulate blood sugar. While glucose (sugar) is a necessary energy source for all cells to function, the hormone insulin, courtesy of the pancreas, does much of the heavy lifting of blood sugar regulation. But when there's not enough, or if it's not up for the job, too much sugar in circulation can be deadly. Over the long term, uncontrolled diabetes can lead to kidney failure, heart attack, blindness, and death. In 2023, diabetes was the seventh leading cause of death globally, responsible for ~1.6 million deaths.

Given the healthy risk factors for heart disease we've already seen, it's not surprising that diabetes was also not a problem for much of human history. As shown in the Appendix (table A9.1), fasting morning blood glucose levels in subsistence populations are low, well below the thresh-old used to diagnose diabetes. Circulating blood sugar levels vary day to day, and so it would be better to look at blood sugar levels averaged over a longer time period. Luckily, the A1C test (a measure of how much oxygen-transporting hemoglobin proteins in your blood are coated with sugar) does exactly that, averaging blood sugar over a few months. Glucose tolerance tests that assess how your body handles sugar after a meal are also informative. Where assessed among traditional-living groups, results of these and other tests, including measures of insulin, are all consistent with minimal diabetes.

I suspect that diabetes is the first deadly chronic disease to creep in after rapid transitions toward more Western lifestyles. Turkana shifting away from herding or farming are three times more likely to have diabetes, and also more likely to be hypertensive, obese, and have higher total cholesterol (table A9.1). In Canada, some First Nations groups saw their diabetes prevalence double in just a single decade![15] Among Tsimane today, a growing number of folks have blood sugar levels that would classify them as "pre-diabetic." At least two Tsimane adults living in town have died from complications due to unmanaged diabetes, and a few others I know are trying to manage their new diagnosis. When I was having dinner in town with members of the Tsimane government in 2016, and discussed what a diabetes prevention plan might look like, a Tsimane official in his mid-30s newly diagnosed with diabetes by our team deliberately left the usual fried white rice on his plate and drank water instead of the obligatory soda. When I saw him in 2023, he was back to white rice and Coke.

Dementia

Alzheimer's disease (AD) and related dementias (as a group, abbreviated ADRD) are well on their way to becoming a major cause of death in the US, and in the world at large. Over 6 million Americans live with AD, and numbers are expected to triple by 2050. Its worldwide prevalence is expected to increase fourfold by the same time. Aging of the global population alone accounts for much of this increase. Lest you think these are limited to rich countries like the UK and Australia, over half of all cases now come from low- and middle-income countries, and this proportion will only increase in the future. The sense that this is a "disease of affluence" holds no truth. Anxiety over the growing stats is compounded by the lack of any cure. By the time it's diagnosed, it's often already advanced.

While the label "dementia" is nonspecific, AD currently reigns as its main expression in the Western world. It shares many of the same risk factors we've seen several times already that would presumably be rare in subsistence populations: cardiovascular disease and its associated risk factors, and sedentary living. Other dementia risk factors, however, are likely to be present in subsistence groups: inflammation, hearing loss, minimal formal schooling, head trauma, air pollution, and genetic susceptibility. And, of course, old age.

Given that ADRD mostly manifest after age 65, and especially after age 75, you can imagine that if it were to exist in small-scale societies, it would be hard to identify with precision. Dementias are also tough to accurately diagnose. There's no simple test like blood pressure or a blood-based biomarker like glucose.[16] Standard assessments are based on urban living, asking questions about the day of the week, the name of the president, and the like. For these reasons, and with other sources of morbidity more prominent, we still have much to learn about dementias and whether they are an inevitable companion to old age.

In most Indigenous languages, there are no words for dementia. The name is derived from the Latin root *demens*, meaning "outside of one's mind," and what we diagnose as dementia may be confused with mental illness. Awareness and understanding of dementia even in contemporary native groups in high-income countries is still poor. Some native groups today associate dementia with memory loss, behavior change, social loss, and unhealthy lifestyles.

Weighing the positive and negative risk factors together, is dementia more or less common in subsistence groups than in postindustrialized settings? Again, teaming up with the HORUS team and with the talented neuropsychologists and neurologists at the University of Southern California (whose daily bread and butter is diagnosing and managing dementia cases), we sought an answer. We adapted typical screening tools to Tsimane life. We combined neurological and clinical assessment with an interview of a spouse or other caregiver, along with years of testing different areas of memory and other cognitive skills to see if folks had been stable or getting worse (of the sort compiled in figure 8.6d). We also conducted head CT scans to look for any lesions, atrophy, or calcification in the brain. The expert USC team and our experienced Bolivian physicians made their independent diagnoses, then ironed out their differences. This is about as good as it gets in a low-resource setting, a working diagnosis without PET scans, spinal taps, or postmortem evaluation of brain tissue.

When all was said and done, we found only five cases of dementia among Tsimane, and one case among Moseten (a population related to the Tsimane but living a more acculturated lifestyle)—all relatively mild cases in adults over age 80. This amounts to roughly 1% of those age 60 and up having dementia.[17] Figure 9.3 compares this low rate with other studies. Much like we found with heart disease, dementias are much rarer in Tsimane and Moseten than elsewhere. Healthy heart, healthy brain. The few dementias we did observe were mild, and not Alzheimer's-like. Demented folks appeared rigid, were slow moving, and had problems with their gait. These and other elements curiously overlap with symptoms of Parkinson's disease, a progressive nervous system disorder that hinders movement and coordination.

If what we found in rural Bolivia is indicative of dementia's occurrence in our past, then most people making it to their seventh and eighth decades wouldn't have needed to worry with existential dread about Alzheimer's. Not only is dementia rare, but brain atrophy also appears to be slower among Tsimane and Moseten than in high-income countries, like Germany and the Netherlands.[18] But while dementias are more common in the United States and Europe, a startling alarm call is that Indigenous populations in high-income countries seem to have the highest rates of all.[19] Figure 9.3 shows that Aboriginal Australians, Chamorro living in Guam, and former fisher-folk, the Melenau of Malaysia, have higher rates than those reported in the United States, Europe, and everywhere else where reliable data exist.

While dementia is rare among Tsimane and Moseten, we did encounter rates of mild cognitive impairment (MCI) similar to those found in the United States. MCI is a catchall tag for early memory or

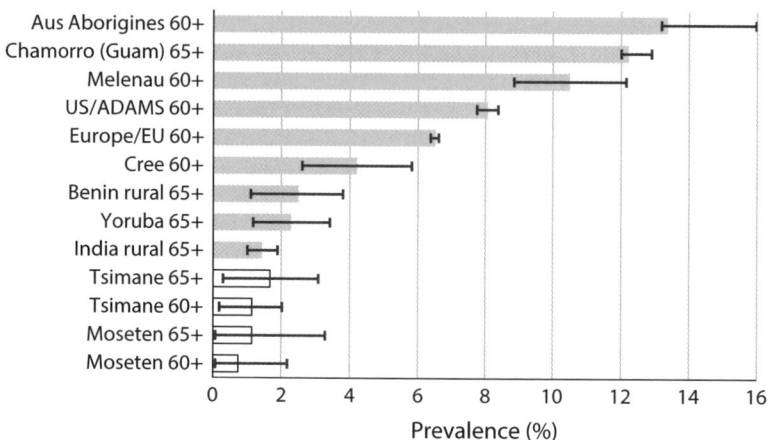

FIGURE 9.3. **Prevalence of dementia.** Modified from Gatz et al. 2022. For additional populations added: Melenau, Pu'un, Othman, and Drahman 2014; Yoruba, Hendrie et al. 1995; Benin, Guerchet et al. 2009.

cognitive loss that doesn't really affect daily living. Whether such impairment is reversible, remains stable, or, worse yet, progresses to dementia remains to be seen. It's likely that some of these cases are due to dietary deficiencies or other temporary comorbidities. Another mystery is that most older Tsimane and Moseten show clear calcification in the arteries feeding the base of the brain. Some degree of brain and cognitive aging is evident in subsistence populations, but why doesn't this calcification lead to pathology? Ongoing study will provide answers.

Cancer

Now that heart disease mortality has been declining in the US and UK, cancer has been vying to overtake the number-one spot as the leading cause of death. At the same time, cancer death rates in the US have been declining over recent decades, but one in two people in the Global North, and one in four people globally, will still get cancer over their lifetimes.[20] All this might sound contradictory. While less smoking and better treatment have helped reduce cancer deaths, earlier detection, the obesity epidemic, and changes in other cancer risk factors have increased the number of new cases in the US since 2020 for six of the ten most common cancers.[21] Cancer isn't going away any time soon. Many experts believe that even if we could extend life by eliminating other common causes of death, many of us would eventually die of cancer. Unlike with infections, cancer strikes when our own cells go rogue, intent to conquer, from lone mutations that accumulate in cells or that deceptively co-opt the usual brakes on regulated cell division to other critical events that favor the local spread to many, often distant, vital organs. It's a standard refrain that cancer is also a mismatch disease, plaguing us now because we live too long, don't use enough sunscreen (or use too much toxic sunscreen), smoke too much, inhale industrial toxins, and eat too much processed meat.

While some modern exposures and behavior surely affect cancer risk, constraints on how cells divide may be relatively fixed. And certain carcinogens, like the sun's radiation and aerosols from cooking fires and biomass burning, have been part of ancestral environments for hundreds of millennia. By its very nature, cancer has been a persistent threat since the dawn of multicellular life. Cancer exists in most animals, and is more common in mammals than other vertebrates, and more common among carnivores.

If cells develop rebellious mutations at a certain rate, then, all else equal, you'd expect larger organisms (having many cells) and longer-lived organisms to have more cancer. But they don't. That contradiction is noteworthy enough to have a name: Peto's Paradox. A classic example of the paradox in action are elephants—huge *and* long-lived, but they have arguably very little cancer for such an enormous species. Large and long-lived species have adaptations to solve the potential cancer problem (or at least keep it at bay). Elephants seem to accomplish this with extra copies of the mighty anti-cancer gene *TP53*. Long-lived species also tend to have mutational clocks that tick slowly. And so the good news is that humans—being reasonably large *and* longevous—are not at a greater risk of cancer than tiny, short-lived mice.[22]

So, cancer may have always been with us, but is it really the case that it's now worse than ever? By one estimation, lifetime cancer risk is believed to top out at 5%, for wild animals and human hunter-gatherers.

Under novel environmental conditions, like under captivity, or in the presence of industrial pollutants, cancer-causing infections, or unhealthy lifestyles, that risk is much higher. Currently, the chance of being diagnosed with cancer during your lifetime is 40%. The median age of first diagnosis is 66 years. This elevated risk comes in part from living longer. A longer life means more time exposed to risk factors, and more time for the mysterious multistep undoing of renegade tumors busting through our cells' restraints. After sharing age profiles of human survival (from chapter 3) with population biologist and cancer researcher Michael Hochberg, he estimated that a hunter-gatherer by age 70 should be a third to half as likely as an American to have cancer.[23]

What we know about cancer comes from looking at the skeletal record (and mummies) throughout ancient history. The same caveats we needed when inferring lifespan from the skeletal record in chapter 4 apply here as well. And there's more: we know something about only ancient cancers that leave traces in bone. Cancers affecting soft tissue will be invisible to us unless they spread and infiltrated bone, or left some evidence on a bone's surface. Much like we saw when comparing registered ages of death with inferences from a skeletal sample among the California Chumash, a clever study by the forensic anthropologist Carina Marques compared registered causes of death among Portuguese from the nineteenth and twentieth centuries with causes assessed using paleopathological methods on the bones. Over a third of the true cancer deaths had no skeletal indications whatsoever to suggest cancer, and only 18% showed definitive confirmation.[24]

So any attempt at coming up with prevalence rates in the past will be lowballing for sure, but we do what we can. Fortunately, some brave paleo-oncologists have recently assembled the largest compilation of malignant neoplasms from over a hundred archaeological sites spanning across almost five millennia, from 3000 BC to AD 1900. As shown in figure 9.4, cancer has been with us throughout the ages. Many of these are bone metastases, multiple myelomas (cancer of blood plasma), osteosarcomas, and nasopharyngeal carcinomas. Where a site of origin can be surmised, it's been breast, prostate, lung, and skin. And, as expected, cancer rates were much higher in adults than in children over the same period.

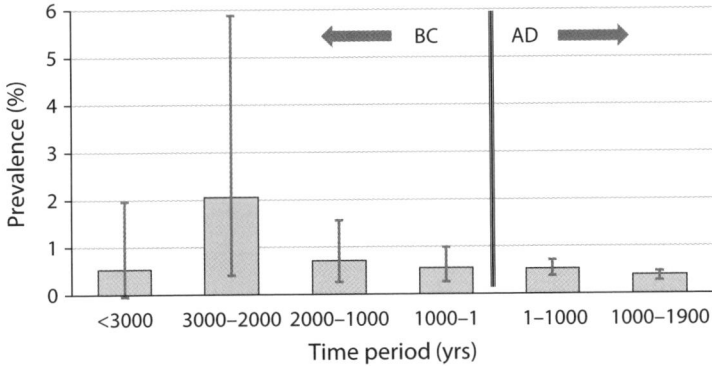

FIGURE 9.4. **Cancer prevalence over human history.** Based on 108 skeletal samples from before 3000 BC to 1900 AD in the Americas, Europe, Asia, and the Middle East. Error bars show 95% confidence intervals. Given the wide confidence intervals due to small sample sizes, and the usual methodological caveats like we saw in chapter 4, precision is elusive. Evidence suggests not much difference in cancer prevalence over time. *Data source:* Marques et al. 2022.

Important for our purposes is that, despite the usual caveats when assessing prevalence from the skeletal record, the percentage of specimens showing signs of cancer doesn't seem to have varied over the past five thousand years (or, at the very least, it certainly doesn't seem to have risen through agrarian transitions to industrialization, and up to the beginning of the twentieth century). This suggests that if there's mismatch, it's not with agricultural transition or even early transition to industry, but that postindustrialized environments of the last century may be more cancer-prone. Remember from chapter 2 that this is the same time when life expectancy increased rapidly throughout the Western world. Even when taking into account the ages of specimens in the sample, and calculating "age-standardized" rates instead of the crude rates (ignoring age) in figure 9.4, there's no rise in cancer in recent history.[25]

What do we know about cancer in contemporary subsistence populations? Mostly anecdotes, cases mentioned here and there. Inuit, for example were believed to have low cancer rates until the nineteenth century. Now they have one of the highest lung cancer rates in the world. Cigarette smoking is five times more common than in the non-Indigenous

Canadian population. But cancer is not caused just by smoking, industrial pollutants, and other by-products of urban life. One in five or six cancers are caused by infection. Human papilloma virus (HPV) is linked to cervical cancer, and *Helicobacter pylori* bacteria with gastric cancers. But cancers of the nasopharynx and salivary glands, like we saw in the skeletal record and among Inuit, are linked to the Epstein-Barr virus. I say "linked" because many people have these viruses and never get cancer. Who's to say whether some of these viruses were more common and destructive in the past than in modern times? Certainly HPV vaccines and antiviral medications have helped reduce incidence of these cancers.

Given the difficulty of diagnosis, my collaborators in Bolivia and I haven't yet studied cancer thoroughly. It's unclear how much cancer we should expect to find. On the one hand, low obesity, high fertility, lower reproductive hormones, high physical activity, and nonprocessed diet are all protective factors for the Tsimane. On the other hand, HPV, *H. pylori*, and other infections are common, and inflammation is elevated. Immune surveillance may be compromised, especially in later ages.

So do Tsimane have more or less cancer than we'd expect? We've seen cases presented to our team, often too advanced to treat easily, like osteosarcomas. Cervical cancers are the most common, followed closely by abdominal cancers. As might be expected given low cumulative exposure to reproductive hormones because of high fertility and intensive breastfeeding (see below), we've rarely seen cases of breast cancer. Lower testosterone may also be protective against prostate cancer.

Between 2008 and 2014, we saw forty cancer cases (thirty were among women). Incidence was four to six times higher among those age 70+ than those in their 40s. Overall, the number of cases per population (incidence rate) before age 50 was similar to that observed in the US, but it's about a third to half of US rates in older adults.[26] Given no early diagnosis because of a lack of screening (except for cervical cancer because of administered PAP smears), only more advanced-stage cases probably come to our attention. In any case, cancer appears to be universal, but postindustrial environments seem to increase the risk in later adulthood.

What about Other Ailments of Aging?

Old age also brings to mind a host of other less fatal conditions, like arthritis and other joint problems, muscle wasting, hearing loss, cataracts, enlarged prostate, depression, and more. The postponing or mitigation of these ailments has been viewed as an important part of "healthy aging." But how inevitable are these other age-related problems?

Here I'll be brief, and grotesquely generalize based on my experiences among Tsimane in order to highlight how at least one well-studied group fares on a number of aging-related conditions. A few complaints typically viewed as normal aging are absent in Tsimane. In men, prostates are normal sized (that is, very minimal benign prostate hyperplasia, BPH), and testosterone doesn't seem to decline with age. Don't expect a market for testosterone shots in rural Bolivia anytime soon.[27]

But other conditions exist among Tsimane, like elsewhere. Osteoporosis is evident in older women, despite being physically active. Cataracts are not uncommon. Osteoarthritis and lower back pain are common, as is depression (which I discuss in chapter 10). And lest you think that all is the same or better in Tsimane-land, some things are worse for older Tsimane. As most Tsimane women have had many births, they are prone to having a prolapsed uterus. While mild prolapse can occur anywhere, Tsimane women are likely to have more severe cases, which, aside from discomfort, can increase risk of infection.

Recall from chapter 3 the role of infections as a primary cause of death. Certain elements of the immune system seem to senesce rapidly because of a lifetime of immune surveillance and combat. Naïve CD4+ helper T cells—the same ones obliterated by HIV—are essential for mobilizing immune defenses against unfamiliar pathogens. Among Tsimane, they're considerably depleted by age 50, perhaps offset by elevations in other immune soldiers—natural killer cells.[28] Given how infections are still the major source of morbidity and mortality among Tsimane, any rapid decline in immune function can be fatal.

Stone Agers in the Fast Lane Revisited

The idea that our hunter-gatherer ancestors were free of many chronic, degenerative diseases of aging is like canon in some circles, set in stark contrast with affluent Westerners tormented by heart attacks, diabetes, osteoporosis, and other mismatch diseases. The notion of mismatch frames much of evolutionary medicine (and fodder for paleo fanatics trying to mimic stereotypes about hunter-gatherer lifeways). Being mismatched suggests that we'd be better off if we reverted to live like hunter-gatherers. A simple idea, somewhat harmless at best, but dangerous at worst. If our lifestyle and habits better matched the dictates of our genes, maybe we'd have less chronic age-related illness—but would we really be better off? As I've repeated throughout this book, we didn't evolve *in order to* live long lives, nor to be disease-free at late ages if such healthy bliss comes at the expense of fertility, or of survival earlier in life. Figuring out whether there's a mismatch and its extent can still be relevant, though, especially when trying to pinpoint potential remedies. So let's break down the idea of mismatch.

A crude version of evolutionary mismatch assumes that most of human genetic evolution occurred during the Pleistocene, then stopped right at the onset of the Holocene some ~12,000 years ago. With this dividing line, agriculture and animal domestication are taken to be the initial watershed killjoy in our long epidemiological history.

But mismatches have occurred throughout our long species history. They occur every time we move from one location to another, or are exposed to new foods, pathogens, and livelihoods.[29] In other words, whenever we're poorly adapted to the novel conditions of the environment we find ourselves in.

Modern humans originated in Africa, itself a diverse continent, and then dispersed across the globe over the last 80,000 years into an astonishing array of tropical, temperate, and even arctic habitats. By 40,000 years ago, human hunter-gatherers were present in almost every part of the globe. The novel pathogens, foods, and weather conditions that people encountered on every continent save Antarctica, coupled with dramatic, rapid climate change during the end of the Pleistocene and the beginning of the Holocene, must have resulted in a large number of mismatches. The spread of cultural innovations helped overcome some of these challenges, but these challenges also provided grist for the natural selection mill to operate, promoting adaptations to novel pathogens, foods, and climactic conditions. Although 85%–90% of human genetic variation occurs within populations, the 10%–15% of genetic variance that differs between populations is evidence that there was ample time for local adaptations to many of these diverse environments.

As described in chapter 3, the shift from nomadic hunting and gathering to include plant and animal domestication and greater sedentism occurred between 4,000 and 12,000 ya in different locations, appearing earliest in the Levant ~12,000 ya, in Southwest Asia ~9,000–10,000 ya, and later in the Americas 4,000–8,000 ya.[30] In the Holocene Neolithic, goats, cattle, and pigs were domesticated, and we began to eat a less diverse range of foods. We came to rely on a limited number of carbohydrate-rich but nutrient-poor staple crops, with greater susceptibility to famines. Our settlements became more permanent and larger, due to intensive irrigation and economic specialization. Unfortunately, such changes also helped support virulent contagious infections, many arising from contact with domesticated animals. As reviewed in the book *The Story of the Human Body* by Dan Lieberman, mismatches between genes and these altered environments included numerous infectious diseases such as measles, smallpox, and influenza, nutritional diseases such as pellagra and rickets, and some chronic health conditions such as tooth cavities and osteoporosis.

The agrarian to urban transition transformed diets, physical activity levels, social and residence structures, and more. Recent innovations like sanitation systems, refrigeration, antibiotics, and modern dentistry largely fixed the mismatches caused by the agricultural revolution, but now Pandora's box is reopened, revealing new mismatches.

As Lieberman nicely summarizes, these new mismatches come in three flavors. Some are caused by *too little* exposure to organisms and activities that were once common, such as nonlethal pathogens, diverse microbial communities, physical activity, and prolonged breastfeeding. Women also have far fewer pregnancies now than in the past. Others result from *too much* of what was once rare. Think sugar, salt, and saturated fat. Lastly, other mismatches arise from exposures that are *too new*, like nicotine, trans fat, high heels, and extended exposure to the humor of prop comic Carrot Top. Together, these contribute to a range of conditions like myopia and flat feet, but also obesity-related ones like we just described: type 2 diabetes, atherosclerosis, and AD.

Contrary to the idea that genetic evolution stopped at the door of the Holocene some 430 generations ago, the pace of genetic change sped up over the past ten millennia, particularly with respect to diet and immune function.[31] For example, working with the talented geneticist Amanda Lea, we found evidence of recent selection in twenty-one regions of the Tsimane genome, affecting immunity and metabolism.[32] While many of these evolutionary shifts may not have been major game changers, they also weren't just icing on the Pleistocene cake. They helped populations adapt to new diets and exposures. Certainly, though, there has been almost no time for selection to cope with the profound mismatches caused by the last couple centuries of industrialization, large-scale urbanization, and modernization. If you had to draw a line and identify the watershed responsible for the major killers today, my money is on recent centuries, not recent millennia. Those changes lie outside the flexible responses our bodies can tolerate, given the vast variety of environments and conditions experienced by our ancestors. For these reasons, hunter-gatherers are not the only nonindustrial groups to be relatively free of many chronic diseases of aging. Traditional-living horticulturalists (like the Tsimane and Kitava of the Trobriand Islands) and pastoralists (like the Turkana and the Yanomamö) show similar patterns upon close examination. Much of the rise in chronic disease morbidity has occurred over the past 150 years, and even the past 50 years.

Time to Play the (Mis)Match Game

If the Industrial Revolution is a better watershed, the even more recent "Western" obesogenic lifestyle behaviors common today are its latest mismatched product. These behaviors are now global, and getting worse. The Global North was "postindustrialized" long before the 1970s, but obesity prevalence in the US and UK has tripled just over the past four decades, increasing the risk of metabolic disease, heart disease, stroke, some cancers, and other causes of morbidity and mortality. But there's a silver lining. While obesity was increasing in the US from the 1960s until about 2010, premature heart disease mortality decreased, due, in part, to major declines in smoking.[33] Despite those sustained improvements, mortality rates have flattened, perhaps as poor air quality in big cities increases risk. Those risks may be offset by the invention of statins, blood pressure medications, coronary stents, and the like. While mortality rates have leveled out, cardiometabolic health continues to falter, largely due to further increases in obesity and diabetes.[34]

As I've suggested, not all changes result in greater health risks. For hundreds of thousands of years, health and well-being have been affected by variation in diet, activity levels, pathogen burden, reproductive behavior, breastfeeding, technology, risk buffering, and the degree of social isolation. Genetic differences reflect some long-term exposures, and plasticity over different time scales can adjust physiological responses to some degree, but opportunities for mismatch to result in health problems lurk everywhere there is rapid, large-scale environmental change. And with the changes coming more rapidly for populations having lived traditional subsistence lifestyles for millennia, we're seeing chronic diseases arise quickly, shown for dementia in figure 9.3, and as described below for Tarahumara and Turkana. Natural selection is still operating, of course, but it is not quick or strong enough to save our kids and grandkids from avoiding the current wave.[35]

Genetic Mismatch or Worse Lifestyle for Everyone?

While chronic diseases of aging have reared their ugly heads almost everywhere, the pace differs widely among groups and regions. Genetic mismatch could be one reason, as the pace of environmental change has varied among groups and because post-Holocene genetic architectures have also evolved. Many genes vary along geographic gradients reflecting climate and infection. Some of these affect the how and when of inflammation, rendering some populations today more susceptible to inflammatory disorders. Recent ancestry from tropical Africa, where infectious burden has arguably been strong, has selected for a number of inflammation-related genes. Mismatch due to migration to a less pathogenic environment could contribute, for example, to differences in autoimmune and chronic disease rates between tropical Africans and African Americans.[36]

Other recent genetic differences come from the timing of when agriculture was first adopted in different populations or, in its absence, from a long history of having a carbohydrate-rich diet. Varied dietary backgrounds could help partly explain some group differences in the propensity for obesity, diabetes, and metabolic syndrome. For example, having more repeats of sections of the genome related to the *AMY1* gene that helps produce amylase, the protein that digests starches, is associated with lower glucose levels after meals, and less insulin resistance—that is, a lower diabetes risk. This pattern is more common in farmers and hunter-gatherers who consume high-starch diets than in low-starch pastoralists and hunter-gatherers living in rain forests and the Arctic.[37] Another gene called *CLTCL1* also differs between hunter-gatherers and carb-friendly farmers. *CLTCL1* directs the production of a protein that regulates glucose levels and glucose storage in muscle and fat.[38]

Here's an example of a potential genetic mismatch with big implications for AD. It involves a risky gene usually considered a harbinger of doom—but whose evil may be evident only in mismatched obesogenic environments. The *APOE4* allele of the apolipoprotein-E gene is touted as the biggest single genetic risk factor for AD. If you have one allele, you're two to three times more likely to get it, and if your 23andMe personalized genetic test unfortunately revealed that you have two of these suckers, then your risk increases eight to twelve times higher above that of having two copies of the safer *APOE3* allele. Europeans carrying two *APOE4* alleles also show an earlier onset of AD, and almost universally have indicators of AD pathology (even in the absence of an AD diagnosis).[39] If that weren't bad enough, *APOE4* is also a risky allele for atherosclerosis. Understandably, *Thor* actor Chris Hemsworth, after hosting a TV series on longevity, was alarmed to discover that he has two *APOE4* alleles. Although *APOE4*'s frequency is variable from place to place, it exists most everywhere, despite all the harms to our health. How is that possible? In chapter 2, we saw that antagonistic pleiotropy can still favor genes that screw us up at later ages if they provide benefits earlier in life. And there's some indication that they might, at least in the more infectious environments of our past. APOE4 carriers are better at clearing certain infections in childhood, like viral hepatitis C, and deadly diarrhea-inducing giardia and cryptosporidium. Among both the Tsimane and the Bimoba of Ghana, APOE4 carriers have higher fertility than those with APOE3. Tsimane women with APOE4 have their first child at earlier ages, and space their births closer together, resulting in a greater number of total births.[40] Though the mechanism is unclear, an experimental study confirms that transgenic mice with expressed human APOE4 are more fertile.[41]

We've found other curious benefits to APOE4. Tsimane APOE4 carriers show lower levels of systemic inflammation. They also maintain higher blood lipids at lower body mass, which can serve to help buffer costs of immune activation during infection.

The last benefit we found is a potential game changer. Tsimane APOE4 carriers were protected against cognitive decline, but only when actively infected with helminths.[42] The mechanisms to explain this relationship still need to be worked out, but the usual destructive effects of inflammation may be altered under the immune-modulating influence of our wormy overlords (see below). How exciting it would be if similar results were found elsewhere! There's some indication that they might be. APOE4 is unrelated to AD risk among the Yoruba of Nigeria. Its harmful effects, once viewed as universal, are instead most visible among those with greater European ancestry.[43] Taken together, the health costs of APOE4 may be expressed only in pathogen-free, obesogenic, and low-fertility settings.

While AMY1, CLTCL1, APOE4 and other genetic variants yet to be discovered may help explain some individual and group differences in vulnerabilities to chronic disease, the largest impact by far has been in our built environment, the so-called Anthropocene. Next, I summarize several large-scale environmental/lifestyle changes, covering what and how we eat, how we move our bodies, and more.

You Are What You (Don't) Eat

We're omnivores. There's no such thing as "the" hunter-gatherer diet. Hunter-gatherer diets vary widely. One survey from the ethnographic record estimates hunter-gatherer diets to be 22%–40% carbs, 28%–58% fat, and 19%–35% protein. A huge range for sure, depending on how much meat versus plants are found in the diet, and how lean the prey. Toward the Earth's poles, diets are mostly fish and meat, with mixed diets closer to the equator. Even with such a wide range, the contrast with Western diets today is clear. The US diet is low on protein and high on carb (roughly 15% protein, 49% carb).

And what may look similar can differ in important ways. Wild game tends to be lean, whereas we breed animals for their fat (and delicious marbling). When cattle feed on grain, their meat will have a different fatty acid composition than when feeding on grass, as they would in the wild. Grass-fed beef has much higher omega-3s, polyunsaturated fats, and select vitamins (for example, E and A). Farm-raised salmon typically has lower calcium and higher fat, though much of that fat is omega-6. Having a larger ratio of omega-6 to omega-3s in your bloodstream has been linked to greater inflammation and higher risk of heart disease and insulin resistance. A similar story can be told about eggs from free-range chickens versus the cheaper eggs from factory-raised chickens that you can also find at the supermarket.

Any meat-fixated paleo-dreamer must recognize that diets among the CAD-spare Kitava and Tsimane (plus centenarian-rich Okinawa) are rich in carbohydrates. The carbs, though, are complex, like those in sweet potatoes, taro, and manioc root. They're not "simple" carbs or refined sugar, which cause bigger spikes in blood glucose. Complex carbs take longer to digest, and supply more long-lasting energy as well as fiber.

Across subsistence societies, no diet seems to be terribly incompatible with a life free from atherosclerosis and diabetes. Even pastoralists like the Turkana, whose traditional diet consists of lean meat and animal blood, don't show evidence of atherosclerosis or diabetes. Instead, the vital contrast with traditional diets is what has been wryly called the Standard American Diet (SAD): heavy on processed foods including meats, red meat, refined grains, high sugar, high salt, high fat (especially saturated fat), and dairy. The SAD is also notable for what it lacks: not enough green and orange vegetables, fruits, and whole grains. While refined sugar and saturated fat have long been identified as dietary renegades, industrial seed oils from processed soybean, rapeseed, and cottonseed have been blamed as more recent villains.

If Twitter wars are any indication, debates over the health(iest) nutrition are a minefield. Rather than wade through on our tippy-toes, we can zoom out, squint our eyes, and attempt a wide-angle lens view.

One ethically dubious study from the 1980s published in the world's top medical journal is worth a brief mention because it demonstrates the direct harm of the worst aspects of SAD on health.[44] In a precursor to Morgan Spurlock's *Super Size Me* film experiment, the Tarahumara Amerindians of Copper Canyon in the Sierra Nevada in Mexico went full-SAD for five weeks. Their traditional diet is a low-fat and high-complex-carb mix of corn, squash, beans, fruit, and eggs. That was swapped out for an "affluent" diet of cheese, butter, lard, flour, soda—in other words, a high-calorie (4,100 kcals/day!), high-fat diet. People were fed in controlled settings so that everything could be carefully measured. Even by the second week into the study, the main changes in the blood panel were clear. Total cholesterol and the "bad" LDL cholesterol increased by over a third, whereas triglycerides increased by 18%. People gained over eight pounds on average, resulting in almost a 10% increase in BMI. Though people varied in how they responded, everyone saw the same general harm to their health. Tarahumara are closely related to the Mountain Pima (Pima Bajo) of Mexico, and the Pima who migrated north to what is now Arizona—all being descendants of the Hohokam. The US Pima lost their agrarian livelihood at the turn of the twentieth century, after losing water access because of the incursion of white settlers in the area. By the time long-term study of Pima health began in the 1960s, the Pima quickly became known for having one of the highest rates of obesity and type 2 diabetes in the world.[45]

It is difficult to recover from the impacts of SAD food intake patterns. For example, consider another interesting experiment. Similar to the Australian one described earlier, a lesser-known quasi-experiment took advantage of an income security program that required the Quebecois Cree to spend three months each year hunting and trapping "on the land" in order to be guaranteed income. By testing diabetic Cree before and after this excursion, we can see again how an active return to a more traditional lifestyle improves health. Fasting blood glucose lowered by 13%, and a more long-term measure of sugar in the blood (glycosylated hemoglobin, or A1C) declined by 11%, though there was no weight loss or change in blood pressure. One reason might be that the Cree took food with them to the bush bought from a store: eggs, lard, butter, cookies, and sugar. While much more physically active during the time in the bush, this study suggests that both diet and activity need to be altered in sync.[46]

Rapid increases in chronic disease morbidity have coincided with easier access to food, due in part to the falling cost of cheap sugar- and fat-rich processed foods designed to have more shelf life than nutrition.[47] The relatively low protein composition of the energy-dense SAD diet might also make us want to eat more until we reach a certain level of protein satisfaction.[48] Food is abundant, available, enticing, and we eat too much of it. Processed foods are tempting to our taste buds because they're high in fat, salt, and sugar—all limited in hunter-gatherer diets. A common refrain is that we've evolved appetites for these whenever available. But even when matched for calories, macro-nutrients, and fiber content, controlled dietary experiments confirm that ultraprocessed foods still make us want to eat more, and make us gain more weight. The compact, energy-dense packages trick our brains used to eating bulkier fare. And so we're still hungry, especially when we wolf down processed foods.[49] Our dopamine-fueled brains are now revved up on ultraprocessed foods, making up over half of what we eat in the United States and, in Europe, ranging from 14% in Italy to 44% in the UK and Sweden.[50]

Many have highlighted other harmful features of the SAD, including its low soluble fiber, high glycemic load and net acid content, its low ratio of potassium to sodium salts, and its knack for compromising the integrity of our intestinal walls. SAD takes a wrecking ball to our gut microbiome, with harmful consequences for our immune system, The full details go well beyond the scope of this book.[51] But the gist is that

SAD promotes chronic inflammation. Regardless of nutrient content, we eat too many calories. Convenience and additives for shelf life and flavor bursting may not be ideal for long-term health. And not just for our hearts, livers, and kidneys. In a recent study in Brazil, getting over one-fifth of your calories from ultraprocessed foods led to greater cognitive decline.[52]

Moving from the Moderate-to-Vigorous Lane to the Sedentary Lane?

A half-marathon is the equivalent of 27,000–32,000 steps (if you were to walk it), and hunter-gatherers don't run that every day. Sure, subsistence populations rely on their bodies for work, play (no video games!), and travel. But hunter-gatherer and horticulturalist activity patterns are not so clear-cut. Sedentary behavior—like sitting in an office chair from 9:00 to 5:00 and then binging Netflix on the couch—is not compatible with working to live. Instead, moderate activity is the norm for subsistence living. Vigorous activity, like CrossFit, occurs when needed: chasing prey, opening trails in thick forest, climbing trees for honey, cutting down trees, and ceremonial dances. Why exert yourself beyond what's necessary? We didn't evolve to reach a target level of activity, like the current guidelines for US adults: at least 150 mins/week of moderately intensive activity, or 75–150 mins/week of vigorous activity. Conserving energy for vital life processes and reproduction when you can makes evolutionary sense.

As you'll recall from chapter 4, hunter-gatherers cover far more territory than our fruit-and-leaf-loving primate relatives. They dig, climb, chop, grind, pound, throw, collect, winnow, slice. And they walk a lot. Sometimes they sprint and chase. Thanks to accelerometers, we know that hunter-gatherers and other subsistence populations walk about two to four times more steps per day than people in industrialized countries like the US and UK. And far more than the great apes. The daily moderate-to-vigorous activity among foragers and farmers is at least ten times more than in Western countries.[53]

Before you throw your plush La-Z-Boy out on the street, it's vital to recognize that people everywhere like and need to rest. The Hadza typically walk 13,000–19,000 steps per day, but Dave Raichlen and colleagues found that Hadza sit around just as much as Americans or Australians—just under ten hours/day.[54] But their version of sitting is "active rest"—squatting, kneeling, or sitting on the ground, for example— which requires some sustained core and lower limb muscle work for stability. The newfound passion for standing desks and exercise ball chairs may be linked to a growing appreciation for this type of "active resting". (I just wrote that sentence while wobbling on a balance board in fact.)

Of course hunter-gatherers balance rest with activity, and move when needed. Activity is a means to an end, and as such, if we're not active, our body adjusts accordingly. A muscular, flexible body is costly to maintain, so why build it if it's not needed. Similarly, our heart will lose capacity if it's not frequently used in a vigorous enough manner. Our body adaptively adjusts, but given that our sedentary lifestyle is coupled with greater longevity, we're likely to have problems at later ages.

Physical inactivity—otherwise known as couch potato-ing—doesn't just lead to weight gain and obesity-related metabolic disease; it affects pretty much everything in the body.[55] Meta-analyses that compile results from many studies show that being physically inactive predicts a 42% higher risk of diabetes, 24% higher risk of coronary heart disease, and 16% more of stroke.[56] Greater activity shows a larger effect on survival—among high-income countries where much activity is recreational weekend warrioring *and* in most low- and middle-income countries as well.[57] Over a quarter of the world's population currently doesn't meet the minimum recommended amounts of physical activity. Because of its direct and indirect links to many forms of morbidity, physical inactivity is now considered the fourth leading cause of death globally.[58]

Much like a prescribed medication, physical activity has dose-like effects that influence most chronic conditions, including diabetes, hypertension, dementia, and arthritic knees. While activity can lead to short-term weight loss, its ability to sustain healthy body weight has been called into question.[59] On the other hand, physical activity has many positive effects on the body that go well beyond its effects on obesity. It is by far the cheapest effective remedy for cardiometabolic disease, with no adverse side effects, and is successful at reducing hypertension, diabetes, an unhealthy blood lipid profile, and heart arrhythmias. Endurance activity improves heart health and results in healthy structural changes throughout the vascular system more broadly. New mechanisms are still being discovered, like how your pumped-up muscles help combat inflammation and insulin resistance through the secretion of small proteins called myokines.[60] Like we saw in chapter 8, sustained physical activity is the best form of prevention against the harms of aging, and it's true too for mismatch diseases.

Its effects extend even into the realm of otherwise uncurable diseases, like dementia. Given that very little seems to prevent or slow the progress of AD, it's encouraging that physical activity is one of the most effective forms of prevention and management for AD. Meta-analyses that compile many prospective studies together show that physical activity can lower later risk of AD, vascular dementia, and cognitive decline more broadly. The effects of activity on a person who is burdened with AD appear less consistent. A slight majority of studies show that exercise interventions can still improve cognitive outcomes at least as well as donepezil, the medication commonly used to treat AD, which has unpleasant side effects.[61]

Different mechanisms might explain the protective effects of physical activity on AD progression. Some are similar to those that impact cardiometabolic health, like lowering inflammation and countering oxidative damage. Others involve greater cerebral blood flow and boost levels of a protein called brain-derived neurotrophic factor (BDNF), which helps maintain the health and plasticity of neurons. BDNF is also associated with improved memory, and the volume of the hippocampus, the brain's memory and navigation center.[62]

Physical activity also generates mechanical loads that impact muscle and bone mass. This is essential for reducing muscle wasting (called sarcopenia) and weakened bones resulting from osteoporosis. These of course affect our ability to carry out daily activities. Figure 9.5 shows how a sedentary existence over decades leads to muscle wasting and loss

40-year-old triathlete

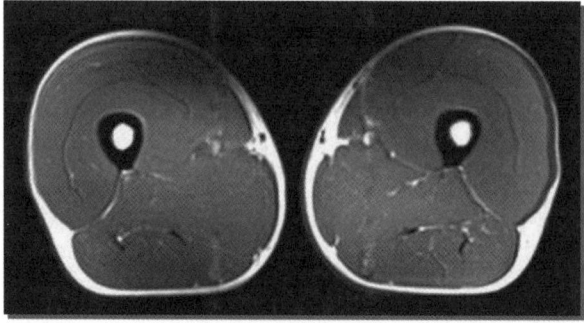

74-year-old sedentary man

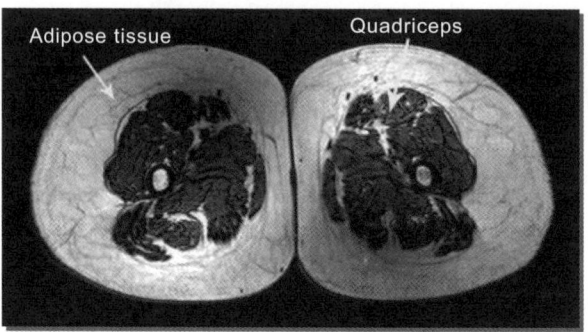

70-year-old triathlete

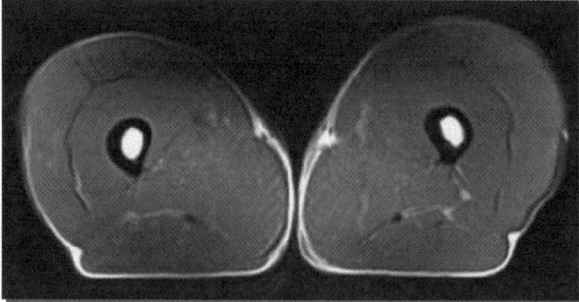

FIGURE 9.5. **Comparison of leg muscles of physically active and sedentary men.** The quadriceps of a 70-year-old triathlete looks like that of a 40-year-old triathlete, but an elderly sedentary man shows a lot of fatty tissue in both the muscle and under the skin. *Source:* A. P. Wroblewski, F. Amati, M. A. Smiley, B. Goodpaster and V. Wright. 2011. "Chronic exercise preserves lean muscle mass in masters athletes." *The Physician and Sportsmedicine,* no. 39 (3):172–178, reprinted by permission of the publisher (Taylor & Francis, Ltd.).

of vitality in your thigh muscles, the quadriceps. But maintaining a physically active lifestyle over the long term, as master triathletes are wont to do, certainly pushes back against what is often considered inevitable decline.[63]

Evidence of Absence, Not Absence of Evidence

Changes in diet and activity, and bad habits like smoking and heavy drinking, reflect behaviors long viewed as risk factors for many chronic diseases. If groups like the Tsimane avoid heart disease just because they're like seasoned athletes who avoid Hostess Snoballs, that would be good to know, but it wouldn't lead to new discoveries. Here's where we delve into more speculative, exciting territory.

Since the 1950s, declines in infectious diseases like measles and tuberculosis coincided with a rapid rise in autoimmune and allergic diseases, including multiple sclerosis, Crohn's disease, asthma, hay fever, and allergies.[64] Cleaning up the water supply and public sanitation at the turn of the twentieth century are largely responsible for reducing infections, along with antibiotics, vaccines, and better hygiene. While these are hailed as great public health achievements, they have come at some cost to our health. Epidemiologists and other health scientists over the past few decades have explored ways in which pathogens, and our immune responses to them, are linked to the recent surge of autoimmune-related diseases. The leading idea is that early and regular exposure to diverse "friendly" microbes, rather than harmful infectious pathogens, helps train the immune system to learn what to attack and tolerate to ensure appropriate immune responses. While nasties like measles may exacerbate autoimmune diseases, diverse commensal bacteria in our gut may help protect us against autoimmunity.[65]

But the human immune system coevolved with a rich diversity of microbial and parasitic species, often dubbed "old friends," that our bodies anticipate to be present. Our current "epidemic of absence" results in immune function mismatches.[66] Antibacterial soap and aerosols, Cesarean-section births, dishwashers, living in cement buildings rather than on farms, and other aspects of contemporary urban life have depleted the types and quantity of exposures to microbes and parasites. Such depletion of exposures can lead to dysregulation of our immune systems (see chapter 8).

Recall that chronic low-grade inflammation is blamed as a primary cause and consequence of many diseases of aging. But inflammation is not naturally evil. It's part of our quick and dirty ("innate") immune response that's our first line of defense against host attack. As with an exploding grenade, inflammation's effects are diffuse rather than precisely targeted. This lack of precision results in collateral damage that, unchecked over time, contributes to atherosclerosis and type 2 diabetes. Insulin resistance provoked by inflammation might be beneficial for fueling immune defenses against bacterial infections, but in the relative absence of pathogens, inflammation in our bodies is "sterile." It comes from obesity, smoking cigarettes, and physical inactivity.[67] Lest you thought the love handles around your middle were just an inert energy storage locker to drain from once the zombie apocalypse arrives, consider that fat (especially visceral fat packed around your organs) is busy sending immune soldiers locked and loaded. Excess fat has long been known to incite inflammation.

Along these lines, another forgotten "old friend" may be important. The squirmy worms that inhabit our intestinal tract are called helminths (hookworm and roundworm fall into this category). Helminths have coexisted with humans for millennia and represent a major feature of early human disease ecology. Throughout human history, helminth burdens surely fluctuated, greater during periods of sedentary living coupled with animal domestication. However, until relatively recently, very few humans were completely helminth-free.

So how do helminths relate to diabetes and heart disease? Helminths exert an energetic cost on our bodies, for teasing and tickling our immune systems. They also like lipids, and absorb these from our guts or from our bloodstream for their own nefarious needs. One unintended benefit, then, is that helminths could reduce our blood cholesterol and stave off the accumulation of plaques in our vasculature. It's not for nothing that tapeworm eggs were once marketed as a dieting aid (though that's likely fake news). As I tell my students who work with me in the

Bolivian tropics where such worms are endemic: eat as much as you want and still lose weight! You're sharing your food with these old friends.

Experimental studies in mice show how helminths improve glucose tolerance and sensitivity to insulin. The absence of helminths, perhaps also with our altered gut microbiome, may also contribute to the dysregulation of our immune system. Helminths can dampen inflammation and help regulate the balance between pro- and anti-inflammatory arms of the immune system.

A majority of Tsimane guts are home to soil-transmitted helminths, primarily hookworm but also roundworm and whipworm. Tsimane immune systems are actively engaged by helminths, showing very high levels of the white blood cells called eosinophils. These are the warriors that help target (but usually tolerate rather than eliminate) the worms. In several studies, we found that Tsimane showing indicators of helminths had lower BMI and lower levels of cholesterol and glucose circulating in their blood. A study of Ende farmers of Indonesia also showed similar findings, and more: those with helminths had lower BMI, LDL, and total cholesterol, and better insulin sensitivity. Though such observations are provocative, they're in the right direction, consistent with the experimental mouse studies. Recent deworming experiments go one step further in showing that eliminating worms can impair insulin sensitivity. A similar but more thorough deworming intervention in Ugandan fishing communities led to worse blood lipid profiles and blood pressure.[68]

In case you're tempted, please don't run to the nearest latrine and dance barefoot any time soon (many worms infect you by piercing their way through the bottom of your feet). Each immune system is different, and there's not a single quick solution that works for everyone. Besides the ick factor, some helminths feast on blood iron and induce intestinal blood loss, increasing your risk of anemia.

Per the usual refrain: more research is needed. It remains to be seen whether the absence of helminths alone or, more likely, in combination with changes in diet, activity, and other lifestyle behaviors will increase diabetes and atherosclerosis risk in populations experiencing transitions in livelihood and environment.

The role of pathogens may also shed light on Alzheimer's disease. The "amyloid beta cascade" hypothesis that dominates most AD work reckons that accumulations of beta-amyloid and tau protein tangles in the brain are the main culprits. But all attempts to eliminate these have been unsuccessful.[69] A novel idea suggests instead that AD is more like an autoimmune disease, caused by microglial-activated toxins influenced by gut and other microbes that cross the blood-brain barrier weakened by chronic inflammation.

According to this newer hypothesis, beta-amyloid is an antimicrobial peptide that helps protect the brain from infection, rather than the criminal cannibalizing healthy tissue.[70] If this were true, it's possible that the different infectious and immune milieu of groups like the Tsimane may account for lower AD risk. Worms, diverse microbiota, and other infectious agents may alter patterns of inflammation in the brain, reducing AD risk. An intriguing meta-analysis shows that many common vaccines, including those against influenza A, tetanus, herpes zoster, and hepatitis, reduce the chances of later getting any dementia, as much as 20%–50%! One likely interpretation is that these vaccines help modulate immune function in ways that protect the brain from infection, inflammation, or both.[71]

But What about All the Other (Insert Differences Here)?

You'd be right to think that there's more to evolutionary mismatch than changes in what we eat, how we move, and what other microorganisms live inside us. Here I'll just mention one other major difference. We have far fewer children now than we used to. Prior to a century ago, a relatively high chance of birthing complication combined with high fertility would have lowered female life expectancy, e_0, considerably, as I discussed in chapter 2. As fertility declined over the twentieth century, the gap in female and male e_0 widened. Female life expectancy e_0 now exceeds male e_0 in all countries.[72] Despite living longer, women are usually portrayed as less healthy than men, a factoid so well established, it has a double hyphenated name: "male-female health-survival paradox." In high-income countries, men endure cardiovascular disease and nonreproductive cancers more than women, while women tend to have inflammatory-related autoimmune diseases like systemic lupus erythematosus, Graves' disease, and rheumatoid arthritis more than men.[73]

The reason for these sex differences in health lie where reproductive biology meets modern conditions. In pregnancy, a woman must strike a balance between upregulating immune responses to combat pathogens and downregulating responses to tolerate her growing fetus. In the balance, immune modulation favors a dampened, anti-inflammatory bias. But in contemporary urban society, low fertility can lead the immune system astray. Combine this with the dysregulation wrought by low pathogen exposure mentioned earlier and, voila—the higher autoimmune disease risks we see today.[74] Comorbidity from immune dysregulation may even extend to mental health. It has long been recognized that depression co-occurs with inflammation. And women report more depression than men in high-income countries. Up to half of women with autoimmune diseases show depression-like symptoms, and depressed women have double the risk of developing lupus.[75]

As I mentioned earlier in the chapter, cancer is affected by reproductive hormones. But energetic stress, active lifestyle, and heavier immune burden in subsistence populations seem to produce lower levels of reproductive hormones like testosterone, estradiol, and progesterone than those in WEIRD (that is, Western, educated, industrialized, rich, democratic) populations. A typical woman in low fertility settings will experience three to four times more menstrual cycles over her lifetime than hunter-gatherer and farmer women (where fertility is high and breastfeeding is intensive). All that extra cycling means more estrogen circulating through the body. Estrogen, despite its perks, is also a carcinogen. In women, high levels of (unopposed) estrogens—from fewer pregnancies, earlier menarche, and later age of first birth—have been linked to greater risk of endometrial and breast cancer. Cancer risk is heightened further by obesity and low physical activity. Among men, it's a similar situation, but with testosterone. Greater long-term testosterone exposure may worsen prostate cancer prognosis. Testosterone supplementation sounds appealing for boosting energy and muscle mass, but nothing comes for free. Blocking testosterone receptors can regress prostate tumors.[76]

A Summary: (Chronic) Disease-Free but Still Aging

Lean high-fiber diets free of processed foods, higher physical activity, minimal smoking, high fertility, and more diverse co-infections are protective factors common to many preindustrial societies against chronic diseases of aging. It's a slippery slope toward chronic disease as nonindustrialized groups "modernize." It would be helpful if a single behavior or factor alone makes the biggest dent in our health, but mismatched environments rarely involve just one changed micronutrient or activity. It's a medley of changes that in combination have synergistic impacts. Unsurprisingly, changes to diet *and* activity improve health more than changes in just diet or activity alone. Rigorous activity won't save you from a terrible diet, and no amount of gorging on kale will protect you from the harms of slug-like immobility.[77]

Herein lies the interesting paradox. Hunter-gatherers and other subsistence populations living traditional lifestyles have a higher chance of dying at all ages than folks in high-income countries, and may even age more rapidly. But they suffer from lower rates of the major killers dogging the world today: heart disease, cancer, diabetes, and Alzheimer's disease. Without those diseases, hunter-gatherer death rates still rise with age, just like they do in every human population. In other words, aging occurs even in the absence of the familiar chronic diseases we associate with elderhood today. Is that healthy aging, though? Hunter-gatherers just die of other causes, many of which are still consequences of the aging body.

By now, I've described ad nauseum many physiological changes that come with age, and diseases of aging. Perhaps you're screaming for something more personal and subjective. What and how does aging *feel*? Enough with VO$_2$max and neuroinflammation! What about psychological well-being and mental health? Turn the page and read on.

10

Foraging Alone?

> I'm very pleased to be here. Let's face it—at my age, I'm very pleased to
> be anywhere.
>
> —GEORGE BURNS, COMEDIAN

"*CHÄTIDYE! SOBAQUI YU'! Jäm'dye mi'?*"

Literally translated as "Kin folk! I am visiting! Good, are you?" these
courtesy salutations announce your arrival when you approach a Tsimane compound. Shouted with excitement and anticipation, they are
often followed by grins and laughter. Daily visits by family and friends
are at the core of Tsimane social life. The small-world networks of tight
communities are the envy of all who mourn the anonymity and isolation of postindustrialized city life. In the immortal words of Joseph
Conrad (and 1980s post-punk band Gang of Four), we live as we dream,
alone.

I've spent several chapters showing how we may be longer-lived and
aging more slowly now than ever before. More of us are also safer, less
poor, and more literate. We're better off on many indicators of well-
being that are tracked over time and compiled for public consumption
by organizations like Our World in Data.[1] But, if this is the case, why are
so many people struggling with depression and anxiety?

In addition, we are witnessing growing concerns about an epidemic
of loneliness and its close sister, social isolation. Whereas loneliness is a

feeling or a perception, isolation reflects the objective lack of relationships, or being separated from the flock. Not all isolated folks are lonely, and not all lonely people are isolated. But they tend to go hand in hand. Solitary living and loneliness affect people of all ages, but especially elders. Currently more than one in four US elders live alone, and one in five reports feeling lonely. Almost half of Americans over 60 report feeling lonely. By age 80, the balance tips to over half. In neighboring Canada, men over age 80 have the highest suicide rates of all age groups, due in part to social isolation, loneliness, and hopelessness.

In Japan, "lonely death" among the elderly, called *kodokushi*, occurs when a person's body is discovered months after death. In one extreme case that made international headlines, a 69-year-old man's corpse was discovered only after his savings were depleted from automatic withdrawals for rent and utilities—a grim three years after he died. In South Korea, this phenomenon of lonely deaths due to living disconnected from others in solitude is called *godoksa* (though observed not just among elderly, but in middle ages too).

As a sign that loneliness and isolation are of growing concern, national governments are increasingly mobilized to address this new global health challenge. For example, in 2018, the UK introduced a Minister of Loneliness, tasked with developing strategies and initiatives to support community connections and improve social integration. Japan quickly followed suit with their own Ministry for Loneliness, to act as a "control tower" for efforts to provide aid to the isolated and lonely. US Surgeon General Vivek Murthy has also addressed the epidemic of loneliness, noting that "human connection is our evolutionary birthright."

One attempt to put isolation in historical perspective shows that living solitarily was relatively rare over most of the past half millennia. Figure 10.1 shows that only from the 1960s onward is solitary living increasing throughout much of the industrialized world. Changing marriage patterns affecting who gets and remains married, childlessness, cultures privileging individualism and privacy, and rising prosperity are some reasons for this trend. A widely cited stat on loneliness claims that its toll on health and longevity is equivalent to smoking half a pack of cigarettes a day, and that it's worse for health than obesity and physical

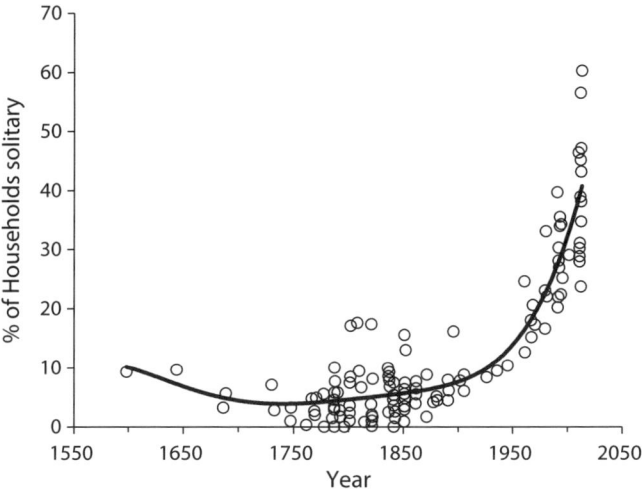

FIGURE 10.1. **Single-person households over four centuries (1564–2012).** Expressed as percentage of all households that were single-person. Solitary living has increased drastically since the 1950s. Sources span UK, Europe, US, Japan. *Data source:* Snell 2017.

inactivity. Loneliness has been linked to higher risks of cognitive decline, dementia, mental health problems, heart disease, stroke, and "all-cause mortality" (a term that refers to deaths from any cause). Its link to death alone has been replicated in hundreds of reliable studies.[2]

Another Mismatch?

With greater social isolation and loneliness at later ages comes the dreary bedfellow, depression.[3] An ever-increasing number of studies and commentaries are linking depression and loneliness to the harmful by-products of capitalism and individualism. For one thing, capitalism may do good for many, but growing inequality and wild materialism is a Faustian bargain. We may land that sweet lucrative job in Silicon Valley, but at a six-hour flight from home in Philly, away from family and friends. And while it is true that our iPhones and social media apps can connect us far and wide, they just make many of us feel more alone.

In a study of 195 countries, cultures emphasizing individualism were associated with having higher levels of depression. As one historian of loneliness remarked, modern loneliness is the "child of capitalism and secularism."[4] This line of thinking parallels the arguments we've seen about physical health and the chronic diseases of aging: we are evolutionarily mismatched to our contemporary urban "stranger-danger" environs. But what is the mismatch really about when it comes to our well-being? Is it about not knowing our neighbors? Is it about the phenomenon of "bowling alone" (popularized by Robert Putnam's influential book about the decline of social capital since 1950 in the United States)? Our growing distrust of and detachment from our local communities? Moving away from family? Separating (and prioritizing) "work life" from "family life"?

There's almost no end to the list of possibilities, but you've probably heard some version of these complaints: our rugged individualist values are colliding with collectivist protections, there are too many choices at the supermarket (of life), and social media makes social comparison a global nightmare. Throw in some rampant economic inequality, loss of trust, climate change, political polarization, and ethnic conflict. Or maybe just too much slug-living, and too many Batman movies.

Many of these proposed mismatches suggest recent changes in our social landscape. We often refer to the "golden years"—a time long ago when life was better than it is today. As the character Andy Bernard from *The Office* (US edition) once rued, "I wish there was a way to know you're in the good old days before you've actually left them." Yesteryear, however, could be 1950 or 1850, or our long preindustrial past. By studying hunter-gatherers and other subsistence societies, we can better understand what life might have been like for past elders as well as today's elders in small-scale societies.

In this chapter, I'll survey what's known about well-being among hunter-gatherers to explore the *quality* of elder lives. Well-being encompasses positive emotions like happiness, contentment, and fulfillment, but it also includes feelings of loneliness and despair. I've established how living to seven decades is a fundamental part of our human story, but now we'll see whether that long life is also well-lived, or whether emotional malaise is an inevitable bedfellow of growing old. Cultures vary in the details, but the gist seems to be that early elder years are pretty good, and later elderhood is not so good. In a twist on the debate about whether elder lives were better or worse in some long-lost golden age, we'll see that modernization has been a mixed bag for elders in subsistence societies. Access to health care combined with living in more permanent villages may improve elder well-being, but other changes might do the opposite. An increase in literacy among younger generations, combined with greater access to books and a changing socioeconomic landscape, can diminish the reliance on aged storytellers, educators, and orators, and in turn many of them feel less useful. But as we'll see, many elders are adapting to changing circumstances, pivoting their experience and perspective in ways that can thwart obsolescence and maintain well-being late in life.

Turn That Frown Upside Down

Putting aside the current alarm bells regarding loneliness, and the concerns over the millions of Paxil and Zoloft prescriptions, there is a glass-half-full version of our collective emotional well-being. Ed Diener, the psychologist known as Dr. Happiness, and his wife, the psychologist Carol Diener, spent much of their careers arguing that most people are happy most of the time.[5] Put a different way, people they studied were at least "mildly happy and satisfied" most of the time. Their conclusions were based on people living in industrialized, mostly democratic, societies. (The important exceptions are people in dire straits, including people experiencing homelessness and people living in the slums of Calcutta.) Maybe industrialized settings effectively meet people's needs and focus attention on self-actualization to foster a positive self-presentation to others.

As we'll see, asking peculiar questions about life satisfaction shows that people outside of the industrialized world are also relatively satisfied with their lives. Maasai herders, Greenlandic Inuit, and Roviana fishers from Solomon Islands, for example, generally report relatively high levels of well-being, as do people across a range of small-scale societies.[6] That being said, the story shifts when we start to look at life satisfaction and well-being across the adult life course.

One surprising result that has captured public attention is that well-being in industrialized countries takes on a distinct U shape over much of the adult life course. That is, well-being starts out pretty high in your early 20s and reaches its lowest point in your late 40s, but then things look up for a bit (until the final years, where they may get rough again). That 40s–50s-ish slump is the infamous midlife crisis.

While there are many ways to conceive of well-being—life satisfaction, happiness, and the like—one of the simplest ways is to just ask someone. The question asked in most studies is akin to, "On the whole, would you say your life is very bad, not good, fair, good, or very good"? Or, "Taking all aspects of your life into consideration, how would you say things are these days? Would you say you're very happy, fairly happy, not too happy?" This type of questioning is admittedly coarse, but so simple that it's been asked almost everywhere to get a sense of well-being.

The U shape that results from this line of questioning is still controversial. Lots of ink has been spilled about it, and a grab bag of methodological concerns is associated with it. Several economists are its most ardent supporters, documenting the U shape in 145 countries and confirming it across hundreds of studies. Even when you take life events that often accompany changes with age into account in the statistical models (events such as marriage and divorce, children fledging the nest), the U shape still remains, sometimes even coming out stronger. The uptick in well-being after midlife is not just statistically significant because of the large samples used in such analyses. It's a meaningful boost equivalent in size to the effect of major life events, like marital separation and unemployment.[7] Improved well-being after midlife has also been confirmed longitudinally. If happier folks are healthier and live longer, then maybe the miserable are not around long enough to answer annoying questions about their happiness. However, the uptick isn't just due to these "selection effects," whereby unhappy people die at higher rates and so are less represented at older ages.[8]

Instead of just relying on cognitive appraisals of well-being, some researchers have also measured "hedonic" well-being, that is, the experiences of joy, happiness, worry, and anger (though also via self-report). Patterns in places like the United States seem to largely reinforce the U-shaped pattern. For example, expressions of negative affect, like stress, anger, and worry, are lower in older adults, while enjoyment and happiness show a U shape over much of adulthood. (Sadness, however, isn't lower among older adults, as might be expected, but at least it's not higher either.)[9]

So robust is the U-shaped pattern that it's argued by many to be a human universal. And the pattern is not limited to humans: even chimpanzees and orangutans seem to show a U-shaped pattern of well-being across their life course.[10]

One explanation for a positive upturn in well-being in older age fits with ideas I laid out in chapter 6. Older adults can be emotionally wiser and have fewer mood swings. They've gained perspective about what counts in life. They adapt to their changing circumstances in realistic and optimistic ways.

The Stanford psychologist Laura Carstensen came up with the idea of "socioemotional selectivity" to describe how older adults are better than younger adults when it comes to regulating their own emotions and maintaining a positive outlook. Her experiments have shown that older adults pay more attention to positive rather than negative information and show a richer balance of emotional states. Part of this, she explains, comes from the recognition of a limited (and diminishing) time horizon that puts things into perspective.[11] Who sweats the small stuff when the end is nigh?[12]

It's Not U, It's Me

From analyses of the famous Gallup poll, the Nobel laureate economist Angus Deacon argued that the U shape only appears in rich countries where the elderly are relatively satisfied with their lives. That's at odds with the arguments that the U shape is a human universal.

If greater well-being after a midlife crisis was universal, where best to look for this pattern than in hunter-gatherers and other subsistence populations? Anecdotes from chapters 6 and 7 offer hints of positive well-being from the many roles that embed elders in kin and community networks. Their knowledge, advice, and other skills provide status and prestige (at least until elders lose their edge and become more of a burden—where death-hastening in one form or another is one cultural solution to rapidly declining elder well-being). In subsistence societies, your physical body is vital for any sustained engagement in daily activities. When health and the ability to function are compromised, life satisfaction is expected to drop. And this is what studies find. But how can we know for sure?

If it feels strange to imagine asking a literate older Belgian, who over a lifetime is accustomed to filling out forms and surveys asking all types of questions, about their global well-being, imagine how awkward it might be to ask a Hadza or Tsimane adult. Lofty abstract questions are bizarre in these cultures. Psychologizing may be normal for some, but not for many nonliterate subsistence populations. Nonetheless, these questions are usually understood, and answers about a person are reasonably consistent when asking a spouse or close friend. Answers also correspond with something more tangible, like how often people smile or laugh. But most surveys even in low-income countries still heavily sample from urban, literate folks. Life satisfaction across the life course among rural Indigenous peoples has only been studied systematically in a handful of groups.

Among Hadza, answers from four related questions were combined to capture overall happiness.[13] Figure 10.2 shows that Hadza women and men report being happy throughout much of adulthood, with no differences in later life. The study, however, included relatively few adults over age 60, so we can't say how happy Hadza elders are at later ages. A comparison with urban Polish using the same questions and same research team is revealing. Poles show less happiness overall, with notable declines in well-being from ages 40 onward. That's a contrast from the rising part of the U shape, but many of the exceptions to the U shape in industrialized countries are in postcommunist Eastern Bloc countries, including Poland. In those countries, there's a "happiness gap" due to perceptions of high corruption, poor government, and economic instability.

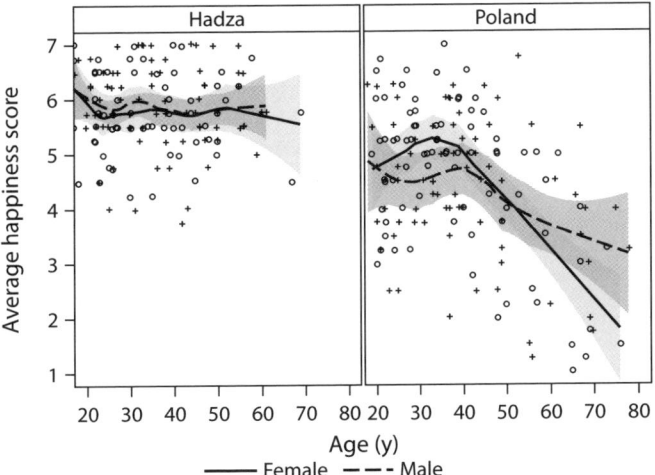

FIGURE 10.2. **Subjective happiness score among Hadza and Polish.** Hadza happiness differs little over the life course, whereas Polish happiness is lower at older ages. *Data source:* Frankowiak et al. 2020.

Under those conditions, life satisfaction scores are lower and often decline with age.[14]

Another study compared global well-being among Baka and Punan foragers, and Tsimane, by using a single question. I first met the lead author, Viki Reyes-Garcia, in Bolivia when I was just starting out as an anthropologist, and I stayed a few nights in the village Yaranda. She had been living there for a year to study how market influences were affecting Tsimane ethnobotanical knowledge. Her team generously shared happiness data with me so that I could look at the age pattern from ages 20 to 80.[15] Unlike the Hadza, all three groups showed differences in overall life satisfaction with age. Average well-being scores were slightly U-shaped among Baka and Punan, although the slump for the Punan was not at midlife but instead around age 60. Among Tsimane, life satisfaction declined across all adulthood.[16] If you think "average" well-being conceals important heterogeneity, then consider the probability that folks reported "low" well-being (that is, life is very bad or not good). Doing so reveals a similar pattern: most older Punan and Tsimane are worse off than younger group members.[17]

The Age of Project AGE

One of the first ambitious comparative projects on elder well-being was called Project AGE (for Age, Generation, and Experience). To complement and update some of the work pioneered by Simmons and others (described in chapters 5 and 6), Project AGE sent experienced ethnographers to study elder well-being among the Ju/'hoansi, Herero, and three more urban, industrialized settings (Clifden and Blessington, Ireland; Hong Kong; Momence, Illinois, and Swarthmore, Pennsylvania in the US).[18] Well-being among !Ju/'hoansi and Herero elders was lower than that experienced in four of the five more-industrialized locales. But for many of the reasons we've already explored, it's hard to compare scores of well-being across different populations with different cultures. Project AGE researchers did try to account for what constituted a very good or very bad life based on local models. But recall the complaint discourse of Ju/'hoansi and Herero elders, a verbal art form aimed at garnering aid and attention from anyone listening. Any public expression of joy might stir onlookers to make requests for assistance! These factors alone may have affected how Ju/'hoansi or Herero responded to questions.

More instructive is what people say. A Ju/'hoansi woman who was over age 60 reporting low well-being explained, "My life is not good. Even you saw me. Last rainy season I slept in this dwelling you see behind me. It has no top, and when the rain fell, it just fell on me. I have no dwelling." A Herero woman in her 50s described why she also reported low well-being: "I have no cattle, no money, no husband, and no children. My cattle have all died in the drought. My children are gone, and I am all alone with no one to support me. Life ten years ago was much better because my son was here to help me. Today I have no one." A Ju/'hoansi man in his 70s reporting moderate well-being said, "I live by my own efforts. I keep myself young and no one else supports me." A Ju/'hoansi woman over age 60, with high well-being, stated matter-of-factly, "I have a good life. Bau [another elder] brings me edible resin

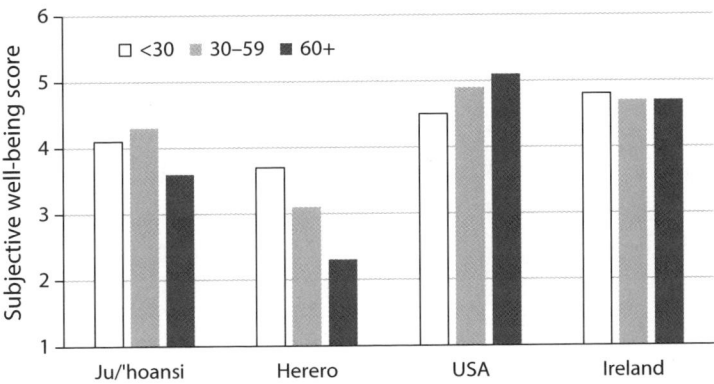

FIGURE 10.3. **Subjective well-being**: Ju/'hoansi, Herero, US (Swarthmore, Pennsylvania), and Ireland (Clifden). Well-being is lower among the 60+ y adults only among !Kung and Herero. *Data source:* Keith et al. 1994, Table 5.3.

and berries. //ushe [her middle-aged daughter] brings me other bush foods. I eat enough to be full, and I have food."

Figure 10.3 shows that, as with other subsistence populations, well-being declines in adulthood among the oldest Ju/'hoansi and Herero—but not so for the Western samples. By the late 1980s when they were studied, the Ju/'hoansi were living in semipermanent villages with schools and a health post, and many had seen livelihood changes (like owning gardens or cattle). There had also just been a severe drought that affected crops and livestock. Ju/'hoansi elders had better objective physical health than in the past and an easier, less mobile lifestyle. But less independence, lower ability to maintain their livelihood, and less cultural value placed on their traditional skills affects elderly the most. Ju/'hoansi elders in worse physical shape were more likely to say that life was worse now than a decade prior, and vice versa for those still healthy and productive.

Madness of "Civilization"?

In an article in the *New York Times*, journalist Erica Goode succinctly summarized a popular view on how depression slows us down when we need to make difficult decisions to help achieve important goals.

According to that view, depression is more common now than in our past, because we have too many lofty goals today. In a romanticized view of hunter-gatherers, she wrote, "One thing that has changed over the eons is the increased pressure people feel to set ever larger goals. Ancestral hominids may have striven to pick enough berries to last for a week; modern humans want to look like supermodels, make a million dollars in the stock market or produce flawless children."[19] More condescending examples abound, such as one from a 1960s study of cardiovascular disease among Maasai herders that claimed, "More than most primitive people, the Maasai find subsistence easy, labor light. Competition is negligible and, some might think, frustrations limited. They have few responsibilities and a quite different attitude toward the world and people about them than do most of us."[20] As discussed in chapter 3, similar myths about the happy-go-lucky hunter-gatherer life abound.

Hunter-gatherers don't use social media to track their popularity, nor do they compare their wins and losses against those of millions. But their social lives are far more complex than the caricatures just quoted. Balancing a diverse social portfolio is tricky. As everyone knows, maintaining social relationships requires work and heavy investments of time, effort, and generosity. But it's worth it. For hunter-gatherers, there's nothing worse than being left alone—much less being abandoned or lost. A "social death," like when being ostracized, may feel as bad as what you imagine death itself would feel like.

In subsistence populations, even the most skilled hunters frequently come back to camp empty-handed, and the most skilled farmers can lose their harvest to pests, animals, or flood. Relying on others is therefore a necessity. And, like we saw in chapter 4, anyone can get sick or hurt in ways that leave them vulnerable. The way we buffer ourselves against all types of risk—including conflicts with other humans—is through our social ties, embedded in networks of reliable relatives and friends. In contemporary urban environments, effective law enforcement, social security, and deliverable groceries are modern substitutes that vastly improve our ability to manage frequent risks. Yet even with many of our needs covered, we still feel isolated and alone when deprived of community.[21]

By comparing the same population over time, or those living in more versus less acculturated settings, we can better assess whether the trappings of modern life have made things better or worse. In some cases, health has clearly improved as a result of greater access to medicine and health care. But, like we saw with the Ju/'hoansi elders, people's perceptions of their own well-being don't always follow suit. Among the Matsigenka of the Peruvian Amazon, physical health had vastly improved over a period of several decades. The Matsigenka had less anemia, fewer intestinal parasites, less malnutrition, and even lower blood pressure. But from people's self-reports, the Matsigenka perceived their health as having severely declined over that time.

How do they reconcile better health with feeling less healthy? Easy. Health is much more than just being free of infection and disease. Matsigenka worry about sorcery and shady dealings with outsiders. An oil company brought some opportunities, but also noise, pollution, and drunkenness. As a result, competition, suspicion, and the loss of trust shattered social relations with other Matsigenka communities. More interactions with townsfolk also induced anxiety and tensions over being told repeatedly about their need for "development." As a 25-year-old Matsigenka woman complained, "Before, we didn't know how to wash dishes, we ate out of *pamocos* [coconut shells or calabashes used as bowls], we used monkey heads as spoons and we didn't have illnesses. We didn't have to boil water and we lived in peace. We were happy without having to wash our hands all the time. Now health personnel come and say we have to wash our hands all the time, boil water for everything, and make latrines to be happy, but before we were happy."[22]

Changes in livelihood and loss of traditional territory can also be devastating. In a comparison of seminomadic Himba herders living a traditional lifestyle with Himba who have moved to town, town-living Himba reported lower life satisfaction than those with a more traditional lifestyle.[23]

From Well-Being to Depression

I've established that in many subsistence societies, well-being declines at older ages. Is a lower level of positive well-being the same as depression? Depression has not been studied much in subsistence populations. This issue is not just due to the methodological problems I've mentioned, but also from a deep-seated sense that it's yet another modern mismatch condition. Many people, scholars included, believe that depression is an affliction of modern civilization. But if sadness, fear, anxiety, and depression are part of our evolved toolkit of psychological defenses, then their expression should not be limited to just the West, or to any one particular time in history.

Two detailed examples among Tsimane and rural Malawians suggest the potential for depression and anxiety to be a human universal—and that these get worse at late ages (or at least acute bouts of low mood, rather than clinical diagnoses of chronic major depression). Not only does depression increase with age among both Tsimane and Malawians, but differences by age make sense in light of the evolutionary function of postreproductive life. Age differences in depression disappear when we take into account changes in physical function, disability, and other indicators of lower productivity with age.

Working with the Tsimane, my colleagues and I adapted scales commonly used in diverse settings for assessing depression.[24] Building on lessons learned from the studies among subsistence populations mentioned earlier, we ran focus groups to help us parse what was relevant and what wasn't. You can't just copy and paste questionnaires developed for urban New Yorkers and think they'll be intelligible or meaningful for folks elsewhere, even if well translated. We asked adults age 40+ whether over the previous month they had experienced sadness, fatigue, changes in sleep or appetite, and other symptoms usually associated with depression. Sadness is captured by the Tsimane word *yoquedye'*. People could respond using a four-point scale ranging from rarely to always. We also noted whether people were smiling and being jovial, a simple but useful way to provide "external validity" that our interview makes sense. From the sixteen questions, we calculate a score,

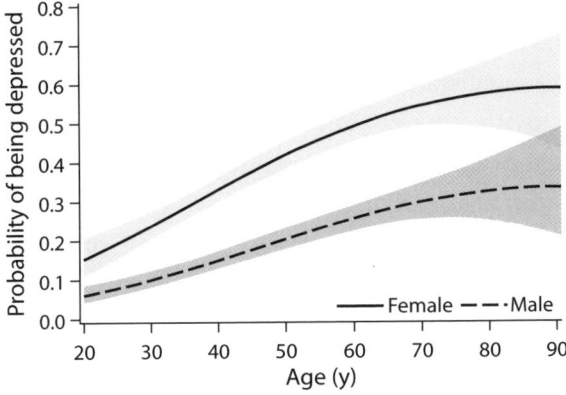

FIGURE 10.4. **Depression and age among Tsimane.**
Curves reflect the probability of having a depression
score in the top quartile. Likelihood of reporting low
mood increases with adult age among Tsimane.

and the values are distributed like a bell curve. This tells us that experience with depressed affect does not reflect a threshold case of having it or not. Higher depression scores did mean fewer smiles, laughs, and eagerness to talk. Overall, about 18% of women, and 7% of men, report regularly experiencing depression-like symptoms. Like patterns observed elsewhere, Tsimane women report experiencing most symptoms more than men. And, like we saw with subjective well-being for other subsistence populations, depression is more common with age. Figure 10.4 shows that the probability of having high depression scores (top 25 percentile) increases with age for both men and women.

I Sing the Body Elastic

But why more depression at older ages? We set out to test whether disability and capacity to do productive work are the linchpins. We calculated a "disability score" from adults performing a series of tasks, like balancing on one leg, bending over to pick up a pencil, walking speed, and other tests borrowed from the groundbreaking MacArthur Aging Study. We also devised a "subsistence score" by tabulating whether men continue to hunt, chop trees, walk long distance, and carry heavy loads, and whether women still weave bags and mats, and whether both sexes can still walk long distances.

Older folks with more disability and less subsistence involvement are more likely to be depressed. As disability and subsistence participation also change with age (see chapter 8), we wondered whether these two factors could account for the age effect. They do. Considering disability and subsistence capacity in a statistical model with age eliminates the age effect in figure 10.4.

This jibes with what Tsimane say. After asking folks, "If you could change one thing in your life to make you happier, what would that be?" most responded about wanting higher economic productivity and better health. These were the top categories for both women and men, accounting for three-fourths of all responses. Some said, "to be young again" and a few said, "I'd die to end my suffering." Others said they would like to be closer to kin, wished there was less social conflict, or desired more modern goods and services. But the majority said some version of "to have better health" or "to work again like before."

Tsimane depression doesn't come from worry over not being able to feed *yourself*. Food anxiety is not uncommon among Tsimane—up to a third of Tsimane adults worry about food security. But that worry is highest in the 20s and 30s, drops to a low in middle age, and stays that way for the rest of life. Older adults don't express any more food anxiety than middle-aged adults. Even if their productivity is lower, they may also have fewer dependents to be concerned about. Instead, the concern is more about others. Tsimane in the privileged position to give away more food than they receive are less depressed than those who receive more than they give.

The story is similar in Malawi. The Malawi Longitudinal Study of Families and Health (MLSFH), run by the demographers Hans-Peter Kohler and Ilana Kohler, studies rural Malawians with a wide range of livelihoods and life experiences. The MLSFH assessed feelings of depression and anxiety using similar questionnaires that they validated with their field team. They also asked about subjective well-being and found that those experiencing more depression or anxiety thought their lives were not going so well. "Umm, duh," you might say. But these types of checks give us confidence that even vague snapshots of well-being capture something real in people's lives.

Like with the Tsimane, depression and anxiety are more common in older adults, and more so among women than men. By age 65, Malawian elders spend a quarter of their remaining lives with moderate to severe depression and anxiety. And like the Tsimane, physical hardship and low productivity are linked to poorer mental health. Roughly two-thirds of the age increase in depression and anxiety is accounted for by changes in health and physical limitations, like pain, on work productivity. As might be expected, other traumatic stressors—like losing your house to fire or flood, poor crop harvest, worry about HIV, and loss of kin— are all linked to higher depression and anxiety. But accounting for these other stressors didn't explain the age differences in mental health.[25] Worse mental health, as it turns out, is directly linked to a compromised ability to accomplish daily activities of use to others, captured in measures of bodily pain and work limitations affecting physical health.

Elder Depression as Evolved Defenses or Maladaptive Mismatch?

It's enough to lose your mojo in later adulthood, so what's the use of feeling bad about it as well? How can such a double whammy make evolutionary sense? Just like those who can't feel physical pain have to be extra careful lest they succumb to cuts, bruises, bone-breaks, and burns, a life without mental pain also leads us astray. Consider congenital insensitivity to pain, which is a very rare genetic anomaly of the nervous system. People born with it have difficulty learning to avoid dangers and rarely live past childhood. Emotions of all sorts, and this includes those related to pain, are not annoying obstacles that just get in the way—they're evolved adaptations that serve to motivate behavior and communicate information to others.[26]

It's tough to think that the emotion of sadness and the broader spectrum of symptoms linked to depression might carry some benefit, but many ideas along those lines have been proposed by a variety of experts. For example, evolution-minded scholars have argued that depression is an honest plea for help to solicit support; a "slow-down" to ruminate and contemplate next steps when things go awry; an appeasement display to avoid social conflict; a strategy to extract resources from others while on strike; and, a form of sickness behavior (akin to what behavioral biologists have reported throughout the animal kingdom), wherein slowing down helps you conserve energy to boost the fighting power of your immune system. Each of these hypothesized functions suggests that there exists a set of circumstances that should elicit depressed affect, which, in turn, should result in some benefit that might not have been realized had there been no depression. Notice that none of these proposals is specific to urban twenty-first-century environs.

Among Tsimane, anthropologist (and fellow codirector of the Tsimane Health and Life History Project) Jon Stieglitz and I found some evidence to support at least one of these ideas. Depressed Tsimane are more likely to have greater immune activation, especially inflammation. We found this to be true both in circulating levels of immune protein messengers in the blood (called cytokines), and when their blood is stimulated in a test tube with bacteria and viral-like antigens. The emotional, physical, and cognitive aspects of depression were all associated with these indicators of higher immune activity. This contradicts the notion that depression leads to a simple suppression of the immune system.[27]

I've known several Tsimane who were depressed when they were faced with making difficult choices, when "two roads diverged in a yellow wood." I'm not sure if the yoquedye they felt really helped them. Among Tsimane, yoquedye is often attributed to "thinking too much" about things, especially things you can't change. Speaking words of wisdom, prior to the Beatles, Tsimane would often say "*paj mo'ya*" (let it be). The equivalent saying in Ache, "*kuaeme*," translates roughly as "don't think about it!" Both ideas seem oriented toward preventing depression in the first place, rather than treating it once it manifests.

A version of the philosophy occurs when mourning the death of relatives. Ache who were thought to cry too much would be beaten "ceremoniously" with sticks by children, while people shouted "kuaeme!" (An amusing story that my former PhD adviser, Kim Hill, shared with me extends kuaeme to the present. While visiting the US in the mid-1990s, an Ache man, Pikygi, stayed with an American in Miami who had lived with the Ache in the past. In Miami, other Americans showed Pikygi how to find pornography on the internet. Pikygi showed little interest, which came as a surprise to the folks who knew him. The Ache are far from prudish. Pikygi said to his American friend, "Those women are desirable, but I can't have sex with them. Kuaeme!")

Depression as a signal to solicit support from other group members is an intriguing idea. If you are surrounded by kin and others who can observe changes in your mood and behavior, being depressed might lead people to attend to your needs. According to this idea, depression is like going on strike, with no fruits of your labor available to others until they meet your "demands." But if you are living alone or far away from those who care most (as many of us are today, like I mentioned in the beginning of the chapter), those signals may go unheeded. Phone calls are nice, but not enough. I suspect chronic major depression is more likely when combining the usual triggers with modern-day social isolation. With no one consistently on the receiving end of our signals, our signals only get louder. Social isolation and loneliness also affect our perceptions of those around us. We become hypervigilant to social threats, and so ironically it can be harder to reach out or be receptive to those who might otherwise help make a difference.

Much of the depression seen among Tsimane and Malawians doesn't persist over time, and it's worth mentioning that they're not using anti-depressants. Rather, depression appears as an acute bout, and it is often due to recent circumstances. Tsimane scoring high in depression one year were not the same folks scoring high in other years. Among older Tsimane, however, depression is more likely to recur, but it is still not as chronic as typically seen in comparable age groups in the United States. The Malawi study assessed depression and anxiety during multiple waves of their work, and over a period of up to eight years. As was the case among the Tsimane, the correlation in the scores from one year to the next was pretty low because different people were depressed from one year to the next. Circumstances change.

While interdependence and communal living help buffer elders, finding yourself growing reliant on others and, even worse, a resource drain, is harrowing (among the caribou-loving Chipewyan of the boreal forest, for example, men unable to hunt report feeling useless and having lost their magical powers).[28]

In industrialized countries, receiving support can also be a mixed blessing.[29] While it feels good to be cared for by others, overwhelming feelings of being a burden can lead elders to depression, anxiety, and a desire to hasten death. Being in the comfortable position of provisioning and being a support giver yields not only status, self-worth, and self-esteem, but also greater well-being, physical health, and higher late-age survival.[30]

While a positive relationship among well-being, health, and functional capacity is found in most places around the world, it's especially tight in subsistence societies where able bodies are needed to serve others. Just as unemployment is a common cause of distress and depression in the industrialized world (even in countries with decent unemployment protection),[31] older adults with functional limitations in subsistence societies have a lower quality of life, just like we saw among Tsimane and rural Malawians.

So are higher rates of depression and lower well-being at older ages maladaptive? What "good" does feeling bad do? The negative valence of bad feelings can push us to compete less for status, to disengage and switch course. Feeling like a burden may motivate a push to increase industriousness, cultivate new skills, or find other ways of being useful—to delay obsolescence. If you're still useful, going on "strike" via depression might pressure others to step up and increase support—and thereby avoid a hastening of death through neglect, abandonment, or, in the extreme, suicide or geronticide (as discussed in chapter 7). If you're not, these paths represent a withdrawal of kin support, channeling resources away from elders and onto others.[32]

Everyone Likes a Good Paradox

The sense of belonging, feeling needed, and being able to provide support to others in interdependent social networks, improves well-being, even in industrialized societies with nucleated households stressing individualistic values like independence and self-sufficiency. Even without the "need" to be buffered, we are still comforted by support, and knowing that others care, and having others to care for. The value of family and community extends beyond the subsistence groups we've seen so far. Yet the average American, who just a few decades ago had roughly three confidants to discuss important matters with, now has none.[33] Stronger social connections improve survival, even when everything else is fine.

The role of social connectedness may help resolve the Hispanic or Latino paradox, first identified in the 1980s by sociologist Kyriakos Markides. This long-studied phenomenon refers to the fact that US Hispanics live a few years longer than US whites, despite having lower average income and education and more risk factors for cardiovascular disease.[34] It's paradoxical because lower socioeconomic standing, more obesity and diabetes risk, and poorer access to health care are all expected to lead to higher mortality. These risk factors pretty much lead to worse health and survival everywhere. But even heart disease mortality is lower among US Hispanics relative to non-Hispanic whites.[35]

Two of the more leading explanations for the paradox so far are inconclusive. The first one, called the "healthy migrant hypothesis," proposes that Hispanic migrants to the US may be extra healthy, while the "salmon bias hypothesis" suggests that Latinos may return to their home country to retire and die, and thereby have not been included in US mortality statistics. Other explanations focus on healthier diets and Hispanics smoking less than whites.

But another strong contender reflects the higher social capital found among many Hispanic populations. That is, the tight family ties, social cohesion, and support networks more common to Latinos in the United States. Embeddedness within a community sharing similar values and identity is protective. Cultural values tend to emphasize the family,

social harmony, and respecting elders. Is that enough to make a difference in elder health and survival? Despite hundreds of studies about the
paradox, no clear all-encompassing explanation yet rises to the top. And
the social capital ideas about connectedness and feeling needed have
not been well tested. More exploration is sorely needed in light of the
loneliness epidemics and the recognition that eating right, exercising,
and the like may not be enough.

Nor is there consensus over explanations for other anomalies called
"mortality crossovers," which refer to when a disadvantaged group experiences higher mortality throughout much of life relative to advantaged groups, but lower mortality at later ages. For decades, it's been
known that the mortality rates of Black people in the US are higher
compared to white people in the US, but only up to about age 85. Thereafter, whites die at higher rates than Blacks. These kinds of mortality
crossovers have also been identified in China, Germany, Canada, and
New Zealand. Usual explanations focus on issues of data quality at late
ages, or on selection bias, whereby higher mortality at younger ages
leads to more robust survivors in the disadvantaged group.

According to these explanations, the crossover is an illusion.[36] But
other factors may come into play. For example, the life exposures and
experiences of the cautious and respectful Silent Generation (born
between late 1920s and mid-1940s) differ from those of Baby Boomers
and Generation X. Likewise, explanations focusing on social capital and
connectedness have not been well tested (and they should be). The
focus is usually on the biological state of the body, rather than its resilience, which comes from how we interact with the social world.

Expectations Shape How You Fare: Chicken and Egg

Ageism is a major issue throughout the world, despite the large number
of older adults alive today and the number of elders in positions of authority. How a society and its inhabitants view elders impacts how elders view themselves, and in turn impacts their own well-being. When
cultural zeitgeist matters, then how people experience aging is inti-
mately linked to their health.[37] Consider that the young participants in

the Baltimore Longitudinal Study of Aging many decades ago who held negative attitudes about being old were twice as likely to have a heart attack three decades later.[38] A related meta-analysis showed that exposure to stereotypical views about aging hurt older adults' performance on cognitive and memory tasks—if you are prompted with ideas that elders are forgetful, you're more likely to forget![39]

In various studies throughout the world, scholars of ageism have suggested that younger generations carry distorted views of their senior citizens/elders. But what do people really think? A comprehensive study in twenty-six countries assessed how younger citizens view older adults within their own country. This included the usual suspects in Europe and Asia, but also South America and Africa. Collectivist countries tend to show lower implicit and explicit age bias against older adults and expressions of greater warmth toward elders.[40] But certain beliefs about elders were common in all countries in the study. On the undesirable end of the spectrum, older people are viewed as less attractive, less able to learn new things, and less capable of doing everyday tasks. On the plus side, being older means that you are considered to be more knowledgeable and wiser than others, and for this, you garner more respect.[41] This is true despite the availability of books and the internet, and the relatively fast pace of cultural change. You'd think that attitudes might be more favorable in countries with a greater share of elders, but in fact, they were more negative: having more old people means worse societal expectations about old age.

In small-scale societies such as Ju/'hoansi, Agta, and Tsimane, we find that ageism is not so focused on age per se, but on physical condition and capacity. Chapter 6 highlighted many reasons that these groups tend to bestow prestige on relatively healthy elders, whereas signs of decrepitude typically warrant pity or disdain. Remember that in these populations, elders are held in high esteem in the absence of books, libraries, TV, and internet. It's remarkable then that in the cross-national study just mentioned above, where all those "substitutes" exist and where skills and knowledge change so rapidly, elders are still valued for their knowledge, wisdom, and experience. The value of elder experience and wisdom transcends what can be easily gleaned from Wikipedia or ChatGPT.

Modernizing influences can alter how elders are viewed. In a comparison of Tsimane, the United States, and Poland, psychologist Piotr Sorokowski and colleagues examined whether valued traits were attributed more to older or younger adults.[42] In this case, populations in Poland and the United States act as a modernizing foil against the more traditional, subsistence-oriented Tsimane. The team asked people to indicate whether a number of traits are better ascribed to an elder or a younger adult. Like the studies I just mentioned, they found that wisdom about how to live and how to resolve conflicts or problems was most ascribed to older adults in all three societies. But other positive traits were ascribed more to elders *only* among Tsimane: learning new information, having a good memory, and being more satisfied with life.

That elders are believed to be more satisfied with their lives seems contradictory to what we saw earlier. I suspect that younger Tsimane were thought of as less satisfied not because they're more miserable on the day-to-day. But instead, because they may be viewed as more ambitious or dissatisfied with the status quo, in light of the rapid pace of socioeconomic and cultural change. Also, Tsimane elders were more likely than not to be considered deserving of respect and authority, something which was not true of the elderly in Poland or the United States. This is all the more surprising because formal leadership roles among Tsimane these days are more often held by men in their 30s through 50s, due in part to their command of Spanish language and greater confidence in dealing with outsiders.

A better approach at exploring how modernization affects perceptions of aging is to do intra-cultural comparisons. Enter the Dani, central highlanders of western New Guinea who traditionally hunted wild pigs and grew magnificent sweet potatoes. Among Dani, Sorokowski again compared how older adults were viewed by asking people to nominate individuals who best exemplify certain traits. But this time he compared Dani living in rural settings with those in urban settings, and compared those who were unable to read or write with those now literate.[43] Ages of adults most worthy of respect were about a decade older in rural settings. Wise advice-givers were eight years older among illiterate Dani nominators than literate raters. Illiterate Dani also thought older adults were more satisfied with life than did literate raters.

Together, these studies, combined with detailed case studies about the effects of modernization on elder well-being, suggest that it's wrong to overgeneralize about modernization's effect on elders. It hasn't been overwhelmingly negative, as originally conceived and normally argued (recall the modernization theory I mentioned in chapter 6). In a rapidly changing environment, some of the elders' knowledge may be obsolete, but life experience, long-range perspective, and deeper appreciation of history can't be replaced. Some status might be lost, but elder lives can still be enjoyable if well integrated, appreciated, and needed. When less physically intensive means of production are made available, elders can be more self-sufficient for a longer period of time. With sedentary living, elders are more easily cared for by others. The shade thrown at elders in more industrialized countries (despite their making up a larger share of the population) may have less to do with youth-obsessed culture than with tense competition over limited resources, like prime real estate and other economic opportunities.

Elder Voices

One of the best ways to learn how elders view their lives is simply to ask. My team and I have asked hundreds of Tsimane adults over age 50 about many aspects of their lives. We learned that loneliness and social isolation exist but are relatively rare. Very few (6%) reported frequently feeling lonely, while three-fourths said they're never lonely. Hardly anyone said they ate a meal by themselves over the past week. Over three-fourths said they enjoyed talking with family or friends three times or more in the past week. And those few who have no one to depend on for food, or who spend more time eating alone, are the few who are more likely to say they're lonely. If you're skeptical about what people say, we have nice checks. In what are called "focal follow" observations, Tsimane trained as anthropologists on our team recorded all instances of conversation and other behavior by a focal elder in random one-hour snapshots. On average, elder women talk to about five different people per hour, and elder men about four. Men's interactions are spread across more categories of people than are women's. The majority of women's interactions are with their daughters, husbands, granddaughters, grandsons, and sons. For men, it's their wives, sons, nephews, grandsons, brothers-in-law, and brothers.

Tsimane elders also feel well supported. Very few said they didn't have someone to care for them when they were sick or feed them when they didn't have food. When it comes to talking to others about their problems, older men are slightly more likely to say they have no one to talk to. Overall, though, Tsimane elders feel needed and useful. A 67-year-old woman said, "They [my kids] always ask me to watch the grandkids when they're working in their fields." Elders also talk about remaining productive in their fields and helping relatives in many farming-related tasks. Some older men still hunt animals like the large rodent-like paca. Many men make or fix tools like bows and arrows, ax handles, and dugout canoes. Elder women mention having *shocdye'* ready to drink at all times, creating a vital venue for conversation and hanging out. Others talk about doing community work, like cleaning the school, clearing trails, and helping women give birth. And a few mentioned how their social security payments help their families purchase necessities like fishing nets and shotgun shells.

Seven Decades Again?

By age 80, Tsimane elders are four times as likely as those under 60 to say they don't feel needed, but many still do. The need to watch over the house and watch grandkids or great-grandkids may be even greater in acculturated villages, where parents are more likely to be visiting town, selling produce, or involved in intermittent wage labor.

Along the lines we saw reflected in the biomarkers of chapter 9, Tsimane elders view their own health decline as accelerating after age 80. Even if surviving to these late ages, they're likely to think that things are only downhill health-wise for their future. However, like we saw with depression, age is unrelated to feelings about your future health once you take other factors into account. Those factors all reflect physical condition and health: physical pain, problems carrying out daily activities, poor eyesight, and ailing health.

The concerns that Tsimane elders often talk about reflect worries over loss of kin support. A 64-year-old woman said, "My daughter married in Cuverene so far away, now she doesn't visit me." A 72-year-old woman said, "I'm worried about my daughter and her husband—they don't live nearby to help me with something to eat, to eat together." And, from a 68-year-old woman, "I just want all my family to be together again"; from one of the oldest Tsimane, a 92-year-old woman, "I want to know how my daughters are doing and my grandkids"; and from an 86-year-old man, "I worry that I'll be left behind. My grandkids and relatives always help me out with food."

Worries also reflect elders' declining ability to keep up. A 79-year-old woman said, "I don't walk like I used to." Or this, from an 81-year-old woman, "When I was younger, I visited each community, when I was healthy and felt good. Now I can't walk like before when I was young and healthy." The majority voiced concern over urgent production needs, like getting help from kids or grandkids during the upcoming rice harvest. Harvesting rice needs to be done in a short time window before the rains come and spoil the harvest. A 78-year-old man couldn't find anyone to help him build a new kitchen.

In a separate study of Tsimane happiness (*majodye*), Viki Reyes-Garcia and colleagues asked people to list what makes them happy.[44] Big surprise—the top reasons emphasize successful subsistence and generous social relations. The five most important things that makes Tsimane happy are "spending time with close family," "to have a good garden plot," "to have good food," and "success in hunting" and "to drink the shocdye' home-brew." Good food according to the Tsimane includes abundant meat, fish, and assorted foods from farming. Just below food and socializing is "to have good health." When considering instead what explains everyday moments of happiness, Tsimane mention these same reasons but with one notable difference. The main reason Tsimane give for being happy in the moment rings true everywhere: "Nothing bad happened."

————

The increase in reported well-being from people between the ages of ~45 to 70 years throughout the industrialized world may be one of the few mismatches where we're better off now than we were in the past. In subsistence societies, there is no upswing in well-being starting in the mid-40s, though there's also no midlife crisis. Declining health and pro-ductivity weigh heavy. Our declines in health instead can be more easily managed in an increasingly service- and information-oriented economy coupled with modern conveniences. We're able to still carry on in our jobs and in our families. Maybe we can't still do triple cartwheels or jump flights of stairs for fun like we used to, but we manage to adjust our expectations and feel good for having done so. And when we can't carry on, we (often) have formal institutions, like pensions and insur-ance, that buffer against losses in income.

Adjustment works best when there are others in your local network who can take on the roles left behind. But eventually our lives shift. By age 60, the average American spends six waking hours per day alone— and almost eight hours by age 70, almost double the experience of a typical 25-year-old. Time spent with children plummets in our 50s, and with our coworkers in our 60s.[45] By our 70s, well-being in many indus-trialized countries either levels out or declines, just like we saw on the flip side of the U-shaped happiness curve.

Strong connections to family, friends, and community may be the best fountain of youth out there, better than any multivitamin or even the best snake oils on the market. Having just our physical needs met doesn't cut it for happiness and well-being. On several occasions, Tsi-mane friends would express sympathy to me for being (at the time) over 30, but with no wife, no children, and only one sister who lives almost three thousand miles away. In their eyes, I was either very poor or just a grown man-child. They were not wrong. Now almost fifty, I can agree with them. Epidemiological studies confirm that weak social networks are comparable in effect to our worst habits, like smoking and drinking, and greater than the harms of obesity.

While there's no magic pill to provide social comfort, artificially in-telligent robot companions are lining up to tackle the loneliness prob-lem. The New York State Office of Aging gifted companion robots named ElliQ to hundreds of tech-savvy lonely elders with Wi-Fi in late 2022. In the words of their director, these talkative robots were de-signed to help them to "focus on what matters to individuals: memories, life validation, interactions with friends and families." Preliminary studies suggest these robots may actually help reduce loneliness.[46] Even without robot friends, some of us may feel closer to fictional families and friends than to our own. These "connections" offer the safe comfort of familiarity and intimacy but at low cost and no risk. All the one-sided drama can be kept at a distance, and if you're feeling unsafe or overwhelmed—click!

Binge-watching *Six Feet Under*, *This Is Us*, and other TV family dramas may lead us to think that our social needs are being met. But these are escapes, a means to forget—at least until the series ends.[47] Heavy TV watching and use of other social media may be more a symptom of depression or loneliness than their cure. Those faux "parasocial" communities don't belong to us; they can't sustainably fulfill our desires for meaningful connection. A sense of purpose comes from being needed and feeling valued.

11

Bringing It Home

In spite of illness, in spite even of the archenemy sorrow, one *can* remain alive long past the usual date of disintegration if one is unafraid of change, insatiable in intellectual curiosity, interested in big things, and happy in small ways.

—EDITH WHARTON, *A BACKWARD GLANCE* (1934)

Aging isn't even one percent as scary as whatever is going on with the people who try not to [age]

—CALEB HEARON (TWITTER OCTOBER 9, 2022, @CALEBSAYSTHINGS)

WE'VE TAKEN a long journey backward through history and sliced sideways across different subsistence societies today to imagine what older age was like prior to our urbanized and mechanized twenty-first-century lives. From this, what conclusions can we make about our increasingly elderly population today? This is an urgent and important issue. The number of people over age 65 in the US is expected to jump from 56 to 73 million by 2050, making it the fastest-growing segment of the population. When this cohort reaches its peak, more than one in five Americans will be of retirement age.

The situation looks similar around the globe. This trend will impact many domains of life: Will there be enough workers? How will we afford skyrocketing health-care costs? Who will provide much-needed elder care? Will social security run out, and what happens then? Will younger generations be obliged to watch the Rolling Stones "Steel Mobility Scooters" tour? These are (almost) all no doubt serious questions. But they're not new. Old age has always been a part of the human life course, and elders have always been around. We were built to last seven decades. The difference lies in degree, not in kind. And the large wave of elders approaching promises opportunities to harness, not just burdens to manage.

————

Despite improvements in survival at the latest ages, you still won't be bumping into many centenarians on the highway of life. Only 0.01% of the world's population are over age 100—a number that matches the population of Baltimore, Maryland. By 2050, the projected eightfold increase raises it enough to match the population of Los Angeles, California. That's an impressive number, but still a tiny blip among the projected population of the world: ~10 billion. Even if we consider only those older than age 80—about 2% of the world population right now—we see that number climbs to just 4% (~400 million) by 2050. In other words, most of the numerical boom will be among elders with ages similar to those encountered in the subsistence societies discussed throughout this book—there will be many more young-old than old-old.

Young-old elders have long held many vital roles in society. We saw how they advise, teach, lead, narrate, entertain, manufacture, care, and cherish. They belong. In contrast, older adults today are cynically viewed as superfluous in a work-obsessed, youth-oriented, active-adult-centric world. Older adults are considered a problem that society has to handle and are often put out to pasture or discarded. Even under the best circumstances, older adults will be honored but still ignored. Maybe they'll be allowed to volunteer here and there, but they are expected to just cruise, sometimes literally on a cruise ship. In her pivotal critique of contemporary aging, *Coming of Age*, the existentialist philosopher-writer-activist Simone de Beauvoir remarked that "old age is a problem on which all of the failures of society converge. And that is why it is so carefully hidden." What may have seemed a hidden, but dehumanizing, menace in capitalist societies of 1970 when her book was first published is now out in the open. Her gloomy one-dimensional view of elders "condemned to poverty, decrepitude, wretchedness and despair" and with "no refuge from the emptiness of their lives" is a sad reflection of our worst fears.[1]

But it's a myth to think that, in our ancestral past, elder lives were paradisiacal. We see that in societies more typical of our past, elders are valued, but only as long as they are deemed useful, needed, and able. Failing that, they might still get respect if they are in control of valuable resources. But few kudos are awarded just for packing seventy candles on a birthday cake. So it's not true, as the physician and author Roger Landry wrote in *Live Long, Die Short: A Guide to Authentic Health and Successful Aging*, that "only during the twentieth century have they [elders] been knocked to the mat" and considered burdens.[2] In traditional societies, elders can also be neglected, abandoned, or see their lives cut tragically short. Nor should we forget that in modern societies, older adults often refuse to head out to pasture. Many still hold a lot of sway; they vote, own property, and hold positions of influence. At the extreme, we see the Margaret Thatchers, Joe Bidens, and Anthony Faucis of the world—people of immense influence well beyond seven decades. They are role models for sure, though their level of achievement is a high bar. If that's what it takes to earn respect as an elderly person, what hope is there for the rest of us?

Rather than returning to a mythical past via a flailing nostalgia, an appreciation for how we've lived helps us redefine and reimagine the roles and status of our third-agers. We can never forget that our success as a species stems from our longevity dividend. Behold our elders—not mere elderly—whose lived experience merits *elderhood*. This is a rebellious pushback against the usual business of just extending adulthood. Slogans like "60 is the new 40" privilege youthful vitality. Our wrinkles can be covered with creams. Or broadcast to the world as battle scars, signatures of laughter, learning, and experience.

The self-proclaimed ambassador of elderhood, geriatrician Bill Thomas, is leading this reimagining of older age.[3] He emphasizes the essential potential role of elders as intergenerational gurus. In promoting a multigenerational society, he proposes an "eldertopia" to describe an imaginable future: "a community that improves the quality of life for people of all ages by strengthening and improving the means by which (1) the community protects, sustains, and nurtures its elders, and (2) the elders contribute to the well-being and foresight of the community."

In this final chapter, I champion this view to position a hopeful future, one where the burden of old age is reframed as the opportunity of elderhood. Luckily, this future doesn't hinge on new breakthroughs in antiaging pharmaceuticals. Nor does it require us all to be extraordinary, like Jo Schoonbroodt, who at age 71 completed a marathon in under three hours and ran twice as far as he drove that year.[4]

I'll first position some lessons from hunter-gatherers about maximizing a joyful healthspan. Then I'll consider new ways of exporting what works in small-scale populations to fit our needs in large-scale societies. Usually this means blaming ourselves for our own behavior. Major improvements will instead require modifying incentives and restructuring our communities—to shift how we use our body in our current environment, and how we interact with others. I start with our physical health and shift to our broader well-being, which reflects our needs for support, community, and purpose.

To truly reap the gains of our evolved seven decades, we need to champion elders as active members of society and harness their strengths in the same way our ancestors have done for millennia, and the way many still do. Our age-segregated society holds us back from what is possible. We need to see more schemes to mix up the ages. Elders integrated in a multigenerational setting where all can thrive. This includes grandparents, parents, and grandchildren and broader cross-generational interactions for communities at large. Elders more tightly hooked into family and community can help address many societal ills: reducing loneliness and disconnection among young and old alike, countering illiteracy in underresourced areas, building resilience and perspective among the young, helping ensure a more secure social security and retirement, normalizing caregiving, and uplifting elderhood. As elders have mostly moved beyond the competitive, profit-seeking status strivings of adulthood, their perspective can help secure a future worth living for everyone.

Serenity to Accept the Things I Cannot Change . . .

As I mentioned in chapter 8, physical aging is not going away anytime soon. So how much effort should we put into extending life further? Or pushing back against our aging soma to stay healthy as long as possible? Life extension is a multibillion-dollar industry, but despite the many prophetic claims and discoveries—at least for yeasts, flatworms, flies, and mice—no one has developed any products that make an ounce of difference for increasing lifespan, much less maximal lifespan. The diabetes medication metformin comes closest as a potential boost to healthspan. Mixing metformin in a cocktail with other common drugs, people in a small one-year study rolled back their biological age by 2.5 years, measured using Horvath's epigenetic clock.[5] Is this reverse aging? Maybe. But is this the direction we should go to improve our lives?

One hopeful almost-billionaire futurist, Bryan Johnson, sold his company for $800 million (as one does) in order to focus on reversing his own aging. Calling his strict regimen "Project Blueprint," he modified his eating, activity, and sleep. He inhaled a daily cocktail of twenty-four supplements, including metformin and rapamycin, an mTOR-inhibiting drug often used to prevent organ transplant rejection. From this rejuvenation Olympics, amid constant testing, he shed 5.1 years of biological age in a mere seven months—a world record! He saw improvements on almost every biomarker of aging. Since that initial self-experiment, he has continued to biohack his way toward rejuvenating his organs, at an alleged cost of $2 million per year. (He receives blood plasma transfusions from his teenage son and claims to have reversed aging in his penis through painful "shockwave therapy.")[6]

But let's say this difference does translate into better sustained health and higher survival—a big if. At what cost does it come? According to Johnson's website, his regimen costs $2,224 per month in food, supplements, and testing. That's four to seven times what a person in the US tends to spend on their food per month. And beyond the cost, it involves a level of obsessive discipline that I'm guessing few of us possess, not to mention adding terms like "nicotinamide riboside" to our vocabulary. I certainly don't have the patience, time, or mental energy to devote to this, and I study this stuff. If some people want to experiment, and rage against the dying of the light, the rest of us may one day benefit. But we're not holding our breaths.

Hunter-Gatherers as Models?

Instead of banking on miracle pills and petri-dished organ transplants, surely knowledge about hunter-gatherer lifestyles can help prevent or delay diseases of aging, right? That's always the first question I get from reporters. Most of the ways we live our lives today have changed since our hunter-gatherer and farmer pasts: what and how we eat, how we move our bodies, the microbes that vacation with us, and much more, changing fundamentally how we experience aging. Media blogs and fanatic podcasts alike rely on "paleofantasies" to give us advice on how to best improve our health and well-being.[7] These hark on how we need to eat a paleo diet, exercise like a triathlete, bathe in dirt to expose ourselves to diverse bacteria, and attend the occasional fecal transplant party, a sure way to make new friends (and new "old friends")—each attempt to mimic conditions that somehow we've lost.

By now you know well that mortality is higher in hunter-gatherers at all ages compared to almost all other living populations, and that we didn't evolve specifically to age slowly or live long lives. But elder foragers and farmers do have far fewer chronic "mismatch" diseases. Other than freegans and back-to-nature enthusiasts,[8] most of us are not in a position to live our lives as foragers. Nor do we want to. Nor is the world set up that way anymore. But what's the essence of forager lives that we can capture? Table 11.1 lists some suggestions. Many are not new, but we can now appreciate why they hang together. They've been field-tested for millennia.

TABLE 11.1. Short list of "to-do" points for successful aging

Physical activity	Diet	Belonging	Inner circle (Family and friends)	Purpose
Take a walk/hike; YouTube workouts!	Minimize soda	Join clubs, interest groups, religious organizations	(Re)connect with old friends, lost relatives	Share ideas, skills with others
Treadmill during that Zoom meeting	Eat slowly, stop when approaching full	Visit/communicate with family more/consider multigenerational living	Forge new (fictive kin) connections in community	Volunteer and/or work in (religious) community
Park some blocks away from destination	More fiber (psyllium husks)	Think of others; put in the effort	Listen and tell more stories; record them!	Do old things in new ways, with open mind

Note: I don't include the usual activities that are beneficial at any age, like getting enough sleep, flossing regularly, or getting annual checkups. I list just a few examples under each of the five categories: physical activity, diet, belonging, inner circle, and purpose.

Bust a Move

First, we know that we should move more. We knew that before study-ing foragers, but it's the one major consistency across all subsistence groups. No wonder it helps strengthen our bones, muscles, heart, and arteries. Exercise focuses our minds and improves blood flow to the brain. Physical activity has been identified as a form of primary preven-tion for at least thirty-five chronic conditions associated with aging and survival.[9] The approach to movement that foragers take can be helpful in guiding us. It's tempting to drive to the grocery store three minutes away, just because we can.

Moving more doesn't mean weekend warrior snowboarding or joining an expensive gym. Mostly it means walking more. Vigorous activity is nice, but not a whole lot may be needed. Remember that hunter-gatherers engage in a lot of light and moderate activity; they're just not so seden-tary.[10] During the height of the COVID-19 pandemic, I took daily walks around the neighborhood, even though, like many areas, there are few sidewalks where I live. Just 21 minutes per day can have substantial ben-efits for heart health and more.[11] I also binged on 30-minute YouTube-video cardio, aerobics, and high intensity interval training. Not just watching them while eating popcorn but participating as a student. Didn't need any equipment, though I enjoy the videos with kettlebells.

People with high levels of activity are usually those with active living, not those exercising every day. Adding movement to daily life whenever possible is a more easily attainable target. If you can walk or bicycle to work, or for errands, even a couple times a week, go for it! If that's not feasible for long-distance commuters, maybe an e-bike could help. Per-haps paradoxically, you can get similar, if not greater, levels of physical activity from e-bikes than you can from regular bicycles (probably because we're more likely to just get on the bike).[12] Even when we drive, parking at a lot distant from work adds more steps. Wearing a Fitbit seems to encourage us to add steps and shed pounds. Meta-analyses show that being accountable to a digital counter increases our activity, decreases our sluggishness, and indeed helps us lose weight. Works even better if you set a target goal![13] Mine is admittedly tame, but not too hard to exceed: 8,000 steps a day (still less than half of the Tsimane daily average).

There can be significant gain from the accumulation of many small actions. For me, instead of meeting with students and colleagues in my comfortable but boring office, I talk and walk with them around campus. I am lucky that the UCSB campus has a nice lagoon and ocean bluffs as outdoor office surroundings. Walking and talking improves mood, and making physical activity social helps ensure that we actually do it. Even taking the steps to my office instead of the elevator to the second floor I estimate adds 104 steps a day. Not much, but over the years I've been at UCSB, that alone has helped keep off three to four pounds.[14] Other tricks, given my reasonably sedentary lifestyle: I use a makeshift standing desk, and play at least 15 minutes of soccer or basketball with my young son every other day. His latest favorite game, just weeks after the World Cup Finals, he calls "Messi versus Mbappé." After dinner, I like to walk around the neighborhood. On weekends, I enjoy the many nearby hikes in the lush Santa Ynez and San Rafael mountains in the Los Padres National Forest, a pleasant stroll in the nearby San Marcos foothills, or a bicycle ride around town with my kids. I try to get my kids hiking or biking outdoors with me at least three times per month.

Eat Like a Forager, or Don't

Many books scream about gut-busting diets, and I don't have much to add here. I like Michael Pollan's motto: "Don't eat anything your great-grandmother wouldn't recognize as food." His other motto is good too: "Eat food. Not too much. Mostly plants." Limit consumption of ultraprocessed foods. Most Americans need more veggies, nuts, and fruit in their diet. Every little bit helps. Though it's contentious to say it, don't quibble over this versus that amount of carbs or protein. As long as you're eating "real" food, the important point is not to overeat.[15] Physical activity is great for many aspects of health but doesn't work terribly well for weight loss.[16] It will be easier to overeat if you're not very active. I like the Confucian concept of *hara hachi bu* practiced in Okinawa: eat until you're about 80% full, when satisfied but not completely full. Don't eat while distracted or watching TV. When scooping fish and green plantain stew out of a common pot with all nine members of my Tsimane host family the first year I visited, I'd burn the roof of my mouth with each bite. If I waited until the stew cooled down, there'd be nothing left. In that anxious and tense eating environment, I'd hardly feel satiated. I was often hungry, despite eating well. Since then, I've come to savor eating slowly and deliberately, what some call "mindful eating."

Other quick, simple rules: Hydrate, though you don't necessarily need eight glasses of water per day. You may if you live in a desert, but tea, coffee, fruits and other foods high in water content can also help you stay hydrated. Eat more fiber, even if in the form of sugar-free Metamucil (psyllium husk). Fiber helps lower cholesterol and control blood sugar. Eliminate soda, including diet soda. Stevia, made from plant leaves, is super sweet. It's an acquired taste, but a good replacement for refined sugar as a sweetener. Even better if you can learn to drink your coffee black, and selectively read only the studies showing how wonderful coffee is for your health.[17]

Put the Style Back in Lifestyle

Lifestyle accounts for the majority of age-related diseases. Much of the difference between "healthy" aging and unhealthy aging is due to our bad habits. I have mentioned diet and activity, but smoking is another big one. It's estimated that one in five deaths in the US is due to cigarette smoking. From a peak 42% smoking in the 1960s, roughly 13% of US adults currently smoke. It's twice as common among Native Americans, and least common among the more educated and wealthy. Globally, smoking is twice as common elsewhere as in the United States. Men smoke more than women in all countries, except the Pacific island of Nauru.[18] Smoking would be the best habit to change in the interest of improving health and survival. The good news is that it's been on the wane for about half a century in most rich countries and has been declining more recently in low-middle income countries too.[19]

At a population level, positive changes in these lifestyle factors will maximize healthspan and lifespan. High BMI, poor diet, smoking, and inactivity combined account for most of the global differences in diabetes morbidity, for example.[20] In a study covering twenty-one countries—the largest of its kind—70% of cardiovascular cases and deaths were due to modifiable lifestyle risk factors.[21] The Big Four risk factors—smoking, physical inactivity, excess weight, and poor diet—account for 80% of deaths from noncommunicable diseases, which amounts to over half of all deaths.

On a smaller scale, exemplar communities reveal how the right mix of factors can lead to a healthy and long lifespan. Seventh-Day Adventists, such as those who live in Loma Linda, California, are considered to inhabit a longevity Blue Zone.[22] They don't smoke, drink alcohol or caffeine, and are physically active. They follow a "biblical diet" that is mostly vegetarian—vegetables, nuts, whole grains, and fruits. Given their lifestyle, Adventists show lower rates of many cancers, diabetes, and heart disease. Men and women following all the low-risk elements of the strict regimen lived until their late 80s, about ten years longer than non-Adventists in California. That represents perhaps what is attainable under the best of circumstances. Adventists, as a whole (who vary more widely in their behavior), lived about four years more than non-Adventists after age 65 (e_{65}) for men, and about two years more for women.[23] Other Blue Zones with healthy and long-living seniors are found in Italy (Sardinia), Greece (Ikaria), Japan (Okinawa), and Costa Rica (Nicoya). All share some of the lifestyle factors just mentioned. Like foragers, all are involved in daily physical activity without pumping iron or hitting the gym. Most appreciate moderate drinking. (During a summer I spent learning to paint murals in Sardinia in my early 20s, I was amazed by how many people made their own unpreserved wine, drinking each night with dinner. I owe my long-term love affair with the dry but noble Grenache, called Cannonau in Sardinia, to those years— long before I knew anything about health-promoting flavonoids.)

Or . . . Do Nothing?

If you're moderate in most things, your expected future health and survival will look pretty good. At least from the perspective of the long arm of history. That's some achievement. I know this flies in the face of hundreds of books, experts, and billion-dollar industries designed to help us reach Platonic perfection. If you're a morbidly obese, immobile chain smoker, there's room for improvement. But for those who are already typical, or even healthier than average, the statistical radar forecasts gains of (just) a few years for all that extra effort. Depending on your perspective, that might be good enough, a sufficient amount to make the effort worthwhile. If counting calories or sticking to special diets or activity regimes leaves you incessantly stressed, you may be better off doing little. Your body responds to chronic stress like an invasive threat, spreading dangerous inflammation throughout your body. Stress-free-eating bacon trumps stress-eating carrots. Or, as Shakespeare's Gratiano declared in the *Merchant of Venice*: "With mirth and laughter let old wrinkles come. And let my liver rather heat with wine, than my heart cool with mortifying groans."

Another reason to improve population healthspan is to save on exorbitant health-care costs. This is chimed not so much as a mantra as an imperative. The concern is that we spend way too much keeping folks alive, especially in the final tumultuous year of life. That concern, though, is misguided. Estimates suggest that the year prior to death accounts for only 8%–11% of total health care costs. Not just in the US, where per capita health expenditures are two to three times that of other high-income countries,[24] but also in other countries varying in late-age health and health-care safety net infrastructure. Instead, health-care costs are largely for expensive procedures throughout life. They're to treat injuries and ailments that people recover from.[25]

It's true that if everyone stopped smoking, or had a BMI of 25, we'd all be healthier. But would we save on health-care costs? Sadly, no. Short-term cost savings from being a healthier nonsmoker are offset by the greater costs spent during an extended life. Beyond the early 70s, nonsmokers soak up more health-care costs because a larger share of those living at the oldest ages are nonsmokers.[26]

Less Precise Medicine

No doubt staving off frailty for as long as possible is the more realistic holy grail we hope to find. Many think that personalized or precision medicine is the path to getting us there. This approach angles prevention, diagnosis, and treatment to the specifics of your (epi)genome. Maybe so, but from a public health standpoint, far more years can be added to the collective lives on our planet with much cheaper, less tech-savvy fanfare. Increasing primary care, sanitation, malarial bed nets and making sure people have a few good friends go a long way toward ensuring that many more of us make it to seven decades. Better daily living conditions like access to clean water, nutritious food, safe neighborhoods, and preventive medical care would vastly improve health at all ages in most low- and middle-income countries, more so than improved precision medicine or even more intense clinical practice.[27]

Incremental improvements in health throughout the life course not only adds years to life expectancy but will further reduce global inequality in healthy lifespan. Even in high-income countries, more attention to the social determinants of health would go a long way toward reducing health disparities. The past century has already witnessed a sustained increase in lifespan equality rising in step with the increase in life expectancy. Eliminating deaths in the first half of life is still the best way to both extend life expectancy *and* increase lifespan equality.[28]

Making further gains in life expectancy for those falling behind doesn't demand new miracles of molecular and regenerative medicine. The US spends more per capita on health care but shows worse health metrics than many other high-income countries. Even as other countries' life expectancies have started to rebound following the worst of COVID-19, life expectancy in the US continues to decline—a full six years lower e_0 than in other high-income countries. Heart disease, diabetes, violence, disability, birth complications, chronic lung disease— all look worse in the US than in peer countries. Even advantaged Americans who aren't overweight or don't smoke often experience worse health in comparison with their peers in other wealthy countries. The US health disadvantage results in higher rates of chronic disease and mortality among adults, and in higher rates of untimely death and injuries among adolescents and small children. Indeed, many of the preventable deaths leading to higher rates in the US occur before age 50, from opioid overdoses, fatal car crashes, and gun violence. High income and wealth inequality, increasing as it has since the 1970s in the US, affects everything, including health. If we don't do something about rising economic inequality, I suspect any future gains in longevity will be fleeting.

Ironically, it's the oldest in the US who fare best: death rates among Americans over age 75 are lower than in other high-income countries. Given limited resources, let's have less regenerative medicine, and more safe environments, and more prevention in early and midlife.

It's Not as Bad as You Fear

Most people have a fear of getting old (FOGO). One survey posits 87% of Americans having FOGO. The main fears are loss of physical ability (23%), memory loss (15%), having a chronic disease (12%), and running out of money (12%). Dying itself was viewed as *less* scary than these other fears. While these fears have grounding in current reality, the fear of aging is worse than the experience of aging itself. Figure 11.1 shows that the actual manifestation of these fears among US elders is much lower than what young and middle-aged adults expected. Even though "old-old" elders age 85+ report more of these problems, very few (1%) say their lives have turned out worse than they expected them to be because of them. Pew Research surveys of older adults suggest that old-old and young-old report similar benefits of older age, including being treated with respect, more time with family, time spent with grandchildren, less stress in their lives, and financial security. Plus, elders report feeling, on average, at least a decade younger than their years.[29]

Upon reaching age 75, the biologist and popular science author Jared Diamond, whose book *The Third Chimpanzee* helped inspire me to go down this whole anthropology path in the first place, wrote that he "regards my 60s and early 70s as the peak of my own life, and old age as likely to start somewhere around 85 or 90 depending on my health."[30] Perhaps consistent with avoiding "old age," the average American says living to age 89 is good enough. Only one in five wants to live into their 90s, and a mere one in twelve is keen to complete a century.

The people have spoken: quality over quantity.

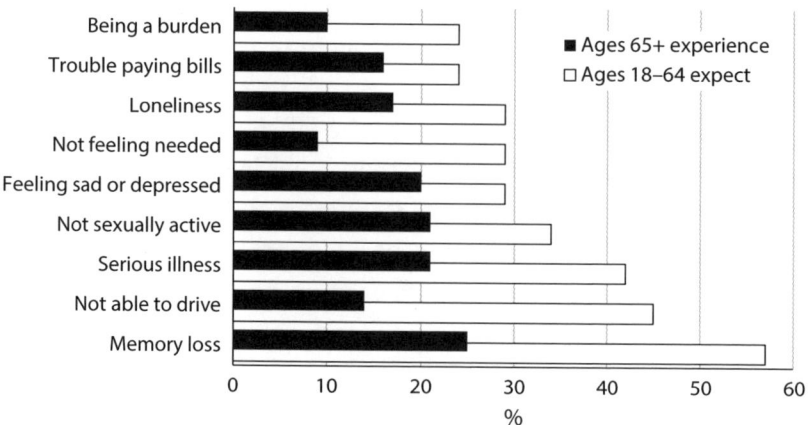

FIGURE 11.1. **Aging is not as bad as most people think.** A comparison between the proportion of adults age 18–64 (n = 1,631) who expect to experience different challenges when older, with the proportion of elders age 65+ (n = 1,332) who report experiencing these challenges. Many more young adults expect hardships with aging than elders actually experience. *Source:* Taylor et al. 2009.

Behavior: Individual to Community

Modifying our health through our behavior is viewed as our personal responsibility—it's up to us to change. Most of our evolutionary history has been spent trying to make sure we get enough food, with nice variety, and not be too exhausted in the process. Only now do we suffer from having too much within arm's reach, and we require superhuman self-control to eat less than we can, and far less of the yummy. Only now do we absurdly walk to nowhere like rats on treadmills, because we've designed our communities to make movement less necessary.

A better, sorely needed approach is to modify our environment so that healthier choices don't require enormous self-discipline. My friend Dan Buettner took this idea to heart by redesigning communities in the US to introduce Blue Zone lifestyles. These community-transformation programs, facilitated with help from local businesses, schools, local government, and community members, aim to increase walkability and increase access to healthy diets. It started with Albert Lea, a small

9,000-person town in Minnesota, which built a walking path around the lake, connected sidewalks between the residential and downtown areas, altered restaurant menus, and placed healthier foods at eye-level height in grocery stores. If you want people to be healthier, these and hundreds of other nudge-like mind tricks work better than making people feel guilty about their weak willpower. And the results in Albert Lea were remarkable: after just eighteen months, the community lost a combined 7,280 pounds, and saw a boost in e_0 of 3.2 years. The icing on the (gluten-free) cake is that these preventive measures save on health-care costs. Many other towns and cities have since committed to becoming Blue Zone communities.[31]

No country has seen a decline in obesity prevalence since its rise. The steepest gains in weight and obesity occurred a mere two decades ago, mainly among adults aged 20–40 years. Several pharmaceuticals show promise for reducing excess weight, like the diabetes drugs semaglutide (Ozempic®, Wegovy®) and tirzepatide (Zepbound®), but prevention is always a better strategy than seeking a cure post facto. Blaming and relying on individuals to resist doing what might seem familiar and natural doesn't work. Top-down and bottom-up approaches can help make neighborhoods more people-centered, positioning our schools and work as part of our healthy living rather than impediments. Certainly there are many obstacles to building health-first communities: attending to high traffic areas, ensuring pedestrian and bike safety, implementing person-centered zoning, fighting against the lower-priced but heavily marketed Big Food Industry—let alone how to make it all work in areas of heavy rain and snow, or extreme heat. During a lucky sabbatical in Toulouse, France, living in the city center with no car, I walked at least 10 km every day, along the Garonne River to work, to the Victor Hugo market to buy groceries, dine, play, or to enjoy the many public parks and areas. No need for the gym. I would've lost weight too, if not for the endless temptation of window dessert porn on my walks home.

In exploring lifespan and aging in subsistence humans, my colleagues and I have found that middle and late ages carry the potential to be free of the main chronic illnesses that currently afflict us. This is the case not just in fluke Blue Zones but could generalize anywhere because we carry what it takes right in our genomes. Groups like the Tsimane, Hadza, and Kitava demonstrate that lifestyle factors have the biggest impact on living a life relatively free of most noncommunicable diseases. That's a reassuring lesson that should fill us with optimism. Indeed, the creative examples like the ones you just read about, that aim to redesign communities around the needs of people rather than of cars have helped improve diet, increase physical activity, reduce smoking, and cut waistlines. The age-friendly community model aims to modify our social and physical environment to facilitate healthy living for people of all ages, but especially among older adults.[32] What foragers and people in Blue Zones really have in common is social: shared, multigenerational communities where elders maintain important roles.

Get Mrs. Palfrey Out of the Claremont!

A half century ago, Margaret Mead complained about the "need to develop a new style of aging" in our society. The US ideals of autonomy and independence may be a source of pride but are also a source of misery (as she noted, "perhaps on the verge of starvation, in time without friends, but we are independent"). She blamed the isolation and irrelevance of elders on older people themselves, thinking that elders chose to withdraw from society. Whether they are living alone or comfortable in assisted living with other elders, the segregation of elders deprives younger generations of role models for how to age and grow older—and of a direct connection with history. As Mead wrote, "They [young people] need to know about their past before they can understand the present and plot the future."[33]

In the 2005 film *Mrs. Palfrey at the Claremont*, a stoic elderly widow aims for independence, checking herself in to a retirement hotel in London. Materially comfortable, she and the hotel's residents are bored, isolated, and lonely. Spurned by her ingrate grandson, Mrs. Palfry finds her life turned around when she meets a frustrated young writer and musician and they become fast friends and kindred spirits. The young writer Ludo becomes her surrogate grandson, and everyone thinks he's her real grandson. Mrs. Palfrey dies after a series of illnesses following a tragic fall. After her death, Ludo is able to write his novel. Outside the confines of the Claremont, fictive kin provide real connection. Both Mrs. Palfrey and Ludo prosper from finding each other.

How could things have turned out differently for Mrs. Palfrey? Let's start with the most obvious. In Simone de Beauvoir's 585-page dark tome on the tragedy of aging, *The Coming of Age*, only three paragraphs are devoted to grandparenthood. Though she acknowledges that grand-parenting brings happiness, in the same breath she laments that a "grandmother's attitude often begins by being markedly ambivalent. If she is hostile to her daughter she is also hostile to the children through whom her daughter asserts herself and escapes from her."[34] De Beauvoir never had grandchildren.

Becoming a grandparent is commonly viewed as one of the primary joys of life. If not, vocal pleas of "When are you going to give me grand-children?!" are in vain.[35] Among the Ju/'hoansi, children were named after their grandparents or other elder relatives, such that their "identities mingle in important ways." Having the same name reinforces the direct link the young have with their forebears, and elders have with the young, to help ensure continuity.[36] Like the Ju/'hoansi, many grandparents throughout the world have nurturing and emotionally close relation-ships with their grandkids.

Just over half of Americans aged 50–64 have grandchildren, while eight out of ten over age 65 have grandchildren. Two-thirds of grand-parents have at least four grandkids. The same Pew Research Survey I mentioned earlier shows that most grandparents value their role and believe they play an important part of their grandkids' lives. More grandparents are living with a grandchild now than a couple of decades ago, but most US grandparents offer just occasional child care. A quarter of grandparents regularly care for their grandkids. Living with a grand-parent is twice as common in Germany and Italy as it is in the United States. Germany and Italy are "old" countries, with over 20% of their adults being age 65+. That's where the US is headed by 2050.

Grandparents also contribute to their descendants in many other ways. According to a representative survey commissioned by the Ameri-can Association of Retired Persons (AARP), US grandparents spend

FIGURE 11.2. **Health benefits of grandchildren.** The majority of adults agree that having grandchildren helps improve their health. From AARP representative study of 2,654 US adults age 38+ with at least one grandchild in 2018. Average age was 66 y; 58% female. *Source:* AARP 2019.

$179 billion each year on their collective grandkids. That's a lot of ice cream, Harry Potter books, and tube socks. But above financial contributions, over half of grandparents consider their wisdom and moral guidance their best assets, more so than babysitting or caregiving. Half report being a valued elder and major source of information on family heritage, culture, and history. Just under half identify playing roles of storyteller, friend, and teacher in their grandchildren's lives.

Above all, grandparenting provides purpose and meaning. Figure 11.2 shows that most US grandparents believe that the relationships they have with their grandchildren perk up their well-being and help them to be more sociable. It also makes them more physically active and feel better about their physical health.[37]

Longer adult life expectancy today provides more opportunities for grandparents to matter in the lives of their grandchildren, no matter the age of the grandchildren. Psychologist and author Mary Pipher, reflecting on one of her patients reminiscing about childhood, wrote, "As one woman puts it, 'my parents were always telling me to hurry up, and my grandparents always told me to slow down.'"[38] In my case, my grandmother on my mother's side was my only grandparent still alive by the time I became an adult. My Grandma Elsie lived to 101 years and was fond and proud of her four grandkids and seven great-grandkids. She was vibrant and smart, a keen observer with a sharp memory for detail. I cherished my visits with her, listening to her stories of family far and close, and the circumstances leading to many in my family migrating from Hungary to the United States, and those early days in New York. As an adult, I had a separate, more profound relationship with her than I ever had with her as a child, growing up as I did in a kids-are-seen-but-not-heard generation. Since most of my family live on the East Coast of the US, my grandmother and my aunt, who lived in Las Vegas at the time, were the closest-living relatives to me in California.

Like with parents of adult children, the grandparent-grandchild relationship has been often regarded as one of unconditional friendship. A longitudinal study of grandparent-adult grandchild relationships over two decades in southern California found that closely bonded grandparents and grandchildren were both less depressed. Like we explored in chapter 10, grandparents who received support without providing any were more likely to be depressed.[39] It's the giving by grandparents in the context of a warm mutual relationship that is rewarding.

But like me and my grandma, many of us don't live in the same house as our children's grandparents, much less the same town or state. In the US, only one in ten grandparents live in the same house as their grandchildren. Distance is indeed the biggest barrier to grandparents being more involved with their descendants. Over half of grandparents have at least one grandchild who lives over 200 miles away; a third live over 50 miles from their closest grandkid. To some extent, technology helps—including phones, video chat, email, and social media. More elders say they like the idea of video chatting than are doing it regularly, which suggests that better user interfaces or extra hand-holding tech support could improve long-distance communication.

Around the corner may be new possibilities. Virtual reality headsets offering opportunities to relive happy experiences from earlier in life may boost well-being for some, as a type of "reminiscence therapy" for those with dementia, but may be best used for conversing in an immersive environment with your relatives and others, especially when the ability to travel is limited. Imagine drinking tea while listening to the

Hamilton soundtrack with your 9-year-old granddaughter despite living thousands of miles away.[40] Hands-on workshops in public libraries, community colleges, and senior community centers could help non-tech-savvy elders, like my mother, get more comfortable with the gadgets and software that help close the physical distance. We could also take a page from Singapore's book, providing "digital ambassadors" to help coach elders and to subsidize low-cost smartphones and phone plans.[41]

Long-distance relationships also offer up the possibility of "skip-gen" travel, wherein grandchildren visit grandparents without their parents tagging along. This is a win-win, especially for folks with few relatives living nearby. We've been lucky to do this a few times now with our own kids. The first time, my in-laws stayed at our place with the kids, while my wife and I took our first much-needed staycation at a hotel about six minutes away.

Win-Win, or Non-Zero-Sum Living

Multigenerational living sounds appealing, though high housing costs, incompatible job locations, limited space, and getting on each other's nerves make it not feasible or desirable for everyone. The Ju/'hoansi and many other groups alter the size and composition of their social groups to meet the needs of dependents, be they children or elders. Such sensible movement of people with group fission and fusion is easier said than done in much of the world today. While flexible multigenerational living has been more common throughout Eastern Europe and Asia, it's often viewed as clashing with the autonomous, nuclear-family-oriented household structures common in Western countries.

Joining together multiple generations is gaining momentum, especially in places with lower housing costs. The number of Americans living in multigenerational households has quadrupled since the 1970s. A major reason for the increase has to do with facilitating elder caregiving and other gains of co-living. But it's also appealing because co-living substantially lowers the financial burden of high housing costs, especially for lower-income folks. The financial savings might even be enough to buffer some families against poverty, as those living in multigenerational households are less likely to be poor than people with other living arrangements. Overall, people of all economic strata find multigenerational living to be far more convenient and rewarding than stressful.[42]

New models of multigenerational living are needed in order to help meet the demands of growing families. One example comes from the California-based Pardee Homes. They launched the GenSmart Suite in ten states, including Nevada, which has one of the highest percentages of multiple generations living together in the US, at 21%. Their homes can accommodate multiple generations, including elderly parents and adult children returning from military service, as well as other caregivers, like nannies, who become part of the family—with separate entries and both private and public spaces. Meeting the needs of multiple generations while maintaining privacy and autonomy is an attempt to maximize the gains of joint living while allaying the common concerns.

Other effective community upgrades include mixed-use spaces, such as multigenerational playgrounds. These have the usual kiddie play areas, but they are intermixed with walking paths and fitness stations for adults. These types of spaces that combine socializing with exercise have long been around in Spain and China, for example, and are now starting to spread around the United States (see figure 11.3).

No Kin? No Problem

Blood may be thicker than water, but being a part of a community that is broader than your immediate kin has always been important to we "groupish" humans. Our need for community has never been greater. Roughly one in six elders don't have any adult children, and therefore no grandkids. The average American today has more parents than children. Once my fellow Generation Xers enter elderhood, they'll have fewer grandchildren than elders have ever had before because of declining fertility rates.

Marc Freedman, the social entrepreneur and author of *How to Live Forever: The Enduring Power of Connecting the Generations,* has a mantra about the future: "The fountain of youth is the fountain with youth." He has helped found several groundbreaking organizations with this

FIGURE 11.3. **Multigenerational fitness park.** Marion Diehl Park outside of Charlotte, North Carolina, includes an electronic climbing game in which children react to flashing lights and a giant rope-climb with several levels that challenge a child's self-confidence. For adults, there are outdoor cardio and resistance machines. The surface of the park has a cushioned rubber surface for "ailing bones and children who fall." Reprinted with permission of Wall Street Journal, Copyright © 2016 Dow Jones & Company, Inc. All Rights Reserved Worldwide. License number 5801620686051.

philosophy in mind. Experience Corps, like its PeaceCorps and Ameri-Corps predecessors, is a call to arms in the name of service. It links elders age 50+ with elementary-school-age children to improve their reading. According to their website, they've linked more than two thousand elder tutors with over 31,000 students across 279 schools. Elders commit at least fifteen hours per week to working closely with kids at school. The impacts of these programs extend beyond the marked improvements to elder mental and physical health.[43] Formal evaluations revealed substantial improvements in student reading comprehension skills, and tutored students saw fewer behavioral problems. And estimates suggest that, even if elder tutors were paid nominal salaries, cost savings could still be substantial for schools.

Similar programs have sprung up in many cities, and hopefully we'll see a lot more of them.[44] New schooling models can help shake things up. Intergenerational schools like those in Cleveland, Ohio, recognize that "tomorrow's discoveries depend on yesterday's understanding." Placing elder "learning partners" in the classroom with students is core to the curriculum.[45] On a broader scale, the AmeriCorps Seniors Foster Grandparents program started in the 1960s as part of the War on Poverty to connect elders (age 55+) with high-needs or so-called at-risk youth. They partner with schools, nonprofits, or other community organizations in each state. Again, the goal is to help provide meaning (and some income) to elders, while also addressing unmet needs of children. In 2018, almost 30,000 elders served more than 170,000 children, with tutoring, mentoring, and daycare.

Where I live, the Council on Alcoholism and Drug Abuse (CADA) has a mentorship program pairing adult mentors with third to eighth graders referred by their counselors for having low self-esteem. The more than fifty mentors are over age 55, "retired folks with a heart and time who want to give back," according to Lisa Gosdschan, CADA's Mentor Program Director. Lisa, a quick-witted, vigorous 67-year-old go-getter, has had four mentees herself, including 15-year-old Beatriz (figure 11.4). "I get more out of this program than they [mentees] do. I enjoy their company immensely. I don't have to be their parent. I'm just an older friend. It's so satisfying watching them just be kids. What we have is a carefree, special relationship." With no kids herself, Lisa has retired twice already, but keeps coming back to help her community. She's also a volunteer long-term care advocate for twenty elder care facilities in the county: "I check in to make sure they're not wearing their breakfast."

I ask Beatriz what she likes about the program. "She [Lisa] really helps me stay on track. She teaches me things I don't learn in school." Lisa proudly tells me that Beatriz is going to college in two years. Lisa states matter-of-factly, "We like each other!"

As with other programs, evaluations show that mentors make a difference in the lives of their mentees: improved mental health, less drug use, better connections with family and friends. Unfortunately, there's a longer list of kids needing mentors than of mentors needing mentees.

FIGURE 11.4. **Child mentor program in Santa Barbara, California.** Executive director of CADA, Lisa Gosdschan (right), her mentee Beatriz (middle), and assistant, Sarah Elkins.

In Lisa's opinion, a big obstacle to recruiting more elders is the lack of transportation. Proximity is key. In many cities, senior centers are being built near schools to help facilitate easier access, especially in areas where public transportation options are lacking. Yeimi Arias, who works for the nonprofit Family Service Agency, directed an earlier version of this elder mentor program in Santa Barbara County with funding from AmeriCorps RSVP. She echoed that it was hard to recruit elder mentors. In addition to access, she said "sometimes [elders] were scared of being paired with teenagers; they preferred younger kids."

Many universities are bringing generations together to co-reside in the same buildings. Student musicians from Drake University in Des Moines, Iowa, act like artists-in-residence, receiving free housing to live, and perform, in Deerfield Senior Living Center. Not only do these types of programs help break down stereotypes, enrich lives, and promote trusting and meaningful relationships, they also address another fundamental need: housing shortages in expensive college towns. Rents are usually discounted for students, in exchange for organizing programs and events, sharing meals, and the like. As Caroline Cicero, director of the Age-Friendly University Initiative at the University of Southern California, where students live in a retirement home, explained, "The younger person could help the older person connect through technology, shopping, basic chores, and one-on-one contact, and the older person offers a lifelong perspective and basic necessities such as food, shelter and affordable housing." One student living with seniors in Minnesota said, "Though there is an age gap, we're not that different from one another. We're all human, we experience challenges and struggles, and we are all learning and growing together."[46]

Marc Freedman and other champions of intergenerational collaborations have long argued that separating the generations impoverishes the lives of both old and young alike. Organizations like Generations United aim to maximize opportunities for mixing up the generations. Given the composition of ages making up contemporary long-lived populations, we now experience roughly similar numbers of up to five generations, from birth up through 70 or so. Mixing generations should not be viewed as charity for the young, or for the old. It is mutually beneficial. Bill Thomas, the geriatrician and elder-care guru, makes the case when he called for a "radical reinterpretation of longevity that makes elders central to the pursuit of happiness and well-being." His idea of an elder-topia hinges on multigenerational interdependence and improved quality of life for people of all ages. Hoping for this kind of world is not a pipe dream. We've already seen how it could work. The foundation for such a potential world is already found in many subsistence populations. The demographic differences between those populations and urban nation-states—more old, fewer young—are ones of degree, not kind.

Getting Creative with Intergenerational Cooperation

Campaigns to recruit elders to get more involved in their communities could be revolutionary. The idea is to shift from occasional volunteering as a form of commodified leisure to routine and pervasive volunteering contributions from the rising tide of current and future elders. By one estimate, US elders over age 55 contributed three billion hours of service in 2018 alone—a value of over $73.5 billion.[47] If unpaid family caregiving is included, that value jumps to over $160 billion. Imagine the cumulative monetary worth from combined efforts of the growing number of elders heading into future decades!

Another arm of the AmeriCorps Seniors is the Retired and Senior Volunteer Program (RSVP), started in 1971. It places elder volunteers to meet a wide range of community needs, including preparing taxes, delivering meals, crisis hotlines, and disaster recovery. In 2018, over 174,000 elder volunteers served 700,000 clients. Efforts to track the impact of these programs show consistently positive effects on physical and mental well-being of elders, and to some extent, the clients as well.[48] This is consistent with many studies attesting to the value that giving and feeling useful by volunteering has on elder health and well-being, life satisfaction, self-esteem, and of course, survival, while reducing functional limitations.

The organization CoGenerate, also cofounded by Marc Freedman, tries to link generations to address a variety of community needs. For example, they brought retired physicians to community health centers in the San Francisco Bay Area to help treat underserved people and to mentor younger practitioners. Their Vaccine Corps mobilized both elders and young volunteers to help provide COVID-19 vaccines to 47,000 folks with limited health-care access in the Bay Area. In addition, they provide incentive grants to any organization proposing to integrate old and young toward the public good.

When I was first working on this book, I thought that if I were more entrepreneurial, I'd develop a local elder/youth buddy system that would be similar to a foster grandparent program, but not just school-based, or for special needs, or to improve literacy. It would be oriented around skills or hobbies, be it yodeling, philately, dancing the Lindy hop, or pinhole camera photography, I imagined. Or could even just allow people to swap stories. A group like this could offer the opportunity to connect and share. For people living in remote or rural areas, or with mobility limitations, an online version could be a nice alternative.

Soon, I found others with greater vision and ambition that have already started down this path. Dana Griffin and Jules Olleon founded an online "virtual village" to bring elders age 60+ in contact with kids outside their community or country. Their company, Eldera, centers on the notion that "every elder has something valuable to give." As Griffin explains, all you need is a "device with Zoom, a generous spirit and a curious mind." Eldera vets elders, and makes the match with kids ages 5 to 18. Meetups on Zoom are more modest than other programs: up to an hour per week. From the testimonial of one pleased mentor, "I'm a mentor myself to a 6 year-old girl from Houston who is terrific. She taught me ballet, shares the books she writes, and I share the one I wrote for kids. We read together, play games, and have just gotten to know each other and care about each other. Along the way, I've become friends with her mom, who has her hands full and is so grateful to have another caring adult in her daughter's life." I signed up my young daughter Evie, who now has a fun new friendship with an animated older woman living in Florida. They talk about their cats, dance, and family.

Big & Mini is another tech-y approach fostering online platonic relationships, but catering to matches with college-age young adults (though people of all ages can be Bigs or Minis). These got a boost during the height of COVID-19 pandemic during lockdowns, when people were feeling particularly isolated. Once virtual reality tech with immersive environments becomes widely available, opportunities will likely boom.[49]

Second (or Third) Careers?

Instead of viewing retirement as an extended, sometimes boring, vacation, longer expected adult lifespans enable us to contemplate what we want out of our second act. One approach, like in subsistence societies, is no formal retirement whatsoever. Instead, elders could adjust their labor to accommodate changes in their physical function. As the poet Kahlil Gibran wrote, "Work is love made visible. . . . It is to charge all things you fashion with a breath of your own spirit, And to know that all the blessed dead are standing about you and watching." As labor patterns shift, even hunter-gatherers have their second acts, specializing in important high-skill activities that harness their wisdom and experience.

Some folks continue working their regular jobs until they can't anymore. In my old eastside neighborhood in Santa Barbara, Roke Fukumura has been working at the same grocery store for the past thirty-two years, across the street from the park where he used to play semi-professional baseball as a teenager. He started that job after supposedly retiring. He comes in at 5 a.m. five days a week, unloads the truck, and stacks the fruits and vegetables. He clearly loves the job: "I enjoy serving the public, and I meet very wonderful people." While I was working on this book in 2022, he turned 100.[50] That's the extreme example, of course. Professions with the most workers over age 66 include tax preparers, clergy, farmers, and bus drivers.[51] I know many academics who, even past 70, have shown no signs of slowing down. Staying employed, or finding a new job at older ages, can be difficult given age discrimination in hiring practices. Some companies do appreciate what elders have to offer and have shown a good track record of hiring older adults. They also offer work policies such as phased retirement and part-time schedules.[52] Companies can even get certified as being elder-friendly.[53]

Some of this extended work life is out of necessity, especially due to concerns about savings for retirement. But it's also about continuing meaningful labor. People don't always have a choice regarding the timing of their retirement, as it is often linked to the nature of their employment. For example, pilots are forced to retire at age 65, and air traffic controllers even earlier, by age 56–61. However, many people in the United States and Europe choose to retire in their early-mid-60s if they are in a position to hold more control over these critical life decisions. Often before this time, many may be facing an empty nest at home and are anxious for change.

A pivot later in life may sound scary, but is full of hope and potential, as many people over the years have asserted. The psychologist Erik Erikson, for example, was a pioneer when it came to his focus on the growth potential of elders, rather than their progressive loss of capacity. In the 1950s he referred to this as *generativity*, a "concern for establishing and guiding the next generation." Another more modern example of a prominent person acknowledging the potential of generation-connecting is the former hotel entrepreneur Chip Conley. When Conley was the oldest non-techie working for Airbnb, he realized his experience made him the sage among his young coworkers, but he also learned from them the tech side of things. He formed the "world's first midlife wisdom school," the Modern Elder Academy, which aims to help people turn their midlife crises into callings for new types of work. Boot camp workshops and sabbaticals, either in person in Baja, California, or online, are designed to harness and champion elderhood.[54]

Modern elderhood is about staying relevant in a rapidly changing world, blending a lifetime of experience with a growth mindset. Defying ageist hiring practices means careful matching of young with old, something networking organizations like Charlotte Japp's Cirkel have been advocating.[55] Employment and training services are also available through state and local organizations like America's Job Center of California, which aims to help older workers transition to a new career. Hopefully we'll see more demand for elder buy-in, and an accessible means to help brush up on new skills and support connecting people together.

One way to learn new skills is from extended learning at local community colleges and universities, often tuition-free. For example, Santa Barbara City College offers free classes, mostly on oil painting, jewelry making, singing, and mindfulness. But the California State University system offers a broader range of courses for all state residents over age 60.[56] Even when not slanted toward job market skills, many universities have lifelong learning programs for both enriching and engaging elders in the community.

A New Kind of Care

Snowballing greater involvement and sense of purpose in our latter years is where it's at. But let's not ignore the obvious. One of the biggest concerns about longer life and an older population is how we manage our frail, post-80s years, where we slow down and start counting our blessings. How to best care for those in need, and how best to attain that care, are two dilemmas where the numbers do matter. The 65+ age group is projected to almost double in size to 95 million by 2060 in the US, and the more frail 85+ age group is expected to almost triple, to 19 million people—an impressive number for sure. But this is under 5% of the 405 million Americans that will be swarming the globe, and just a fifth of the over-65 crowd. Large, but still manageable if we make the effort.

In *The Age of Dignity*, the director of the National Domestic Workers Alliance, Ai-Jen Poo, demands a rethink about how we approach elder caregiving. She pushes for a "New Deal of care" to help accommodate the elder boom. Building a "Care Grid" would require legislation to help support care of elders in their homes. It would protect and secure employment for currently unpaid or underpaid adult caregivers. Employers would more flexibly accommodate employees who choose to be caregivers for elders, just like many do now for childcare. Most Americans don't want to end up in nursing homes, which are also quite expensive. Though there are more nursing homes than Starbucks in the US, fewer elders end up in nursing homes than most Americans suspect or fear.[57] A third of those with serious illness said they'd prefer to die sooner rather than live their remaining years in a nursing home.

Navigating an ideal caregiving situation for elders and their caretakers will be wildly complicated, requiring learning, sacrifice, and anguish. Large-scale investment in the "Care Grid" is what's needed to support the large numbers of frail old-old expected in the coming decades.[58] Poo is hopeful about prospects for the future: "Rather than each of us struggling through in lonely isolation, we will have a shared base of support and structure on which to build our individual lives."[59]

We need a shift in societal values, and then we need to creatively bring in all the bells and whistles of the free market to amplify those values. In Japan, where the proportion of elders is already where the United States will be in another few decades, people of any age receive e-credits called *fureai kippu* for helping elders in their community. These are time credits stored in an online bank that can be exchanged for other services, and you can use them for yourself or transfer them to other people. You can even accumulate them for when you may need eldercare. Such a system is well-suited to settings where money is scarce but time is not. Though fureai kippu and money are both convertible currencies, Japanese elders say they prefer the time-banking barter system over yen because it's more personal. According to interviews with elders, people who worked for fureai kippu were believed to be more motivated by compassion than those who worked for money. It suggests cooperation over competition. The system took off, then lagged when Japan introduced a national system of compulsory care in the early 2000s. The new welfare system was viewed by elders as a reasonable alternative, as it was affordable, reliable, and helped destigmatize the use of caregivers. However, recent problems arising from tightened eligibility criteria have led to a renewed interest in the fureai kippu system.[60] Other countries, like China, Switzerland, and Malaysia, are now exploring similar time-banking options to improve elder care.

Geronticide Revisited?

As the surgeon and author Atul Gawande points out in *Being Mortal*, maximizing survival often conflicts with maximizing well-being. Our tendency to over-medicalize and prolong life at all costs is increasingly being called into question. The philosopher-theologian Ivan Ilich championed resistance against the "medical nemesis" back in the 1970s, wherein the "managed maintenance of life on high levels of sublethal illness" is the "ultimate evil of medical progress" (I remember seeing his lectures from the time he visited Penn State in the 1990s, and he had a large cancerous tumor growing on his face that he refused to treat). Decisions about end-of-life care are complicated without knowing the implications of going through different medical procedures. Given our personal values about how we hope to live out our remaining years, extending life weighs against the effects of a strained, dangling survival on our family. The Choosing Wisely campaign, for example, provides information to help consumers evaluate which procedures are necessary and which may not be worth it.[61]

When evidence reveals that some procedures may have low efficacy, they often still become recommended treatments. This is because of the fee-for-service financial incentives pushing what can seem like an endless schedule of diagnostic tests and procedures. Fortunately, alternative options are improving. Palliative care, or holistic medical attention focusing on providing comfort and quality of life for both patient and family, is an option for people with serious chronic illness, and it is also increasingly covered by health insurance. Evidence from randomized control trials shows that palliative care improves patient quality of life and burden of symptoms, in addition to caregiver satisfaction, relative to usual care.[62] With this approach, a patient's symptoms and pain are managed, without curative treatment necessarily being a target. When side effects outweigh the benefits, no cure or treatment may be sought. In that case, palliative care is similar in some ways to hospice care.[63] In-patient palliative care programs have quadrupled over the past couple of decades and are now available in over two-thirds of US hospitals. Outpatient and community-based systems are also becoming more common. Programs of all-inclusive care for the elderly, or PACE, is another patient-centered integrative care plan under Medicare. Emphasizing care over cure, it typically includes a mix of hospice, often at home, with the help of coordinated teams of specialists. In her powerful memoir *Knocking on Heaven's Door*, the journalist Katy Butler describes the turmoil surrounding these types of last-ditch medical care decisions, and the ill-fated consequences. After her once-vigorous 79-year-old father suffered a stroke, he received a pacemaker that kept his heart going well beyond the rest of his body. Frailty and dementia soon overshadowed his former vigor. "I'm living too long," he lamented to Katy's mother. Katy and her mother grappled with the decisions many of us fear.

As Katy struggled with what to do, she confessed, "I did not want him to die because I thought he was useless or because I was ageist. . . . I wanted him to die because I loved him. I wanted to stop our family's suffering." The 12% of Americans aged 70 and older caring for living parents are struggling with the same dilemma.[64] While it may be the case that, as surgeon-author Sherwin Nuland advised a suicidal elder woman who thought she was of no use to her family anymore, "our very lives benefit others, simply by the security of their knowing that we are there"—there comes a point where the balance is tipped. Just like members of the Tiwi, Sirionó, and many others, Katy realized that she "would understand that things that look heartless to outsiders must sometimes be done out of love."[65]

By the end of the long and winding road, many elderly have exhausted their savings, for caregiving, nursing homes, and out-of-pocket medical expenses. Draining finances, home value, and other assets to prolong life as much as possible cuts into the nest egg we hope to pass to our children and grandchildren. Material inheritance is the logical extension of our evolutionary drive to move beyond ensuring early life survival, to helping our descendants make a successful toehold in competitive labor markets of love, work, and shelter. The average inheritance in the US from parents, grandparents, and other benefactors is about $46,200, a sizable chunk that can make a huge difference for rent, a down payment on a house, a new car.[66] It's ugly and makes us feel guilty to even think it, but there may very well be a point where we're worth more to our living loved ones dead than alive.

Indigenous Futures

As we strive to make the best of our longevity in the postindustrialized world, we must also work toward ensuring that the lives of elders in the tribal societies we've learned so much from continue to thrive. In the popular media, there's strong interest in harnessing insights, wisdom, even the fecal microbiota of hunter-gatherers to serve the health and well-being interests of people in the more affluent postindustrialized world. But I'd be doing a disservice if not addressing the prospects of current and future elders in subsistence-oriented populations.

Maintaining traditional livelihoods is a challenge. Insecure land tenure, land and marine use restrictions, and encroachment by colonizers, gold miners, poachers, and other incursions to native lands all chip away at the ability to be self-sufficient and plan for a future. Without adequate alternatives, lifestyles change. Cheap processed market foods often fill the gastronomic void. Younger folks in many groups leave to find jobs in towns. Alcohol and other vices sometimes become a sanctuary for those failing to thrive amid globalization. Conditions of course vary around the world, and constant vigilance and support are necessary to ensure a better way of life, one filled with dignity and confidence.

As we saw in chapter 3, survival has improved for many Indigenous populations. Raiding and warfare are largely eliminated in most areas. Several deadly illnesses are now more likely to be prevented or treated. Over the past two decades, I've seen more immunization campaigns reach remote villages in tropical Bolivia and help control a few lethal childhood diseases. More Tsimane families are becoming part of the cash economy, and younger generations are learning to speak Spanish and how to write. An Ache teen I befriended twenty-five years ago contacted me recently on Facebook. He's now a religious father of three children and raises pigs. Several dozen Tsimane have active Facebook profiles, and many more use WhatsApp. Out-migration of Tsimane is still limited, and many of the folks I know who left the territory in search of new opportunities later came back. Aside from mercury poisoning and other ailments coming from polluted water and food, a worsening of the diet leading to diabetes and hypertensive heart disease are probably the biggest health threats on the horizon.

More disruptive and visible are the broader cultural changes afoot. Food and livelihood are core elements of culture and identity. As one Sirionó elder told me when I visited his village in 2018, "The young [Sirionó] don't know how to eat anymore. They don't know how to drink. They'd rather eat chicken than armadillo. They like [white granulated] sugar more than honey." In many groups, this current young generation may be the last to speak their native tongue. While we explored in chapter 10 the dwindling relevance of elders in the midst of these changes, the reality is that tribal elders are needed now more than ever. As COVID-19 took the lives of many Native American elders, for example, one Muscogee leader described the toll as a "cultural book-burning." A Lakota man who lost both of his parents remarked, "It takes your breath away. The amount of knowledge they held, and connection to our past."[67]

Even with books, internet, and schools, there is no good substitute for the practical wisdom and perspectives that elders provide. Donna Butts, director of the intergenerational living nonprofit Generations United, cuts to the chase: "Those in their early twenties don't have the experience to know that life will go on. . . . We've survived disasters before. We've survived diseases before. We've survived recessions before."[68] Elders connect people of all ages with their roots and affirm cultural identity through direct links with past traditions, rituals, and ceremonies. These cannot be outsourced, looked up on Wikipedia, or gleaned from a phone app.

We can marvel at the Tsimane, Hadza, Ache, and other Indigenous groups you've read about in this book for resisting the strong tides of cultural transformation. For so long, they've maintained their traditional livelihoods and staved off the not-so-inevitable chronic diseases. As their lives and livelihoods are now shifting, the need for jointly partnered collaborations couldn't be stronger. We continue to learn from Tsimane, Hadza, Turkana, and other groups, not despite but *because* of these socioeconomic changes happening. They are important because they may help us to discover novel gene variants, methylation patterns, microbiomes, and other potential breakthroughs that might shed new light on why diseases of aging, much less aging itself, may look different than what we're used to. Despite the changes afoot, not everyone will be affected the same way, and any knowledge that can help mitigate new health threats would be a blessing.

We also have much to learn about how styles of living affect well-being, connecting and navigating social worlds in ways that build resilience and give meaning to life. Many groups are tempering the balance between self-reliance and interdependence amid this phase of change. To do this effectively, we need to ensure that the research model moving forward is collaborative, based on trust, and conducted jointly with Indigenous leaders and communities throughout the process. This departs from older, more extractive practices where "they" exist as exotic others to benefit "us."

The standard ethical credo "do no harm" is a too-low bar, an obsolete relic. It's not acceptable to leave Indigenous populations no better off than they were before researchers from high-income countries arrived on scene. There are many ways to make research more inclusive, and participatory. Many Tsimane and Moseten have expressed curiosity and concern about their changing epidemiological landscape and a desire to learn the tools to help navigate the new terrain. New initiatives are starting to help make that happen.[69] Despite the obstacles of uncertain political and environmental futures, I'm hopeful that collaborative research can help improve the health, well-being, and legacy of transitioning populations.

Final Thoughts

Our evolved seven-decade lifespans owe much to our cooperativeness, learning, intelligence, intensive parenting, and grandparenting. A long lifespan is not just a consequence of our success as a species but is one of the factors that helps explain our long-term success—our global spread and persistence over time, especially during periods of recovery after crises. Before "successful aging" was ever a thing, targeted for the best of us, our intellectual seer Leo Simmons of chapter 6 had the insight over a half century ago that many subsistence societies "solved the problems of successful aging for a few persons long before they could assure any old age at all for the many."[70] Now that we have the many who will live up to and beyond seven decades, our future depends on how we harness the powers of our elders. Less burden, more advantage.

Sherwin Nuland captured the sentiment beautifully: "We must study to be old, in somewhat the same way that we, when growing up, studied to be adults. . . . Aging is an art form—in itself a type of creativity."[71] In this chapter, I have just scratched the surface of the different ways in which aging can be generative. It's exciting to think of the new creative possibilities for aging-as-art. Imagine teams of geriatric superheroes. The Elders, an international group of a dozen "independent global leaders working together for peace and human rights," is one such team. Nelson Mandela helped shape this organization's mission, believing that their "almost 1,000 years of collective experience" could help reduce poverty, climate change, and intractable conflicts. The current chair of the Elders is cofounder Mary Robinson (80 years old), former president of Ireland and former UN High Commissioner for Human Rights. Their current efforts are to strengthen the United Nations, work toward universal health coverage, and foster tolerance toward refugees. This is just one extreme example, and at a global scale, but imagine the possibilities for local versions of *Avengers*-like Elders in each of our communities.

Making this paradigm shift happen will require better integration, a truly multigenerational world, scaled up from the small-scale societies we came from. "To learn from the old we must love them," wrote author Mary Pipher, "and not just in the abstract but in the flesh, beside us in our homes, businesses, churches and schools."

I'm reminded of my first night on an Ache trek in the Mbaracayú Forest. I was with five Ache families. In a clearing, everyone slept with their limbs intertwined, a mass of bodies with campfires on the perimeter. Children draped on adults, elders in the mix, never abstract. A defense against angry jungle threats, and the comfort of knowing who's got your back. I slept soundly those nights and with intense dreams.

Last night, with heavy rains beating down on the house, my son Ollie begged me to sleep in his bed. Again, limbs intertwined. This time I didn't sleep as soundly. In just over eight out of ten universes, I'm lucky to make it to age 70, when he'll be 28. My daughter, 31. Will they live nearby or across the country? Will we still be close? Will I be in good enough shape to still go on hikes with them? Since having kids, my resolve to survive pulls stronger. This urge to stay connected and help them thrive is consuming. It sometimes feels like a weakness or a vulnerability. I take comfort in thinking that anxiety over caring may have been a driving psychological force behind our ancestors pushing to work a little harder, share a little more, and live a little longer, so many millennia ago.

ACKNOWLEDGMENTS

THIS BOOK has been slow cooking for years. I am grateful to all the people who kept asking me when my book was coming out, including those who didn't know I was even writing one (but assumed I should be). As this is my first book, my gratitude will be unabashedly excessive.

I stand on the shoulders of giants who helped pave the way so that this book could be written. That includes two grandparents and other ancestors who arrived at Ellis Island roughly a century ago to escape persecution and pursue a better life. On my intellectual journey, Kim Hill helped transform my mere interest in hunter-gatherers into bona fide reality, and helped whet my appetite for the Amazon, and field-friendly empiricism. Martin Daly, Margo Wilson, Hilly Kaplan, Magdi Hurtado, Steve Gangestad, Eric Smith, Bruce Winterhalder, Ray Hames, and Jim Brown contributed to my intellectual growth while I was in graduate school at the University of New Mexico. Before that, as an undergraduate at Penn State, the late Jeff Kurland ("Dr. Jazz") and War-ren Morrill ignited the spark that led me to question everything through an evolutionary and ecological lens, then to abandon physics and to pursue an actual career in anthropology. I thank Jim Wood and Henry Harpending for showing me by example that a background in math was great preparation for social science.

Special thanks to the Ache of Arroyo Bandera, especially Enrique Tikuarangi, Gabriel Krajagi, and Ruben Chachugi. My adventures with the Ache fueled a love for fieldwork. Although I dabbled in archaeology as an undergraduate, for me there was no substitute for being able to ask people questions and to become a part of their lives (and they a part of mine). My early work in Paraguay with the Ache got me hooked. I was fortunate enough to find a second home in Bolivia, with strong ties and good friends. I thank Rebecca Ellis, Ricardo Godoy, Viki Reyes-Garcia, and Eddy Perez for helping house and situate me during my first visit to Tsimaneland in 1999. I've come to know people from more than seventy villages, but the people of Cosincho, Munday, Aperecito, and Cuverene will always have a special place in my heart. I'm immensely appreciative of all the Tsimane, Moseten, Ache, Roviana, Turkana, and Sirionó who have opened their homes to me, and shared their lives.

Hilly Kaplan and I started the Tsimane Health and Life History Project (THLHP) in 2002, and its scope has blossomed over the years. More than eighty Tsimane, Borjanos, Paceños, Cruzeños, and Orureños have helped build the project and establish the THLHP as a multinational and multiethnic institution. Many of the young workers who started a couple decades ago are now impressive, talented leaders of their communities. Living and sharing experiences with Maguin Gutierrez Cayuba, Juana Bani Cuata, Bacilio Vie Tayo, Arnulfo Cari Ista, Genaro Roca Moye, Marino Lero Vie, Lorgio Canchi Tayo, Agustina Bani Cuata, Jesus Bani Cuata, Ervin Gutierrez Cayuba, Leonarda Maito Moye, Limbert Hiza Tayo, María Dyersin Lero Cari, and Alfredo Zelada Supa always made my trips interesting and enjoyable.

As a co-director of the THLHP, I've been privileged to learn from and work alongside the best. Hilly Kaplan is a leader in many domains, whose drive to inject a heavy dose of theory to illuminate any topic has been a model for how I try to address any research question. His notion of "embodied capital theory" creatively linked ideas about human longevity to economic productivity and resource transfers, which I discuss in this book in chapter 5. This theoretical framework has inspired many people within the evolutionary community and informed the early years of the THLHP. I've tried to model myself (sometimes disastrously) after Hilly's calm demeanor in the face of adversity, his ability to engage deeply and successfully with whatever problem is at hand (from trying to determine better ways of scoring calcified arterial plaques to digging out a field vehicle stuck in the mud), and his ease at talking with anyone.

Former postdocs Ben Trumble and Jon Stieglitz became new co-directors as Hilly and I expanded our efforts to study health and aging in more detail. My co-directors are outstanding colleagues and cherished friends. Communicating and working with them almost every day has been a joyful career highlight.

Since its inception, THLHP has always combined research with routine medical care to serve a broad range of health needs. I've been fortunate enough to have worked with many courageous and hilarious Bolivian physicians who often represent the public face of the THLHP in Tsimane villages as part of our roving medical team. Daniel Eid Rodriguez is not just a smart physician but also one of my closest friends. After many years working in the communities, Raul Quispe Gutierrez knows almost all 17,000 Tsimane by name or by appearance. Edhitt Cortez Linares was one of the first physicians to work in Tsimaneland and is also a close friend. Special thanks to many of the other dedicated physicians and biochemists who have worked with me: Karen Arce Ardaya, Ofelia Foronda, Juan Copajira, Marco Cartagena, Ivan Maldonano Suarez, Jhon Aguilar Garcia, and Nohemi Zabala Crespo. I also thank Carmen Mavis Ardaya and Roberta Mendez for their years of dedication and careful accounting—we'd be lost without them.

I've learned as much from my students and postdocs as I hope they've learned from me. Their work is peppered throughout this book. They're largely the reason that I enjoy collaborative research—they make the process more pleasurable than the product, and I'm proud to consider them part of my extended family: Raziel Davison, Tom Kraft, Adrian Jaeggi, Ben Trumble, Aaron Blackwell, Cristina Gomes, Bret Beheim, Dan Cummings, Katie Sayre, Stacey Rucas, Eric Schniter, Chris von Rueden, Melanie Martin, Anne Pisor, Jeff Winking, Lisa McAllister, Angela García, Sarah Alami, Amy Anderson, Helen Davis, Carolyn Hodges, Paul Hooper, Emily Miner, Yoann Buoro, Emily Cobb, Carmen Hové, Ronnie Bailey-Steinitz, and Tianyu Cao.

I've also been blessed with amazing colleagues near and far. Over a beer, glasses of wine, or a meal, they have all helped shape how I think about many of the topics covered in this book: Alan Cohen, Julien Ayroles, Amanda Lea, Tuck Finch, Ron Lee, Dan Lieberman, Kim Hill, Steve Gaulin, the late John Tooby, Leda Cosmides, Sam Pavard, Herman Pontzer, Brian Wood, Dave Raichlen, the late Frank Marlowe, Randy Nesse, Claudia Valeggia, Thom McDade, Chris Kuzawa, Brooke Scelza,

Kristen Hawkes, Amy Boddy, David Lawson, Nicole Thompson González, Monique Borgerhoff Mulder, Rob Boyd, Pete Richerson, Eileen Crimmins, Steve Horvath, Shripad Tuljapurkar, Jamie Holland Jones, Joel Rothman, Nicole Albada, Clark Barrett, Dan Fessler, Alyssa Crittenden, Bill Leonard, Rob Walker, Joe Henrich, Sam Urlacher, Sam Bowles, Margy Gatz, Greg Thomas, Randy Thompson, and the HORUS group.

I'm grateful to friends and colleagues who gave me feedback on parts of the book, including Lisa Baum, Amy Boddy, Leda Cosmides, Sharon DeWitte, Tuck Finch, Kim Hill, Jeff Hoelle, Kaye Reed, Jim Rilling, Jon Stieglitz, Ben Trumble, and two anonymous reviewers. Thanks to Tom Kraft, Tianyu Cao, and Katie Sayre, who helped polish up a few of my clunky graphs. Thanks also to my kind colleagues who shared their precious field photos with me. I would've loved to have included many more.

Thanks to my editor at Princeton University Press, Alison Kalett, who saw a book in me before I did and swooped in at just the right time. She also showed me that good breweries exist even in Yardley, Pennsylvania. I'm grateful to Hallie Schaeffer for patiently fielding my many questions about figures and permissions. I was very lucky to work with Amanda Moon and Thomas LeBien of Moon & Company to help ensure that my message was clear, my metaphors less mixed, and my digressions few(er). My copyeditor, Susan Campbell, helped further sculpt my message, and my production editor Angela Piliouras swiftly helped make it whole. I take full blame for any errors or murky prose that still linger.

Lastly, I want to thank my wife, Lisa, for her kind heart, her unwavering support, and for keeping me perpetually grounded during the simultaneously thrilling and agonizing task of writing and revising. I owe biggest thanks to my children, Evie and Oliver, whose distractions are always welcome, and who all of this is for anyway.

———

Several nonprofit organizations aim to improve the well-being of Indigenous populations, including those appearing in this book. The following list provides you with some information about the needs of various Indigenous communities and how to make a donation:

The Hadza Fund supports health relief and related research among the Hadza (https://www.hadzafund.org/).

Kalahari Peoples Fund supports rights of San speakers for health, education, land, livelihood. and language (https://www.kalaharipeoples .org/).

One Pencil Project provides school supplies and higher-education scholarships for Tsimane, Himba, Twa, and other groups of the Kunene region of Namibia and Angola (https://www.onepencilproject.org/).

Organizations with a more global focus on Indigenous well-being include Cultural Survival (https://www.culturalsurvival.org/), Survival International (https://www.survivalinternational.org/), International Work Group for Indigenous Affairs (https://www.iwgia.org), Forest Peoples Programme (https://www.forestpeoples.org/), and Center for World Indigenous Studies (https://www.cwis.org/).

APPENDIX

TABLE A3.1. Modal lifespans for hunter-gatherers and other populations

Population	Modal adult lifespan (modal age of death)	Standard deviation	% of adult deaths at mode	% adult deaths at and above mode
Hadza	73	8.1	2.2	29.1
Hiwi	70	7.4	2.5	10.8
Ache	69	8.9	2.1	28.1
Yanomamö Xilixana	75	7.6	1.7	22.3
Tsimane	69	7.8	2.5	28.9
!Kung 1963–74	70	7.6	2.6	29.5
Ache reservation	77	6.3	3.1	32.4
Aborigines	71	8.8	2.6	35.8
Sweden 1751–59	72	7.4	2.3	24.3
United States 2002	85	1.7	3.5	35.3
Wild chimpanzees	15	16.8	3.9	100.0
Captive chimpanzees	45	7.9	2.7	31.9

TABLE A3.2. Causes of death by age group among hunter-gatherers and horticulturalists

	Ache (forest)	Ache (settled)	Ju/'hoansi	Tsimane	Agta	Hiwi	Machiguenga	Northern Territory	Gainj	Total	
	%	%	%	%	%	%	%	%	%	#	%
(a) <15 yrs	(n=230)	(n=84)	(n=164)	(n=423)	(n=112)	(n=94)	(n=82)	(n=74)			
all illness	22.2	65.5	87.8	79.9	95.5	44.8	63.8	67.6		825	65.3
degenerative	8.3	20.2	3.7	10.4		10.4	9.6	24.3		120	9.5
accidents	6.1	3.6		9.7	1.8	15.6	11.7	6.8		102	8.1
violence	63.5	10.7	8.5	7.4	2.7	27.1	2.1	1.4		216	17.1
(b) 15–59 yrs	(n=125)	(n=22)	(n=127)	(n=192)	(n=77)	(n=31)	(n=19)	(n=68)			
all illness	28.0	86.4	79.5	74.7	69.5	35.3	33.3	61.8		400	60.5
degenerative	3.2	13.6	3.1	16.5	4.9	2.9	14.3	25.0		61	9.2
accidents	23.2			8.8	4.9	8.8	42.9	0.0		85	12.9
violence	45.6		17.3	12.9	14.6	44.1	0.0	13.2		115	17.4
(c) 60+ yrs	(n=27)		(n=52)	(n=60)			(n=2)	(n=33)			
all illness	18.5		51.9	66.1				72.7		95	54.6
degenerative	22.2		40.4	25.4				21.2		49	28.2
accidents	25.9			8.5				6.1		18	10.3
violence	33.3		7.7	1.7			100.0	0.0		12	6.9
(d) All ages	(n=382)	(n=104)	(n=343)	(n=690)	(n=364)	(n=139)	(n=117)	(n=175)	(n=44)		
all illness	23.8	67.9	79.3	69.6	86.7	41.0	57.3	66.3	79.0	2333	70.1
degenerative	7.6	16.0	9.0	12.2	7.6	7.9	10.3	24.0	7.0	306	9.2
accidental	13.1	2.8		8.4	2.7	12.9	17.1	4.0		166	8.1
all violence	55.5	4.2		7.5	3.0	30.2	3.4	5.7		354	12.5

Source: Adapted from Gurven and Kaplan 2007. See text for more details.

TABLE A4.1. Life expectancies at birth and in adulthood: A global survey over the past two millennia

Country/sample	Time period	e_0	e_{15}	e_{25}	e_{40}	e_{45}	e_{60}	Source (see below table)
India	1872–1881	24.7	31.3	26.2	18.9	16.4	9.3	1
	1881–1891	25.1	29.9	24.0	18.8	16.6	10.1	1
	1891–1901	23.8	31.7	25.8	17.9	15.8	9.5	1
	1901–1911	23.0	30.5	25.0	18.0	16.0	10.0	1
	1911–1921	20.2	26.1	23.6	17.9	16.1	10.7	1
	1921–1931	26.8	33.0	26.6	18.6	16.5	10.3	1
	1931–1940	31.8	38.1	32.0	23.3	20.6	12.6	1
Egypt	1927–1937	31.0						1
Mexico	1930	33.3						1
Chile	1930–1931	39.3						1
Brazil	1890–1920	39.4						1
Korea	1930–1931	33.7						1
Grenada (slaves)	1818	25.9	29.5	23.7	16.5	14.6	7.9	2
Poland (Breslau [Wroclaw])	1687–1691	33.5	36.9	30.4	21.8	19.2	12.1	3
British peers (both sexes)	1550–1574	38.0	35.9	27.9	18.0	14.2	10.2	4
	1575–1599	37.2	29.9	27.7	19.0	16.6	10.2	4
	1600–1624	34.7	33.2	26.8	19.0	16.5	10.6	4
	1625–1649	33.0	33.0	26.8	19.2	16.7	11.2	4
	1650–1674	31.9	34.1	28.4	21.7	18.8	11.8	4
	1675–1699	34.2	34.8	29.2	22.0	18.8	11.4	4
	1700–1724	36.2	37.1	30.8	23.3	20.4	12.1	4
	1725–1749	38.1	40.6	34.6	25.6	22.3	13.5	4
British	1541–1799	35.5		32.8				5, 6
Medieval English men	<1276	35.3	32.7	25.6	17.8	15.5	9.4	7
	1276–1300	31.3	28.6	23.5	16.6	14.8	8.3	7
	1301–1325	29.8	26.8	21.4	15.7	13.9	9.3	7
	1326–1347	30.2	25.1	21	17.7	16.1	10.8	7
	1348–1375	17.3	25.2	22.9	18.1	16.8	10.9	7
	1376–1400	20.5	22.9	22.5	19.2	16.1	10	7
	1401–1425	23.8	29.4	27.5	19.3	16.6	10.5	7
	1426–1450	32.8	30.9	25.5	20.4	18.1	13.7	7

(continued)

Country/sample	Time period	e_0	e_{15}	e_{25}	e_{40}	e_{45}	e_{60}	Source (see below table)
Children of English kings [William I to Henry VII]	11th–15th c	30.7	27.5	21.4	14.7	13.3	7	7
English Monks (Canterbury, Durham, Westminster)	Cohorts 1395			29.2				8
	1405			29.8				8
	1415			30.1				8
	1425			28.8				8
	1435			27.3				8
	1445			24.3				8
	1455			23.2				8
	1465			21.9				8
	1475			19.1				8
	1485			19.9				8
	1495			25.1				8
	1505			28.4				8
China (Wang clan)	0–499				22.3	19.7	12.4	9
	500–999				23.3	20.2	11.1	9
	1000–1199				24.5	21.2	12.6	9
	1200–1399				25.5	22.1	13.0	9
	1400–1499				25.3	21.9	12.8	9
	1500–1599				24.4	21.4	13.1	9
	1600–1699				24.3	21.1	12.5	9
	1700–1799				22.7	19.5	12.2	9
Ancient Rome	300–400	21.8	17.5	17.0	16.5	16.5	13.3	10
Hispania, Lusitania	300–400	37.5	25.5	22.5	19.8	18.5	12.3	10
Africa	300–400	46.5	37.0	33.0	27.5	25.0	17.5	10
Roman Egypt	12–259	23.9	33.4	27.6	19.9	17.4	10.0	11
Roman Egypt	100 BC?	22	24	24	17	16	11	12
Hunter-Gatherers (in book)	Average	31	38	33	24	21	13	13
	Range	21–42	29–43	23–37	14–28	12–25	6–15	

Sources: [1]Davis 1951; [2]Koplan 1983; [3]Dublin and Lotka 1949; [4]Hollingsworth 1977; [5]Wrigley and Schofield 1981; [6]Wrigley et al. 2005; [7]Russell 1948; [8]Hatcher et al. 2006; [9]Zhao 1977; [10]Macdonell 1913; [11]Bagnall and Frier 1994; [12]Pearson 1902; [13]Gurven and Kaplan 2007.

TABLE A4.2. Maximum lifespan among mammals. A regression of ln(maximum lifespan), adjusting for body size and study sample size

Parameter	Estimate	Standard error	t Value	Pr > |t|
Intercept	1.419	0.036	39.06	<.0001
ln(body size, g)	0.140	0.005	30.77	<.0001
Group				
Monkeys (vs. non-primate mammals)	0.647	0.040	16.31	<.0001
Apes (vs. non-primate mammals)	0.885	0.221	4.01	<.0001
H. sapiens (vs. non-primate mammals)	1.557	0.439	3.55	0.0004
Sample size				
Large (vs. small)	0.290	0.044	6.62	<.0001
Medium (vs. small)	0.342	0.031	10.9	<.0001

Notes: Maximum lifespan is the highest reported value for a species. This simple statistical model explains two-thirds of the variation in longevity among 976 mammals. Model: $R^2 = 0.65$, n = 976, F = 299.17, $p < 0.0001$. See text for more details.

TABLE A6.1. Effects of co-resident elders on village mortality

Effect	Estimate	Standard Error	DF	t Value	Pr > |t|
Intercept	0.6053	0.2459	10	2.46	0.0336
Distance to town (km)	0.0054	0.0024	25	2.23	0.0348
Percentage age 60+	−0.0483	0.0288	25	−1.68	0.1058
Village Size	−0.0002	0.0006	25	−0.33	0.7431

Notes: A mixed effects regression model that adjusts for village size and proximity to town, and accounts for villages from the same geographic region sharing unrelated risk factors. Based on the results of the model, the effect of elders on annual village mortality lies outside conventional statistical standards of significance, but its effect size (−0.048) translates to a difference of about one life saved over a five-year period when comparing villages with high vs. low proportions of elders. The percentage of elders age 60+ is 1.8% for the low (bottom 10th percentile) and 6.7% for the high (top 90th percentile).

TABLE A7.1. Predictors of death-hastening behaviors

#	Outcome	Predictor	OR	Sig?	N	Source (see below table)
1	Abandonment	Complaints about physical weakness (yes/no)	3.91	yes	95	1
2	Killing	Complaints about physical weakness (yes/no)	3.65	yes	95	1
3	Complaints about physical weakness	Agriculturalist vs. non-agriculturalist	0.45	yes	89	1
4	Complaints about physical weakness	Harsh vs. moderate/mild climate	0.80	no	68	1
5	Geronticide	Harsh vs. moderate/mild climate	2.08	no	68	2
6	Geronticide	Permanently settled (yes/no)	0.46	no	95	2
7	Geronticide	Constant vs. irregular food supply	1.48	no	91	2
8	Geronticide	Forager/fisher/pastoral vs. farmer	3.25	yes	89	2
9	Geronticide	Absent/informal vs. present social hierarchy	3.02	yes	95	2
10	Geronticide	Bilateral descent (yes/no)	4.28	yes	91	2
11	Geronticide	Low vs. high social rigidity	2.25	yes	95	2

Sources: [1]Maxwell et al. 1982, Tables 1–4; [2]Maxwell and Silverman 1990, Tables 1–7.

Notes: Reanalysis from (Maxwell, Silverman, and Maxwell 1982; Maxwell and Silverman 1989). OR, or "odds ratio," shows the effect of the presence or absence of a trait on an outcome. For example, row 10 shows that geronticide is 4.3 times more likely to be reported in societies with bilateral descent versus those with other forms of descent. "Sig" shows whether the result in that row is statistically significant at conventional levels ($p < 0.05$). "N" refers to the number of societies included in the analysis.

TABLE A9.1. Cardiovascular risk factors among subsistence populations.

Population	Type[a]	Obesity (%)	BMI (kg/m²)	Blood lipids (mg/dL)				Blood pressure				Diabetes	Immune	Source (see below table)	Notes
				Total	LDL	HDL	Trig	Hyper (%)	SBP	DBP	Increase w/ age?	FG (mg/dL)	CRP (mg/L) mean/median		
Ju/'hoansi	HG	0	18.8	119	60	39	101	0	119	73	No			1	
Hadza	HG	0	20.2	111	59	36	80	13	120	71	Modest	83	2.6/1.5	2,3,4	
Congo-Brazzaville	HG		20.1	119	89	35	80	19	128	77	Modest			5,6	
Tsimane (<2004)	H	3	23.2	138	75	37	129	1	113	70	Minimal		9.2/3.2	7,8	
Tsimane (2015)	H	6	24.1	151	91	40	102	5	116	73	Minimal	79	3.7	9	
Shuar (Cutucu)	H		24.5	153	89	41	131		113	77		78	?/0.5	10	
(market integrated)	H		25.0	177	96	55	129		116	74		78		10	
Yanomamö	H	0	21.2	132	73	37	111	0	95	63	No			11,12	
PNG (Tukisenta)	H	0	21.2	150			127	5	124	81	No	76		13	Males, heavy smokers
Kitava	H	0	19.3	202	137	45	106	7	119	71	Modest	68	?/0.5	14,15	cocounts
Turkana (rural)	P	0	20.0	154	61	73	100	6	116	77	Modest	100		16,17	
Turkana (urban)	P/U	13	21.9	182	63	75	117	17	126	80	Modest	95		16,17	
Masai	P	3	20.7	150	85	42	121	4	118	71	No			18,19	
USA (NHANES 2006)	U	27	28.6	196	113	54	139	29	119	69	High	102	4.2/2.1	20	ages 18–65

Sources: [1]Truswell and Hansen 1976; [2]Raichlen et al. 2016; [3]Raichlen et al. 2020; [4]Sherry and Marlowe 2007; [5]Mann et al. 1962; [6]Moussouami et al. 2016; [7]Vasunilashorn et al. 2010; [8]Gurven et al. 2009; [9]Kaplan et al. 2017; [10]Liebert et al. 2013; [11]Mueller et al. 2018; [12]Mancilha-Carvalho and Crews 1990; [13]Sinnett and Whyte 1973; [14]Lindeberg et al. 1994; [15]Carrera-Bastos et al. 2020; [16]Lea et al. 2020; [17]Lea et al. 2021; [18]Mbalilaki et al. 2010; [19]Mann et al. 1964; [20]NHANES 2006.

Note: BMI = body mass index; CRP = C-reactive protein; DBP = diastolic blood pressure; FG = fasting glucose; HDL = high density lipoproteins; LDL = low density lipoproteins; SBP = systolic blood pressure.

[a] H = horticulturalist; HG = hunter-gatherer; P = pastoralist; U = urban.

NOTES

Introduction

1. If you've ever had Guayakí yerba mate (a staple beverage in Paraguay), you indirectly know of the Ache, as Guayakí is a pejorative term used by neighboring Paraguayan populations to describe the Ache. Many hunter-gatherer groups have been more widely known by (often pejorative) terms given by outsiders, and so wherever possible, I use the preferred terms that group members use to refer to themselves.

2. 2019 Revision of World Population Prospects. United Nations Department of Economic and Social Affairs (DESA) Population Division.

3. Dobzhansky 1973.

4. See, for example, Sinclair and LaPlante 2019; Finch 2007; Levitin 2020.

Chapter 1. Defining the Age of Aging

1. See 50+ World, "Seniors USA Museum Discounts," May 10, 2019, https://50plusworld.com /seniors-usa-museum-discounts.

2. See Finnish Centre for Pensions, "International Comparisons," accessed September 19, 2024, https://www.etk.fi/en/work-and-pensions-abroad/international-comparisons/retirement -ages/, https://www.oecd-ilibrary.org/sites/b6d3dcfc-en/index.html?itemId=/content /publication/b6d3dcfc-en, https://en.wikipedia.org/wiki/Retirement_age.

3. Increasing retirement age makes sense for the many who don't retire until after age 65 anyway. But low-earning, less-educated workers with cumulative disadvantage and disability from a lifetime of physically demanding jobs have poorer late life health prospects and lower life expectancy. For these workers, increasing retirement age can be a burden, if not even more detrimental to health. For a progressive proposal on retirement in light of push-pull cultural values (e.g., egalitarianism, universal social security, and the "myth of [already] purchased benefits"), see Alstott 2016.

4. AARP 2019.

5. Tsimane people indicate gender through different forms of the same noun root. For example, *nanas* refers to a girl, whereas *nanaty* refers to a boy. To avoid annoying duplication, I use just one form of gendered nouns when using Tsimane expressions.

6. Van Arsdale 1981.

7. The ! and other symbols reflect mouth-popping click sounds in their language.

8. Biesele and Howell 1981.

9. Glascock and Feinman 1981.

10. UNDESA 2019.

11. Suzman, Willis, and Manton 1995.

12. Data on Finnish population size, births, and mortality rates downloaded from Human Mortality Database (May 5, 2021). Using a common method for projecting implications of

changing demographic rates (Leslie matrices), I determine what the new stable age distribution would be holding survivorship rates constant but reducing fertility from 4.7 in 1878 to 1.4 in 2018.

13. In 1949, the highest e_0 was in Norway (71.5 years). By 1950, Norway's e_0 was 72.3. It wasn't until a half century ago—1969—that a dozen countries had e_0 over 72 y! Data source: Our World in Data, https://ourworldindata.org/life-expectancy.

14. Drefahl, Ahlbom, and Modig 2014. All Swedish demographic data were downloaded from the Human Mortality Database website, https://www.mortality.org/, January 2021.

15. For an entertaining and up-to-date review of "our species' greatest achievement," see Johnson 2021.

16. My editor made me give you his full name: Wilhelm Hector Richard Albrecht Lexis.

17. Lexis 1878.

18. Väinö Kannisto, the late Finnish statistician and demographer, deserves credit for bringing Lexis's obscure measure, the modal age of death, to greater public attention. As we saw for Sweden, he showed that as the mode got bigger over time in Finland, the variation around that mode declined (Kannisto 2001; Cheung et al. 2005). Kannisto showed that later modal ages of death were met with a shorter lifetime remaining after the mode, in a study comparing sixteen countries, and four countries over time (England, Switzerland, Finland, Netherlands). He also showed that as the mode increased, the dispersion around that mode decreased. Dispersion is captured by assessing the standard deviation around the mode.

To explore the relationships among modal age of death, life expectancy, expected years of life remaining at the mode, and standard deviation around the mode, I compared these metrics for Sweden using period lifetables from 1751 to 2019. All data were accessed from the Human Mortality Database. The Pearson correlation between life expectancy at birth (ranging from 38 to 83), e_0, and modal age of death (ranging from 64 to 90) was 0.92 (p < 0.0001). While the modal age of death was also associated with greater life expectancy at older ages, like, say, 80 y (r = 0.73, p = 0.002), gains in e_{80} were more modest: an increase of four years, going from five to nine years over a 270-year period. The modal age of death was inversely related to the expected number of years remaining from that mode (r = −0.87, p < 0.0001, ranging from twelve years when the mode was 64 to four years when the mode was 90). The variation around the mode reduced by about a third over the 270-year period.

19. Canudas-Romo 2008.

20. Sanderson and Scherbov 2008, 2010.

21. Table 3: Sanderson and Scherbov 2010.

22. Whitney 1997.

23. See Joseph Frankel, "Zoo Animal Little Mama, World's Oldest Known Chimpanzee, Dies at 79 Years Old," *Newsweek*, November 15, 2017, https://www.newsweek.com/zoo-animal-little-mama-worlds-oldest-known-chimpanzee-dies-79-years-old-712622

24. Wilmoth 1995.

25. See Matt Purdy, "The First Person to Live to 150 Has Already Been Born," *The World*, August 7, 2015, https://theworld.org/stories/2015/08/07/pushing-limits-human-lifespan, and John Nosta, "The First Person to Live to 150 Has Already Been Born—Revisited!" *Forbes*, February 3, 2013, https://www.forbes.com/sites/johnnosta/2013/02/03/the-first-person-to-live-to-150-has-already-been-born-revisited.

26. Since 1990, the Gerontology Research Group (GRG), founded in 1990 by Stephen Coles and Steven Kaye, has been publishing up-to-date lists of all supercentenarians whose vital information is verified through multiple sources, such as birth certificates, baptism records, and census records. The GRG offers a global list of verified supercentenarians organized by year of birth. Much effort goes into validating ages of birth and death. These lists are routinely updated and available to the public. Access is available at https://gerontology.wikia.org/wiki

/Category:Lists_of_supercentenarians_by_year_of_birth, last accessed June 24, 2021. As of that date, this tally includes 2,380 individuals born from 1800 until 1902; after 1902, we run into what demographers call right-censoring: an awkward way of saying that some supercentenarians on the list are still alive. Ignoring living cases could bias estimation of ages of death downward, and so I examined only complete cases through 1902.

27. The average age of death was 110.8 y. Overall, there was only a 0.35-year increase in the average age of death of supercentenarians born before 1800 to 1902; this increase was not a statistically significant trend. On average, women survived two and a half months more than men.

See also Gavrilova and Gavrilov 2020. They make a similar conclusion using the GRG dataset and also the International Database on Longevity. They show that further growth in longevity among supercentenarians leveled out for those born after 1879. The Gavrilovs also show that mortality rate accelerates after age 113, which is inconsistent with claims of decelerated mortality rates at the oldest ages. Because fewer than one per million reach age 110, statistics on mortality rates at such late ages are uncertain. Mortality rates per year may approach a plateau, but do not exceed 0.5 See Finch, Beltrán-Sánchez, and Crimmins 2014.

28. Garmany, Yamada, and Terzic 2021.

29. Fries 1980, 1989. As Richard Suzman, the director of the Division of Behavioral and Social Research at the National Institute on Aging, remarked, "The compression of morbidity was prophetic in the sense that Jim looked at the reduction of morbidity and disability at a time when most gerontologists and epidemiologists thought we would see a pandemic of disability." From Swartz 2008.

30. Olshansky, Carnes, and Grahn 1998.

31. Jagger et al. 2008.

32. Crimmins, Zhang, and Saito 2016.

33. James et al. 2018.

Chapter 2. Evolution and Aging

Epigraph source: Minois 1989, pp. 14–15

1. For a nice overview of different aging trajectories, see Jones et al. 2014.

2. Birks and Evans 2009; DeKosky et al. 2008.

3. Given the information given in Genesis 5, Adam is believed to have lived to 930 y! See also "Living to 1,000: The Man Who Says Science Will Soon Defeat Ageing," *Cambridge Independent*, June 17, 2018, https://www.cambridgeindependent.co.uk/business/living-to-1-000-the-man -who-says-science-will-soon-defeat-ageing-9050845/.

4. Wang et al. 2020.

5. From *De Rerum Natura* (On the Nature of Things, Book 3), as mentioned in Fabian and Flatt 2011:

"But if an older man, more advanced in years, in his misery should complain of it, wailing about death beyond all reason, would nature with more justice not call out and in a sharp voice chastise him. . . .

She would be right, in my view, to say this—right to rebuke and criticize the man. For old things, driven out by what is new, always yield, and one must renew one thing with something else. So no one is sent down into the abyss and black Tartarus. Material is needed for the growth of later generations—yet all of them, once their life is over, will follow you. Men have died before and will die again, just like you. Thus, one thing will never cease being born from something else. Life is given to no man as a permanent possession—instead all men receive it as a loan."

6. I joyfully digress to mention semelparity takes its name from Semele, the Greek princess of Thebes, lover to womanizing Zeus, who died after urging Zeus to show her the full extent of

his divine power. Mortals apparently can't witness the gods without spontaneously incinerating. Never to let one of his precious spawn go to waste, Zeus removed and sewed her unborn child into his thigh, giving birth months later to Dionysus, the "twice-born."

7. Weismann's idea of programmed death, though, presaged one of the most exciting discoveries of twentieth-century cell biology. In 1961, Leonard Hayflick and Paul Moorhead discovered that normal human cells (embryo fibroblasts) undergo routine cell division a finite number of times, and then age and die, a process called apoptosis. This countered the leading idea at the time, that cells could replicate indefinitely. The limit was about forty to sixty, in their experiments. Many others have since confirmed that this "Hayflick limit" to cell division exists in almost all somatic cell types, whether adult or fetal cells. See Hayflick and Moorhead 1961; Hayflick 1994.

8. Remolina and Hughes 2008.

9. Medawar 1952; Haldane 1942; Fisher 1930; Hamilton 1966.

10. Woody's mother, Nora Belle Guthrie, also died of Huntington's disease, after having spent many years stigmatized because of "bizarre" behavior (she was believed to have caused a house fire, and Woody's sister, Clara, died of burns after setting herself on fire in an argument with her mother), spending her final years in the Oklahoma Hospital for the Insane. Two of his daughters from his first marriage also succumbed to Huntington's disease in their early 40s. A real hero in this sad tale is Marjorie Guthrie, Woody's wife, whose herculean efforts around the world, and in advising government panels, helped spread awareness about Huntington's disease.

11. Ringman 2007.

12. Conrad et al. 2011.

13. The deceptive reporting of results may overstate the cases: a 30% higher risk of heart attack if you take calcium supplements sounds scary, but the absolute difference is 5.8% chance of a heart attack versus 5.5% if you took a placebo. Safe to say you should probably still make sure you get your calcium. See Bolland et al. 2010.

14. Cannata-Andia, Roman-Garcia, and Hruska 2011.

15. Austad and Hoffman 2018; Byars and Voskarides 2019.

16. For an excellent recent review of antagonistic pleiotropy in humans, see Byars and Voskarides 2020; Voskarides 2018; Byars et al. 2017.

17. Lanikova et al. 2017; Voskarides 2018.

18. Zhao et al. 2013.

19. Williams 1957.

20. Kirkwood and Rose 1991; Kirkwood 1977.

21. See Tomczyk et al. 2015; Jones et al. 2014.

22. Hamilton 1996, p. 90.

23. Finch 2009.

24. Kirkwood 1977, p. 301.

25. Vilchez, Saez, and Dillin 2014; Kirkwood 2011; Kaushik and Cuervo 2015.

26. López-Otín et al. 2013; López-Otín et al. 2016.

27. See "Opossums, Hydras and Hummingbirds: What We're Learning about Aging from Animals," NPR, *Invisibilia*, April 6, 2020, https://www.npr.org/sections/health-shots/2020/04/06/828350701/possums-hydras-and-humming-birds-what-were-learning-about-aging-from-animals.

28. Austad 1993.

29. Ricklefs 1998.

30. Stearns et al. 2000; Kirkwood and Austad 2000.

31. Borgerhoff Mulder et al. 2012.

32. Temple 1987.

33. Maybe you're thinking, but what about the bubonic plague?! Known as the Great Mortality, it wiped out at least a third of Europe between the fourteenth and eighteenth centuries, and

also wreaked havoc in China and North Africa. As a monk observer noted, "there was no dispar-ity in sex or age, taking men, women, the old, the young, plebs and nobles, paupers, the rich and powerful, priests and the laity." Sounds about as indiscriminate as you can get. Except we know that elite nobles fared better than commoners, and even among nobles, women and the elderly were hit harder. Quote and analysis of noble mortality from plague appear in Cummins 2017.

34. See Abrams 1993; Williams et al. 2006; Caswell 2007.

35. Reznick et al. 2004.

36. Medawar 1952. Quoted and assessment of aging in wild animal populations in Nussey et al. 2013.

37. Nussey et al. 2008; Nussey et al. 2013; Gaillard, Garratt, and Lemaître 2017.

38. Sherratt et al. 2011.

39. Walker et al. 2000; Jenkins, McColl, and Lithgow 2004; Van Voorhies, Fuchs, and Thomas 2005.

40. See Table 1 in Nussey et al. 2013, which summarizes information on the effects of four-teen mutations to extend life in genetically modified model organisms. In each case, there are costs borne earlier in life.

41. Hamilton and Mestler 1969.

42. Nieschlag, Nieschlag, and Behre 1993; Min, Lee, and Park 2012. Castration prior to pu-berty helps preserve a boy's voice as soprano, mezzo-soprano, or contralto. This, combined with adult lung capacity, gave a castrato's singing voice a unique quality and large range. Castration to alter vocal quality in males was not uncommon throughout Europe before the end of the eighteenth century.

43. Beyond mere trade-offs between reproduction and maintenance of our soma, one recent study even suggests that the effects of castration may be DNA-deep, at least in sheep (Sugrue et al. 2021). They report slower epigenetic aging in castrated sheep compared to intact males. Some of the sites affected by castration bind to androgen receptors, which regulate male hor-mone expression.

44. The original logic of having thirty-one flavors was so that ice cream lovers could have a different flavor each day of the month, https://news.baskinrobbins.com/about; about the three hundred theories of aging, see Medvedev 1990.

45. Carnes, Staats, and Sonntag 2008.

46. Franceschi et al. 2018, p. 61.

47. Burstein and Finch 2018.

48. Perlman 1954.

49. See Bulterijs et al. 2015.

50. Kennedy et al. 2014; Leng and Kennedy 2019.

51. Gompertz 1871.

52. Finch, Pike, and Whitten 1990; Sacher 1977.

53. Juckett and Rosenberg 1993; Sasson 2021.

54. Gompertz 1825. For a nice overview of Gompertz's contributions to biogerontology, see Olshansky and Carnes 1997; Kirkwood 2015.

55. Olshansky and Carnes 1997, p. 4.

56. For background on Pearl's historical efforts, see Olshansky and Carnes 1997. More on biological laws of mortality: Carnes, Olshansky, and Grahn 1996. For comparison of *Drosophila* and US male mortality profiles: Pearl and Parker 1922; Pearl and Miner 1935.

57. Heligman and Pollard 1980.

58. I feel strange kinship with Strehler, not just because like me he's a Pennsylvania native who later moved to California, but he also helped form the National Institute on Aging, which has generously funded my research for the past two decades. He also helped found the University of Southern California's Gerontology Center, a fount of many wonderful collaborators. (I also

dearly miss the fireflies of my childhood in Philadelphia. Strehler even discovered the light-emitting compound of the firefly!)

59. For example, I model mortality rates for female Swedish birth cohorts (from 1751–54 through 1925–29) using the Gompertz-Makeham equation described in the text, $h(x) = c + A \times e^{b \times x}$, where x is age, c is a constant risk independent of age, A is the initial mortality rate, and b is the slope parameter describing the rate of exponential mortality increase with age. Each data point reflects the estimates of these parameters from a single birth cohort. The scatter of data points fall almost exactly on a line, forming the negative association between A and b known as the Strehler-Mildvan correlation. The correlation is −0.96 (where −1 is the maximum possible).

60. Recent studies show deviations from the Strehler-Mildvan correlation, especially after the mid-twentieth century. Instead of supporting the notion that aging is "nearly constant regardless of the environment," favorable features of the material and ecological environment explain some variability in changing survival rates among different nations with age. There are different interpretations of this phenomenon, some more interesting than others. For example, in analyses tracking people born at the same time ("cohorts"), it could be that differences in some underlying fragility leads to the population becoming increasingly robust over time as the frail die out sooner. Those robust survivors may show a slower rate of mortality increase than if those frailer folks had survived and were now part of the mix at later ages. This, recall, was one of the explanations behind potential "negligible senescence" at late ages.

Background on these ideas: For the Strehler-Mildvan theory of aging and mortality, see Strehler and Mildvan 1960. For discussion of empirical support for the Strehler-Mildvan correlation, see Yashin et al. 2001, 2002; Li and Anderson 2015; Zheng, Yang, and Land 2011. For compensation law of mortality, see Gavrilov and Gavrilova 1991.

Chapter 3. Catching Up to the Present

1. For a review on the bioarchaeology of care, see Tilley 2022.

2. It may seem strange to define a group based on its mode of food production—e.g., hunter-gatherer, pastoralist, and the label "subsistence" population suggests that all food is obtained from the land and sea, and not purchased or traded in markets, and no mail delivery of ready-made meals. Many so-called subsistence populations do currently get some of their food from these other means, though the majority is still obtained through more traditional production activities. Subsistence need not mean that a group is fully responsible for acquiring every calorie consumed. Many subsistence populations are also interdependent with neighboring folks, trading valued commodities and such. Subsistence doesn't need to mean self-sufficient.

3. See Kaplan 2000; Sahlins 1968.

4. This includes the use of a stable age distribution based on one of the model life tables (West) of Coale and Demeny (see Coale and Demeny 1966). Model life tables are used when vital registration information is incomplete or unreliable. Based on actual data from groups of countries varying in mortality levels and shape of mortality curves with age, they allow you to make a best guess about age schedules of mortality, with just partial information (e.g., infant mortality rate). The "West" family of model life tables reflected high mortality countries that didn't quite fit into the other three families (North, South, East). They're based on real life tables mostly from European countries.

5. Wilmsen 1989. See also Solway and Lee 1990. The comments and response at the end of the article are especially entertaining and give a sense for how high the stakes can be when the stakes are so low.

6. Schrire 1984; Stiles 1992.

7. See Headland and Bailey 1991.

8. For a wonderful overview of the behavioral ecology of hunter-gatherers, see Kelly 2013.

9. Blurton Jones 1991.

10. *Naked and Afraid* is a US reality TV show, in which two naked survivalists (usually one woman and one man) must survive in the wilderness for three weeks. Each person is allowed to bring one useful item with them, often matches or a knife. The camera crew intervenes only when there is a medical emergency. Apparently this show has survived for seventeen seasons so far.

11. The late Lew Binford, an influential US archaeologist who wrote the world's largest and densest tome on hunter-gatherers that I doubt anyone, including his editor, have ever read, summed it up nicely: "It's obvious there are no pristine hunter-gatherers, but to say you cannot generalize in any way to the past because modern behavior is unique is, in essence, an attack on science."

12. See Henrich 2017.

13. Kelly 2013, p. 2.

14. For an up-to-date discussion about using contemporary hunter-gatherers to learn about past demography, see Page and French 2020.

15. The anthropologist Richard Lee and primatologist Irven Devore cofounded the Harvard Kalahari Project and were core organizers of the 1966 Man the Hunter conference in Chicago on hunter-gatherers. In the 1968 volume that came out of that conference, Lee and Devore wrote, "We cannot avoid the suspicion that many of us were led to live and work among hunters because of a feeling that the human condition was likely to be more clearly drawn here than among other kinds of societies" (68).

For popular depiction of urban "hunter-gatherers" living in the Great Basin area of the United States, see "Meet the Contemporary Hunter-Gatherers Who Live Off the Land in the American West," *Business Insider,* December 14, 2017, https://www.businessinsider.com/adrain-chessers -photos-of-hunter-gatherer-americans-2014-5, and "Going Native," *Salt Lake City Weekly,* August 9, 2001, https://www.cityweekly.net/utah/going-native/Content?oid=2129158.

16. Here are the key texts laying out the demography of the five foraging populations: Ju/'hoansi: Howell 1979, 2010. Hadza: Blurton Jones et al. 1992; Blurton Jones 2016; Dyson 1977. Ache: Hill and Hurtado 1996. Hiwi: Hill, Hurtado, and Walker 2007. Agta: Early and Headland 1998; Headland 1986.

17. Weiss 1973, p. 77.

18. Neel and Weiss 1975.

19. Early and Peters 2000.

20. Gurven, Kaplan, and Zelada Supa 2007.

21. Isendahl 2011; Watling et al. 2018.

22. Cohen and Armelagos 1984.

23. See Furuse, Suzuki, and Oshitani 2010; Babkin and Babkina 2015.

24. Rindos 2013.

25. Hawks et al. 2007; Cochran and Harpending 2009.

26. Thornton 2005.

27. US Center for Disease Control, https://www.cdc.gov/nchs/data/vsrr/vsrr015-508.pdf.

28. Chagnon 1968; Hart, Pilling, and Goodale 1988.

29. Hill and Hurtado applied regression analysis to estimate ages of Ache hunter-gatherers in their relative age list, after knowing some real ages, their relative age ranks, and some sense of the age gaps between individuals (see Hill and Hurtado 1996). Newer methods based on Bayesian analysis are nice improvements on Hill and Hurtado's method. These newer methods allow you to estimate the uncertainty in age gaps more directly, resulting in a posterior distribution of possible ages, rather than a single point-estimate (see Diekmann et al. 2017).

30. Horvath 2013; Levine et al. 2015; Chen et al. 2016. For a practical user-friendly guide to epigenetic clocks in this fast-moving subfield, see Ryan 2021.

31. Horvath et al. 2016.

32. For a description of Agta women's hunting, see Estioko-Griffin 1985. For an overview of some controversies about the sexual division of labor in hunter-gatherers, see Gurven and Hill 2009.

33. Lancaster Jones 1963, 1965.

34. Johnson 1981; Wood and Smouse 1982.

35. Pennington and Harpending 1993.

36. Pennington and Harpending 1991.

37. Mortality rate ages 0 to 15 (per 1,000 births) in 2019: Central African Republic (125 per 1,000), Lesotho (94 per 1,000), and Chad (138 per 1,000). Source World Health Organization Global Health Observatory website.

38. Volk and Atkinson 2013. This paper builds on an earlier sample compiled by Barry Hewlett (see Hewlett 1991).

39. Infant mortality rate was at least 11% of births in 1965, around when Scheper-Hughes first started working in Brazil. See Scheper-Hughes 1993.

40. Based on reproductive histories I conducted from 2002 to 2005. Of 356 women who had at least one birth, 207 (58%) had lost at least one child. Of the 109 women over age 40 who had given birth, 94 (86%) had lost at least one child. This is close to what you'd expect for a Tsimane woman who has given birth to nine babies (the average total fertility for the Tsimane): probability of at least one child dying $= 1-(1-p)^9 = 0.92$ where $p =$ probability that a child dies before age 15, in this case $p = 0.25$.

41. In addition to the groups already mentioned, I include the Aka of Central African Republic and the Bakairi of Brazil (Hewlett, van de Koppel, and van de Koppel 1986; Picchi 1994). For additional sources and more information on causes of mortality, see Gurven and Kaplan 2007.

42. For more about mortality and conditions of Tsimane infancy, see Gurven 2012.

43. In 2015, a young Tsimane woman in one of the more acculturated communities located near town gave birth to triplets. She ended up giving one away to a Borjana woman who was working in the local municipal government in San Borja. The woman and her family thought this was giving that child the best chance in life, given the difficulty of caring for three newborns at the same time. The young Tsimane woman is raising the other two children.

44. Bugos and McCarthy 1983.

45. On ritual child homicide among Ache, see Hill and Hurtado 1996.

46. Hill and Hurtado 1996, p. 436.

47. Howell 1979, p. 58.

48. Gurven and Kaplan 2007; Finch 1990.

49. Maternal mortality rate is defined as the number of maternal deaths per 100,000 live births. The World Health Organization qualifies a maternal death as "the death of a woman while pregnant or within 42 days of termination of pregnancy, irrespective of the duration and site of the pregnancy, from any cause related to or aggravated by the pregnancy or its management but not from accidental or incidental causes." Data from 2020 downloaded from World Bank website, https://data.worldbank.org/.

50. Sadly, Semmelweis was ridiculed and criticized for his steadfast belief that cleanliness could make all the difference. The idea that physicians were unclean, or that childbed fever didn't have multiple causes, involving an appropriate balance of the four humours in the body, was a radical one at the time. Unfortunately, Semmelweis was committed to an asylum and died after a severe beating by the guards.

51. See Keeley 1996; Gat 2008, and for an entertaining tour-de-force on how violence steeply declined with modernity, see Pinker 2012.

52. Source for national data on homicide rates: Global Burden of Disease Collaborative Network, Global Burden of Disease Study 2017 (GBD 2017). For US homicide rates, FBI Crime Data Explorer, https://cde.ucr.cjis.gov/LATEST/webapp/#/pages/explorer/crime/crime-trend.

53. Blurton Jones 2016, p. 144.

54. See also Wrangham, Wilson, and Muller 2006.

55. As each group's life table is based on relatively few deaths and risk-years (e.g., Ache life table is based on a total of 353 deaths out of 16,105 person-years at risk), I smoothed mortality hazards using a simple and tractable Siler model. This model allows us to use the available data to make the best estimate of what mortality patterns look like at all ages, including late ages. The Siler model includes three components of mortality: declining mortality from birth through childhood, a constant mortality hazard across the life span, and an increasing component in older ages. Infant and child mortality are thus modeled with a negative Gompertz function. The final component is the familiar Gompertz exponential described in chapter 2, and the second term is an age-independent constant term that has been referred to as the Makeham parameter. For the statistically inclined, the mortality hazard, $h(x)$, is defined as $h(\text{age}) = a_1 \exp(-b_1{}^*\text{age}) + a_2 + a_3 \exp(-b_3{}^*\text{age})$. The parameter a_1 describes the initial infant mortality rate, and b_1 describes the rate of mortality decline. The proportion of deaths due to juvenile mortality is captured by the first component as $\exp(-a_1/b_1)$. The parameter a_2 describes age-independent mortality, which is usually interpreted as exogenous mortality due to environmental conditions. The parameter a_3 is the initial adult mortality rate, and b_3 describes the rate of mortality increase. Mortality hazards can be computed from the survivorship curve (l_x), and then a technique called nonlinear regression can be used to estimate the five parameters of the model. With these, we can get smooth functions of survivorship, mortality hazards, and the age distribution of deaths. It's the latter that we analyze to look for modes and spread in the duration of adult lifespans.

56. For example, see Cohen 2004.

57. See Levitis, Burger, and Lackey 2013. Postreproductive representation, or PrR, can be readily calculated from the usual life table variables: $\text{PrR} = l_m/l_b {}^* e_m/e_b$, where l_m is survivorship to post-fertile period (m), e_m is remaining life expectancy at age m, age b is the age at onset of reproduction in adulthood. By considering the proportion of adult years spent postreproductive, PrR incorporates adult survival to age at menopause, and not just years remaining after menopause.

58. On chimpanzee menopause, Wood et al. 2023. On Ngogo chimpanzee survival, see Wood et al. 2017.

59. See Croft et al. 2015; Ellis et al. 2018.

60. Weiss 1975, p. 95; see also Neel and Weiss 1975.

61. Early 1990, p. 137.

62. These numbers come from the Extended UN Model Life Tables, https://www.un.org/en/development/desa/population/publications/mortality/model-life-tables.asp. These are one-parameter model life tables that require only a single mortality parameter to find a fit, as each age-specific probability of dying is linked like a chain to all others in the life table through statistical modeling (quadratic function). Their predecessors, the Coale-Demeny life tables, were based primarily on observed mortality patterns in different areas of Europe (e.g., northern vs. southern vs. eastern European countries). The UN life tables reflect a broader geographic range, including South Asia and Latin America. Each model life table is believed to reflect a combination of particular cause-of-death structure and cohort history. Ten models are included in their set (four Coale-Demeny and six UN model life tables).

63. The term "pygmy" has been arbitrarily ascribed to any human population where adult men average less than 5 feet tall.

64. To read the sad story about Ota Benga, and how he at least got to throw a chair at Florence Guggenheim's head while living at the American Museum of Natural History, see Bradford and Blume 1992.

65. As reported in Migliano, Vinicius, and Lahr 2007.

66. Eder 1987.

67. The average age of death may resemble the life expectancy at birth when fertility and mortality are relatively unchanged, and there is no population growth. Thus, the average age of death from eighty-three deaths during a tumultuous time in the 1970s cannot be reliably equated with the average life expectancy. See chapter 4 for more about how violated assumptions can lead to wild (and wrong) estimates of life expectancy.

68. Eder 1987, p. 115, Table 13.

69. For a lengthy description on why life expectancy and other life table information about Asian and African pygmy populations may not be reliable, see Becker et al. 2010.

70. Bailey 1985.

71. Fix 1977. From regional censuses, albeit with made-up ages, Fix reported that 10% of Semai men and 7% of women were over age 50. This is at odds with what he found using model life tables (taking Semai infant mortality rate of 24% and child survival to age 15 of 65%, and fertility as key indicators): 30% of the "stable" population should be over 50. Most likely Semai demographic rates were changing rapidly.

72. Stefansson 1960, pp. 124–128.

73. van den Eerenbeemt 1985.

74. Brainard 1986.

75. If an infant has an average chance of dying of 20%, then in statistical parlance, we could say that observations of infant mortality rates should follow what's called a binomial distribution. We can calculate the variance of this distribution, in this case $0.2*0.8/N$ where N is the number of births. The more births in the sample, the smaller the variance and the tighter the confidence intervals will be around our estimate of 20%.

76. Headland 1990.

77. See Black 1975.

78. Porter and Marlowe 2007; Cunningham et al. 2019.

79. The method for exploring the effects of eliminating specific causes of mortality on the life table is called a "decrement lifetable." A good primer can be found in the demographic bible (Preston, Heuveline, and Guillot 2001). For an introduction for the more biologically oriented, see Carey 1989.

80. It is possible, however, that I have overestimated the effects of acculturation on !Kung survivorship because of gaps in the prospective life table created by Howell (1979, Table 4.6). Howell (1979) shows that the estimated life expectancy of acculturated !Kung based on that life table is about 50 years, which is ten years higher than the national estimate of Botswana during the same time period.

81. Lower child mortality with the shift to more sedentary living: Ghanzi !Kung, see Harpending and Wandsnider 1982; Kutchin Athapaskans, see Roth 1981; Turkana pastoralists of northwest Kenya, see Brainard 1981; and Nunamiut Iñupiat of Alaska, see Binford and Chasko 1976. Among Adavasi Juang farmers of India, sedentism led to a slight increase in mortality (see Roth and Ray 1985).

Chapter 4. The Long Road to Longevity

1. Malthus 2013, ch. 9, para. 7, lines 1–4.

2. Clark 2007, p. 96.

3. The view that lifespan was limited pervades older works based on the skeletal record, and was then popularized as truth (e.g., Vallois 1961; Acsádi, Nemeskéri, and Balás 1970; Weiss 1973, 1981; Washburn 1981). Even when acknowledging the occasional older adult, the general perspective portrayed is that there were very few older postreproductive-age adults alive at any one time (e.g., Kennedy 2003).

4. Weiss 1973, p. 78.

5. Halley did assume stationarity and a closed population (i.e., no migration). Amusingly, Halley had no experience with demography, but as an editor of the *Philosophical Transactions of the Royal Society* in London, he wanted more quality publications, and by writing the paper himself, he helped ensure the quality. See Ciecka 2008. Also, it's hard not to appreciate the run-on titles of scientific papers way back when: Halley's 1693 publication was called "An Estimate of the Degrees of Mortality of Mankind, Drawn from the Curious Tables of the Births and Funerals at the City of Breslaw, with an Attempt to Ascertain the Price of Annuities upon Lives."

6. Due to some confusion over terminology by Raymond Pearl when he first calculated life expectancy from the Halley table, life expectancy at birth has been incorrectly described as 33.5.

7. For additional commentary on some limitations of the Halley life table, and a figure of the age distribution of deaths from the source data used by Halley, see Bellhouse 2011. In a postscript to his 1693 publication, Halley opined about the "shortness of our lives, and think ourselves wronged if we attain not old age," as only half survived to age 17. Nonetheless, he argued that such a winnowing early in life made survival into older age that much more reward-ing. He advised against us "murmur[ing] about what we call an untimely death," and to consider ourselves lucky if we "have survived, perhaps by many year, that period of life, whereat the one half of the whole race of mankind does not arrive."

8. Wrigley and Schofield 1989.

9. Wrigley et al. 1997.

10. Hollingsworth 1977.

11. Russell 1948.

12. The statistician William Guy in 1845 also calculated life expectancies at different ages among English aristocrats between 1200 and 1750, based on assembled genealogies. He also excluded those who died prematurely from violence, poisoning, or accidents. Though some of his methods were not standard, he showed $e_{21} > 40$ years throughout the five centuries (except during plague).

13. Harvey 1993; Hatcher, Piper, and Stone 2006; Hatcher 1986.

14. DeWitte, Boulware, and Redfern 2013.

15. As mentioned earlier, baptizing the dead into the faith is a Mormon Church doctrine started by founder Joseph Smith. Not only is it a way to ensure immortal afterlife with your saved loved ones (and ancestors you never knew about), but, lucky for us, the effort to do so motivated a massive effort to collect genealogies worldwide. Billions of family histories are stored in a protected vault in the Wasatch mountains in southeast Utah, capable of withstanding a nuclear blast.

16. Cummins 2017.

17. Given that the maximum age reported in Table 3 of Cummins 2017 is 123 years, it is likely that some ages are off.

18. Zhao 1997.

19. Macdonell 1913.

20. Pearson 1902.

21. Another more comprehensive study of Egypt, at the height of the Roman Empire from the first through third centuries, uses papyri censuses and model life tables to show comparable patterns of survivorship. See Bagnall and Frier 1994.

22. Note: the criteria for being famous didn't change over the sample period. YouTube stars and other "famous for being famous" folks are not included. Study on longevity of 300,000 celebrities, see De la Croix and Licandro 2015. For an earlier treatment on the topic, which first recognized the role of "differential selection factors" in generating such lists of celebrities, whereby eminence is achieved at different ages in different professions, see Lehman 1943.

23. Stelter, De la Croix, and Myrskylä 2021.

24. These are the same problematic assumptions we saw in chapter 3 when discussing the life tables assembled for anthropological populations by Ken Weiss.

25. Bocquet-Appel and Masset 1982, 1996; Konigsberg and Frankenberg 1994. For a more updated review tailored to the study of human longevity, see Konigsberg and Herrmann 2006.

26. Howell 1976, p. 25.

27. Gage 1998.

28. Lovejoy et al. 1977; Meindl, Mensforth, and Lovejoy 2008.

29. Walker, Johnson, and Lambert 1988.

30. Adding to this, traditional age estimates are heavily biased by relying on the age distribution of particular "reference samples," a problem referred to as "age mimicry."

31. Meindl, Mensforth, and Lovejoy 2008.

32. Transition analysis: Boldsen et al. 2002. For a recent update on the use of transition analysis for aging skeletons, see Getz 2020.

33. One such newer method is called tooth cementum annulation counting, which calibrates age estimation to the incremental lines of cementum on the teeth. Preliminary correlations with known ages in a few archaeological samples seemed promising, though the method doesn't seem to have caught on. One reason may be because it's somewhat invasive to the bone. Another might be that cementum deposition may be affected by environmental factors unrelated to age. For the latest on this "cementochronology," see Bertrand et al. 2019. Skeletons and wrist watches quoted in Konigsberg and Herrmann 2006, p. 297.

34. Bullock et al. 2013.

35. These newer methods don't have a cute short name, but have been referred to as "proportion-odds probit regression with age on the natural log scale," employing maximum likelihood methods and hazards analysis. For a pitch for these newer methods, and summary of their application to reevaluating Loisy-en-Brie, Averbuch, and Indian Knoll, see Konigsberg and Herrmann 2006.

36. Sasaki and Kondo 2016.

37. The first systematic treatment relating ages-at-death in skeletal samples to fertility was by Sattenspiel and Harpending 1983. Adjustments to life expectancy estimates that allow for nonstationary population growth are given in Horowitz, Armelagos, and Wachter 1988.

38. Consistent with the fertility change exercise discussed in the text, it turns out that mean age at death is instead related to the birth rate (to be more exact, it's equal to the reciprocal, or 1/birth rate). When a population is stationary, or not growing or shrinking, e_0 equals the mean age of death. As mentioned in chapter 3, the assumption of stationarity is a bold one—it makes the guesswork a lot easier, but that doesn't make the guesswork right. When a population is instead growing, the inferred life expectancy from mean age at death will be too low.

39. See Meindl, Mensforth, and Lovejoy 2008.

40. The AnAge database was downloaded in June 2021 from https://genomics.senescence .info/. For information about the database, see De Magalhaes and Costa 2009. Using these data, I performed a linear regression of maximum lifespan (logged) as a function of group (monkeys vs. apes vs. humans) relative to nonprimate mammals. The model adjusts for body size and study sample size. Even after adjusting for body size and how well species are studied, humans are longer-lived than apes, apes are longer-lived than monkeys, and monkeys longer-lived than other mammals. See Appendix table A4.2 for regression results.

41. Neanderthal genes seem to make up roughly 2% of DNA of modern-day folks of Eurasian descent. Among other things (like effects on skin development, perhaps in lower ultraviolet light regions), Neanderthal genes in modern humans seem to affect gene expression more than the coding of proteins. See Silvert, Quintana-Murci, and Rotival 2019.

42. Hoffmann et al. 2018.

43. Trinkaus 1995.

44. Trinkaus 2011.

45. Caspari and Lee 2004. For critiques about interpretation of older to younger (OY) ratios, see Hawkes and O'Connell 2005.

46. Sacher 1975.

47. Source data were compiled from Will et al. 2021; Hammer and Foley 1996; Judge and Carey 2000. *Ardipithecus ramidus* data come from Suwa et al. 2009. *A. anamensis* data come from Haile-Selassie et al. 2019. I estimated missing values of *Homo* brain or body size from regression equations relating known values from the Will et al. dataset: log(body mass, kg) = 0.622 + 0.390*log(brain size, kg). To estimate maximum lifespan, I used regression equations from both Hammer and Foley, and Judge and Carey, relating brain and body size to maximum lifespan, and present both the lowest and highest estimates to give a broader range.

48. O'Connell, Hawkes, and Blurton Jones 1999.

49. Hublin et al. 2017.

50. Primate information comes from Colchero et al. 2016. Table S1. These include sifakas, muriquis, white-faced capuchins, yellow baboons, chimpanzees, gorillas. I added Japanese macaques from McDonald Pavelka and Fedigan 1999. For these seven primates, maximum lifespan is taken to be the age when adult life expectancy is < 5%. The curve fits in the figure are lowess smooths.

51. Vallois 1961, p. 222.

52. Williams 1957, p. 407.

53. See "Faces from the Ice Age," *BBC News,* May 28, 2002, http://news.bbc.co.uk/2/hi/science/nature/2012385.stm. For a recent analysis of the figure depictions, see Chisena and Delage 2018. The elaborate etchings of people were on the floor, not the walls. As cave floors are often destroyed during excavation, it is unclear whether other caves might have had similar lifelike etchings. In any case, the La Marche discoverers did not find similar etchings at other sites, plus other detailed etchings have been found more recently by other teams both at La Marche and in other caves dated to the same Magdalenian time period ~15,000 years ago, such as the Réseau Guy-Martin cave. See Delage 2016. Even if some of the depictions were indeed fakes, at least the orchestrator(s) seemed to think older adults lived in the Paleolithic!

54. For a recent, thorough treatment of the problems associated with inferring the number of older adults in the "bone demography," see Blurton Jones 2016, Supplementary Information 8.7, https://www.cambridge.org/files/4314/4767/5780/SI_for_ch8__Mortality.pdf.

55. Blurton Jones 2016, p. 17.

56. Blurton Jones 2016, p. 22.

57. Goodall 1986; Finch and Stanford 2004, p. 4.

58. Under exponential growth a population at time t, $N_t = N_0 \exp(rt)$, where r is the annual population growth rate. A population doubles when $N_t = 2*N_0$, which occurs when $t = \ln(2)/r$.

59. See "Fact or Fiction? Living People Outnumber the Dead," *Scientific American*, March 1, 2007, https://www.scientificamerican.com/article/fact-or-fiction-living-outnumber-dead/.

60. Walker, Sattenspiel, and Hill 2015.

61. Gurven and Davison 2019. See also, for earlier treatments: Boone 2002; Hill and Hurtado 1996; Blurton Jones 2016.

62. Richerson, Bettinger, and Boyd 2005.

63. Aburto et al. 2021.

64. Even under brief bursts of high mortality, e_0 might drop in ways that could affect the bone profile recovered millennia later. A catastrophe like the 1918 influenza pandemic, where death from organ failure and fatal pneumonia disproportionately affected adults ages 15–44, might result in an age-at-death distribution not unlike the classic Libben pattern, if victims are more likely to be uncovered at the same site. Age profiles found even under the best of conditions may not represent the age profiles of living populations, especially over longer periods of

time. As explained in chapter 3, adult lifespan potential in the living population doesn't change much, even in these catastrophic cases.

65. The effects of the measles epidemic were brutal: "All the windows were wide, because the patients could not stand the light, and when they entered the houses, they could hear nothing but moans of illness in the twilight." Quote from: Sandra Gunnarsdottir, Haraldur Briem, and Magnus Gottfredsson, "Extent and Impact of the Measles Epidemics of 1846 and 1882 in Iceland." *Laeknabladid,* 100 (2014). Life table data from Iceland 1881–1883 used to ascertain life expectancies downloaded from Human Mortality Database.

66. The big differences in e_0 stem from reliance on period (cross-sectional) rather than cohort life tables. Only cohort life tables describe the experiences of all people born in the same year. Period life tables will reflect the higher mortality from the measles epidemic by people of all ages in that same year, whereas cohort lifetables show members of a cohort each affected by only one year (e.g., those age 5 in 1882 are represented in the 1877 cohort, those age 20 in 1882 in the 1862 cohort, etc.). Indeed, whereas period e_{20} for 1881, 1882, and 1883 (representing one year before the epidemic, year of the epidemic and one year after) was 40, 29, and 36 y, the equivalent cohort e_{20} for those who were 20 during the same years (i.e., those born in 1861, 1862, and 1863), was very similar: 43, 45, and 46. Data source: Human Mortality Database.

67. Kennedy 2003.

Chapter 5. Why Long Lives?

1. See "Extended Adolescence: When 25 Is the New 18," *SciAm,* September 19, 2017, https:// www.scientificamerican.com/article/extended-adolescence-when-25-is-the-new-181/. This idea about extended adolescence in recent decades was promoted in the popular book in the 1980s (Littwin 1986). For a more recent and global perspective, with the proposal to increase the age range of adolescence, see Sawyer et al. 2018.

2. The cost that popped up on the cash register was $847.63, estimated to be the monthly cost of raising a child in the US in 1989.

3. Gurven and Walker 2006.

4. Kaplan et al. 2000, from Table 3.

5. See Cordain et al. 2000.

6. Butter, churned usually from the cream of cow's milk, is mostly fat, though it does have some vitamins (A, D, E, K2), and other nutritious elements (e.g., butyrate). A kilogram of butter, on average, would contain about 845 g of fat, over half of which is saturated fat. That's more than thirty-five times the daily recommended amount for a healthy diet!

7. On tree-climbing by human hunter-gatherers, see Kraft, Venkataraman, and Dominy 2014.

8. On honey and human evolution, see Marlowe et al. 2014. On Hadza and reliance on honeyguides, see Spottiswoode and Wood 2023.

9. Wood et al. 2021.

10. Hill, Barton, and Hurtado 2009; Gurven, Stieglitz, et al. 2012.

11. For overview of meat hunger in tribal societies, see Simoons 1994. For Maori quote, see Firth 1929, p. 276.

12. Men alone hunt in the majority of hunter-gatherer societies, while men and women both regularly hunt only in a small number of societies. In no society ever reported do only women hunt. Women are more likely to hunt small game, often with dogs, near camp and in groups. See Gurven and Hill 2009; Hoffman, Farquharson, and Venkataraman 2023.

13. Chimpanzees, on the other hand, don't willingly share as much, mostly because there's little need to. Sure, there's feeding competition and dominance hierarchies that affect how food is circulated, but everyone more or less can usually access the foods they like to eat.

14. For an exploration of the bridges between biological kinship and human social structure, see Chapais 2009.

15. Horstman and Kurtz 1979.

16. Fortier 2001.

17. Bertoni 1941, p. 39.

18. Gurven and Kaplan 2008.

19. Kaplan et al. 2012.

20. Ache and Hiwi food production and consumption profiles by age come from Kaplan et al. 2000. Ju/'hoansi data come from Howell 2010. Machiguenga and Piro data come from Gurven and Kaplan 2006. Tsimane data come from Gurven, Stieglitz, et al. 2012.

21. For more recent reviews of age and productivity in a variety of fields, see Simonton 1997 and Skirbekk 2004. For review on ages of peak performance in elite athletes, see Allen and Hopkins 2015. A model and data describing how productivity and time vary with age in work activities among subsistence populations is described in Gurven and Kaplan 2006.

22. See Kaplan et al. 2000; Lee 2003.

23. Rilling 2006.

24. Bering and Povinelli 2003.

25. On cerebellum evolution and "technical intelligence," see Barton and Venditti 2014. On connectivity in the human brain, see Ardesch et al. 2019; Wei et al. 2019.

26. On cognitive niche, see Tooby and DeVore 1987. On collective intelligence, see Muthukrishna et al. 2018. On "secret to our success," see Henrich 2017.

27. Gurven, Kaplan, and Gutierrez 2006; Koster et al. 2020.

28. The animal tracker, and researcher of !Xò trackers of the Kalahari, Louis Liebenberg, once proposed the fascinating idea that the art of tracking animals itself is like creative protoscience—wherein hunters collect information on spoor and other signs of prey behavior, then make and test hypotheses about prey movement and motivations. In attempting to out-think their prey, hunter-gatherers demonstrate full expression of our evolved "cognitive niche." See Liebenberg 1990.

29. Meehan 1982; Bock 2002.

30. Hawkes, O'Connell, and Blurton Jones 1997; Hawkes 2003.

31. In addition, only one of the observed Hadza women was actually a grandmother. The rest of the older women studied were aunts or more distant relatives. The larger point though is that older adults can boost the fitness of younger kin.

32. Marshall 1976, p. 97

33. Hawkes, O'Connell, and Blurton Jones 1997; Blurton Jones 2016, p. 367.

34. Hill and Hurtado 1996, p. 424.

35. Sear, Mace, and McGregor 2000, 2003.

36. Jamison et al. 2002.

37. Voland and Beise 2002.

38. Peccei 2001.

39. Muller et al. 2020.

40. Tuljapurkar, Puleston, and Gurven 2007.

41. Gurven, Winking, et al. 2009.

42. On arranged marriage in hunter-gatherers, and consideration of ancient marriage practices, see Walker et al. 2011.

43. Feng and Ren 2022; Carmichael 2011.

44. An elegant approach to understanding why we age, but without assuming any aging in production or mortality, combines elements of embodied capital theory with disposable soma theory. This approach shows that the optimal level of investment needed to repair the soma and

reduce mortality faces important trade-offs, in light of the costs of maintaining both the quantity and quality of our somatic cells. See Kaplan and Robson 2009. Their model also explains why mortality declines early in life: as growth imbues an organism with increasing quality, its expected future value increases, thereby motivating an optimal reduction in mortality with age in early life.

45. Hrdy 2005a; Kramer 2005.

46. Kramer and Ellison 2010.

47. See Sear 2016.

48. Zhu et al. 2023.

49. Ruby, Smith, and Buffenstein 2018; Buffenstein 2008.

50. Sear and Mace 2008.

51. Strassmann and Garrard 2011.

52. Meehan, Helfrecht, and Quinlan 2014.

53. Coall and Hertwig 2010. Many of the groups in which there is no effect of paternal grandparents, but a positive effect of maternal grandparents, tend to live patrilocally, where a couple resides in the village of the man's family, or are patrilineal, where inheritance of land and livestock goes to sons. Anthropologist and myth debunker Beverly Strassman calls this investment in a woman's children in patrilineal systems "covert matriliny." See Strassmann and Garrard 2011. My favorite Strassman myth debunking is her take-down of "menstrual synchrony," wherein groups of women, often co-resident like in a dorm, synchronize their menstrual cycles, aligned perhaps through pheromones. She does this based on first principles, methods, and through failure to replicate findings in natural fertility settings, including among the Dogon of Mali, where she has worked extensively.

54. Mulder 2007.

55. Local needs will also vary, and sometimes help is given to those who need it the most. In a nationally representative dataset, anthropologist (and academic half-sib) Kristin Snopkowski found that Indonesian grandparents provided the most financial support to their poorer adult children, and more domestic household support to their daughters who worked outside the home. See Snopkowski, Moya, and Sear 2014.

Knowing this to be the case, studies linking caretaker help to a variety of outcomes may lead to misleading conclusions, the same way that the best surgeons sometimes have the highest patient death rates. Reputable surgeons often receive the most at-risk patients, and so while skill and experience help save patient lives, it may be misleading to compare patient survival rates among surgeons who receive cases varying in complexity, risk, and other characteristics. See Hartz, Kuhn, and Pulido 1999.

This is another case in which what seems so obvious can sometimes be hard to show empirically. A nice research design would follow a random shock that led to changes in available caretakers, either through death or migration. One could see how changes in caretaker availability and composition led to any number of relevant health outcomes among other family members, as well as changes in who helps and how much. A careful analysis needs to examine the effects of multiple adjustments and labor substitution on the welfare of multiple people, not just a single target child within a family.

56. Davison and Gurven 2022.

57. Allman et al. 1998.

58. Pavard and Coste 2021.

59. The most comprehensive test to date is not among hunter-gatherers, but among Utah Mormon pioneers during the mid-to-late nineteenth century that includes a huge number of people across four generations. Its approach borrows from the field of quantitative genetics and sought to estimate different pathways by which fitness is affected by longer female and male lifespan. Overall, it found little support for grandmother, mother or patriarch hypotheses, or for antagonistic pleiotropy. It didn't test the embodied capital model. Before dismissing any model,

though, it's possible that grandparents weren't traveling together with their families during the disruptive pioneer period of Mormon history, and so maybe opportunities for care were limited. Survival and fertility were also both high during this time as well. See Moorad and Walling 2017.

60. Schuppli, Isler, and Van Schaik 2012.

61. Schuppli et al. 2016.

62. A marine mammal may be the exemplar human. Killer whales (*Orcinus orca*), you'll recall, have prominent postreproductive lifespans (see figure 3.17). I was lucky to see groups of resident orcas near the San Juan Islands a few years ago and talk to a few researchers setting up fancy aerial camera drones. Seeing a pod of black-and-white foraging orcas gliding together in the Pacific Ocean was awe-inspiring. Local enthusiasts can identify every member of the J, K, and L pods, and communicate by walkie-talkie whenever there are shore sightings. Orca modal adult lifespans are similar to those of humans, about six decades. Consistent with the mother and grandmother hypotheses, postreproductive females lead their offspring and sometimes their grandchildren or younger siblings in collective foraging bouts, especially during times when their main food source, salmon, is hard to find. Living in tight family units, the matriarch's expertise both helps provide food to juveniles and indirectly helps teach them important skills when the food quest is tough. Older females tend to lead sons more than daughters, and their aid seems to help sons more as well. While resident orcas are known to stubbornly rely on Chinook salmon, orcas residing more offshore have a more varied diet. Lastly, orcas are apex ocean predators, a title even humans don't hold. In early 2022, a group of orcas were observed collectively hunting the largest animal in the world, a 70-foot blue whale. They've been observed attacking blue whales before, but never adults, and never successfully in an elaborate coordinated way, wearing out the beast bite by terrifying bite until a few jumped on top to drown it. Ironically, the blue whale is also long-lived. It's no wonder that even great white sharks tip-fin around a pod of orcas. See Brent et al. 2015; Totterdell et al. 2022. To witness a video of orcas hunting blue whale, see https://www.theguardian.com/environment /video/2022/jan/27/orcas-recorded-killing-and-feeding-on-blue-whales-in-2019-video.

63. Heldstab et al. 2019.

64. See Reed 1997; Potts 1998; Antón, Potts, and Aiello 2014.

65. From Kraft et al. 2021: For a 50 kg human moving about 14 km, the daily distance traveled by Hadza and Ache, but with the locomotor economy of a bipedal chimpanzee (1.06 kcal kg^{-1} km^{-1}), daily travel costs alone would be ~750 kcal/d, which would be almost a third of total daily energetic expenditure. That's nearly three times the cost of traveling that distance with a standard human locomotor economy (~270 kcal/day; 0.39 kcal kg^{-1} km^{-1}).

66. Kraft et al. 2021: Caloric return rates for both sexes combined: 222 kcals/hr (chimpanzees), 337 (gorillas), 193 (orangutans).

67. See Wrangham 2009.

68. As food technology expert Eric Schulze aptly summarized Maillard reactions: "With the right amount of heat, moisture, and time, those specific sugars and proteins will act like a couple of lust-drunk lovers making out in the back of a Chevy, rapidly becoming a tangled, hot mess, until, nine months later, a whole new creation emerges. Except that with the proteins and sugars, it takes minutes, not months, and instead of a child, the result is an increasingly complex array of flavor and aroma molecules," from "An Introduction to the Maillard Reaction: The Science of Browning, Aroma, and Flavor," *Serious Eats*, May 13, 2023, https://www.seriouseats.com/what -is-maillard-reaction-cooking-science.

69. Thompson et al. 2019. A fascinating re-analysis of older materials provides convincing evidence of deliberate hunting among other Pleistocene-era hominin ancestors, like *Homo erectus*: Domínguez-Rodrigo et al. 2021.

70. Marlowe 2010; Wood and Marlowe 2013; Hawkes et al. 1991.

71. Byers and Ugan 2005.

72. Stiner 2002; Stiner et al. 1999.

73. See Ungar 2017 for a fun overview of what we can learn from teeth about human evolution.

74. Evans et al. 2016.

75. See Larsen 2003.

Chapter 6. To Be of Use

Epigraph 2: Biesele 1993, p. 20

1. Gurven, Winking, et al. 2009. This insane 2009 paper of ours is way too long. It should've been broken down into three papers. What was I thinking? But the Appendix shows a nice summary table of time budgets of Tsimane women and men.

2. The study compared time spent in various work activities among Efe hunter-gatherers, Yekwana and Yukpa horticulturalists, Machiguenga and Mekranoti forager-farmers, and Kipsigis herders to six OECD countries. See Bhui, Chudek, and Henrich 2019 for details. For an earlier treatment of the relationship between labor time and industrialization, see Minge-Klevana et al. 1980.

3. Winterhalder et al. 1988, p. 323.

4. Apparently the 1934 German volume, *Die Behandlung der Alten und Kranken bei den Naturvolkern* (Treatment of the Old and the Sick in Primitive Peoples) by John Koty was not on anyone's bestseller list.

5. Simmons was aware of some of these problems, and even humbly points them out: "Although considerable care has been exercised in weighing the evidence of different authorities, the extent to which scientific objectivity is jeopardized by imperfect information, conflicting reports, lack of preciseness in the sources, oversimplified and perhaps arbitrary classifications, and inescapable subjective judgments is, at this point, woefully apparent" (p. 13).

6. Simmons and Wolff 1954.

7. His calculation of "correlations" also differs from conventional practice. Where I report correlations between cultural traits, I use Spearman rank-order correlations. I also combine his codes for a trait's absence ("0" and "−") versus its presence ("+" and "++").

8. Cumming and Henry 1961.

9. Davison and Gurven 2022.

10. Lee 1979; Silberbauer 1981.

11. Marlowe 2010.

12. Maxwell and Silverman 1970, see Table 1.

13. Simmons 1945, p. 135.

14. Nason 1981.

15. Sharp 1981, p. 107.

16. Marshall Thomas 1959.

17. As primatologist Sarah Hrdy pointed out, despite C.W.M. Hart's important ethnographic contributions, he shamelessly ignored older women and viewed them contemptuously: "a terrible nuisance," "physically quite revolting," and "I rather enjoyed being rude to her." Given his biases, it's not surprising that he complained about being left behind in camp with elderly women and children, believing that he was missing out on all the action: Hrdy 2005b.

18. Hart, Pilling, and Goodale 1988.

19. For a good overview, see Scalise Sugiyama 2017.

20. Biesele 1993, pp. 18–20.

21. Wiessner 2014.

22. This story comes from Huanca 2008. That book is a wonderful overview of Tsimane folklore, interpreted by Tomás Huanca, a Bolivian anthropologist who has worked with the Tsimane even longer than I have.

23. For a fun book-length treatment on how stories can improve group cohesion, see Boyd 2010.

24. Smith et al. 2017, p. 2.

25. From Daniel Smith (personal communication): median age of skilled Agta storytellers was 38 (women) and 48 (men); median age of unskilled storytellers was 25.5 (women) and 34 (men).

Anecdotally, older adults are often described as the best storytellers. Among the Asmat of Papua, New Guinea, old men would have large audiences of boys of all ages with rapt attention in the men's longhouse, as they told many tales about animals who helped the ancestors gain mastery in making a living in the jungle and in managing natural resources (Vanarsdale 2021). Among Coast Salish of the Pacific Northwest, even among the old, not all are great storytellers: "Every old person probably knew all the myths, not everyone was an equally skilled raconteur" (Amoss 1981, p. 232).

26. Schniter et al. 2018.

27. For good overviews of teaching and pedagogy in hunter-gatherers, see Terashima and Hewlett 2016.

28. Schniter, Kaplan, and Gurven 2023.

29. Reyes-García et al. 2009.

30. Amoss 1981, pp. 232–233.

31. Turnbull 1983, p. 55.

32. See "As Biden and Trump Seek Reelection, Who Are the Oldest—and Youngest—World Leaders?" Pew Research Center, May 1, 2024, https://www.pewresearch.org/short-reads/2024/05/01/as-biden-and-trump-seek-reelection-who-are-the-oldest-and-youngest-current-world-leaders/.

33. Garfield, Syme, and Hagen 2020.

34. Endicott 1988, p. 123.

35. Among Yaghan hunter-gatherers of Tierra del Fuego, the Austrian priest/ethnographer Martin Gusinde insists that "certain men, as a result of their advanced age and blameless character, their long experience and mental superiority, have a moral influence of such importance that it amounts to positive control. . . . Members of a wider circle of kinsfolk listen to the word, submit to his instructions, bow to his judgment, and even accept his warning. Like a sincere friend . . . and like a venerable patriarch in a larger circle, he guides and leads by means of good words. . . . He is a conscientious promoter of the old order and a steadfast champion of the good customs of former times. . . . He makes his opinion known with careful consideration and with the best intentions for the welfare of the individual as well as the harmony of the group as a whole" (Gusinde 1937, p. 635). The same Martin Gusinde some years earlier said the same for the Yaghan's neighbors, the Selk'nam, who were earlier encountered by a young 23-year-old Charles Darwin when touring on the Beagle: "Everyone submits when [an elder] gives good advice, when he settles disputes, when [he] mediates between two hostile groups. . . . One follows his judgment in moving the camp, in communal hunts, in calling many together for competitive events or for celebration" (from eHRAF, Gusinde 1931, p. 599).

36. Brown 1922.

37. Quote from Paul Schebesta cited in Turnbull 1965.

38. Huntingford 1954, p. 126.

39. From the Simmons 1945 sample, correlations are found between groups having permanent settlement, and chiefs ($r = 0.41$, $p < .001$, $n = 69$) or councils ($r = 0.25$, $p = 0.058$, $n = 58$).

40. Amoss 1981, p. 233.

41. For a classic ethnography on all the ramifications of a gerontocratic system, see Spencer 1965.

42. Mead 1928, pp. 36–37.

43. Murphy and Murphy 1974, p. 202.

44. Gurven, Stieglitz, et al. 2012.

45. Biesele and Howell 1981, pp. 84–85.

46. Schebesta and Schütze 1954, p. 205.

47. Here are just two more interesting examples:

(1) During the period when native lifeways were being radically transformed by the intro-duction of horses in the southern plains of the United States, adult Comanche men were expected to be aggressive warriors. But the elder men were supposed to be "wise.... It was his task to work for the welfare of the tribe, giving sound advice, smoothing down quarrels, and even preventing his tribe from making new enemies" (Hoebel and Wallace 1952, pp. 146–147).

(2) Among island horticulturalists like those living on Etal Island, land ownership is key. Older folks know everything about the long history of how each piece of land came to be owned, and the history of fights over land. As you might imagine in confined areas of limited territory, land disputes are frequent and can be intense. In a system where "precedent and a detailed knowledge of the past are vital to dispute settlement," the greater knowledge and life-long experience of elders is critical (Nason 1981).

48. For a review of religious concepts in hunter-gatherers, see Peoples, Duda, and Marlowe 2016. Shamans were also present in almost all of the societies studied by Simmons.

49. Winkelman 1990; Singh 2018.

50. Simmons 1945.

51. Rasmussen quoted in Simmons.

52. Heinen 1972.

53. Simmons 1945, p. 160.

54. Vanarsdale 2021, p. 116.

55. Fernández-Llamazares, Díaz-Reviriego, and Reyes-García 2017.

56. For a nice overview of the cultural importance of cocojsi, see Huanca 2008, ch. 7.

57. Ohnuki-Tierney 1974.

58. Rasmussen 1908, p. 124.

59. Gusinde 1937, p. 910.

60. A recent study showed that even a modest increase in midwife services in low- and middle-income countries could prevent about a quarter of maternal and infant deaths (Nove et al. 2021). Though these high survival gains require that midwives are qualified according to International Confederation of Midwives standards, it is likely that experienced older women have improved many lives. In the US, doulas and other trusted labor support improve a wide range of birth outcomes beyond just survival. They also increase satisfaction of mothers' birth experience. See Gruber, Cupito, and Dobson 2013 and citations within Burgess 2014.

61. Biesele and Howell 1981, p. 89.

62. Amoss 1981, p. 231.

63. Simmons 1945, p. 129.

64. Wiessner 1982; Smith et al. 2010.

65. Simmons 1945, p. 57.

66. For an overview of how status was measured in Tsimane, and great work on the deter-minants and consequences of status for men and women, see von Rueden, Gurven, and Kaplan 2008, 2011; Von Rueden et al. 2018; Alami et al. 2020.

67. Sattler, Kaiser, and Hittner 2000.

68. Ashida et al. 2016.

69. Acierno et al. 2006.

70. Hrdy 1981.

71. Diamond 2013.

72. Heinen and Ruddle 1974.

73. These results are from a presentation given by Benjamin Trumble at the California Workshop for the Evolutionary Social Sciences in 2016: "The Impact of a Natural Disaster on Physical and Mental Health in a Small-Scale Subsistence Population." Controlling for respondent age and village size, those living in villages with a greater number of older (age 80+) adults reported fewer lost crops (29% lower crop loss, $p = 0.029$), amounting to 1.1 fewer acres lost per person ($p = 0.001$), and less child illness (Odds Ratio = 0.36, $p = 0.014$).

74. I compared forty villages with updated censuses, each tracked for about nine years, to examine whether a higher proportion of elders (age 60+) in the village was associated with a lower death rate (number per year) among those younger than 60 y. The analysis is a mixed effects linear regression that takes into account the fact that villages in different geographic locations may be more similar in their exposures or death rates. The model also adjusts for village proximity to town and village size (Model $R^2 = 0.19$, $F = 2.66$, $p = 0.063$). Though suggestive, the effect of elders is just barely marginally significant, when going by classical statistical standards. As these results are not published elsewhere, I showcase the main result in the Appendix (table A6.1).

75. Heinen 1972, p. 564.

76. Simmons 1945.

77. Simmons 1945, p. 86.

78. Simmons 1945, p. 73.

79. In general, Hadza grandmas were more likely to be co-resident in camps with their adult daughters than with their adult sons. However, if her son's wife's mother had already died, she would reside with her adult son. See Blurton Jones, Hawkes, and O'Connell 2005.

80. Steyn 1994.

81. In the hot, dry Bay Region of southern Somalia, a group of agropastoralists farmed sorghum and cowpeas and raised camels, cattle, sheep, and goats for milk and meat. Abiding by the three p's (polygynous, patrilineal, and patrilocal), older men traditionally owned most of the land and animals, which gave them primary influence on the decisions of daily life. That didn't sit well with the younger men, who would have to wait until they were at least 30 to gain access to the resources that would allow them to marry. Modernization provided new options for these young men. They could seek wages from opportunities elsewhere, set up shops, or sell sorghum at high prices in new markets. This way they can buy what they need, or migrate out. Anthropologist Anthony Glascock describes how, with this new leverage, they're challenging the old men's control over local resources by demanding land and livestock. They didn't do this before because those outside options didn't exist. Getting land and livestock from the older men helps the young men marry at earlier ages, but reduces the wealth and authority of their fathers. Parent-offspring conflict at its finest. See Glascock 1991.

82. Vanarsdale 2021.

83. Amoss 1981, p. 227.

Chapter 7. Help Others and You Will Help Yourself

1. Tale #78: The Old Man and His Grandson (Zipes 2003). Leo Tolstoy also popularized a version of this story.

2. Logue 1990, p. 346.

3. Simmons 1960, p. 39.

4. Draper and Buchanan 1992.

5. Relying on distant relatives or strangers was believed by the Yakut to be awful: "They begrudge you food, they drive you away from the fire, they begin to curse you for every trifle, they do not care for your illnesses . . . and so hiding in the corner, you die slowly from cold, from hunger, not like a man, but like cattle . . . in silence. . . . The only hope for us Yakut is children." Quote by Sieroshevski 1896 in Simmons 1945, p. 198.

6. Wilbert 1972, pp. 107–108.

7. Sharp 1977, pp. 382–383.

8. Flannery 1953, p. 197.

9. Glascock and Feinman 1980, Table 8.

10. Rosenberg 2020.

11. Turnbull 1965, pp. 228, 284.

12. Endicott 1988, p. 118.

13. Huntingford 1951, p. 34.

14. Hearne 1958, p. 221.

15. Simmons 1946, p. 74.

16. McArdle and Yeracaris 1981. I reanalyzed data that the authors provided in an appendix, in order to examine the relationship between "valued activities" and "high respect." Correlation between valued activity and high respect is 0.2 (Spearman's rho, $p = 0.02$, $n = 135$ societies).

17. In a logistic regression analysis, I explored determinants of whether elders earn high respect in their society. I include whether elders participate in valued activities, and other societal features in the McArdle and Yeracaris dataset: whether households tend to be extended or nuclear, whether there are "high gods" present in spiritual life, and whether some form of slavery exists. Participating in valued activities increases the probability of earning high respect from 50% to 75% (odds ratio (OR) = 2.4 with a 95% confidence interval between 1.1 and 4.9). Of the other factors, only institutionalized slavery is related to elders earning high respect. Where slavery is present, elders earn more respect. One possible interpretation that the authors give is that enslaved people did many of the menial, unskilled tasks that don't earn prestige. And so where slaves are present, elders can concentrate on other, more valued activities.

18. From a reanalysis of Simmons' data, conjectured associations that relate elder prestige to other ecological or societal characteristics show support: Prestige for old men and women was more common where the food supply was deemed more constant and predictable (men: Spearman's correlation, $r = 0.353$, $p = 0.006$, $n = 60$; women: $r = 0.503$, $p < 0.001$, $n = 49$). Having property rights was also associated with prestige (men: $r = 0.442$, $p < 0.001$, $n = 48$; women: $r = 0.465$, $p = 0.013$, $n = 28$). Living with a woman's family (matrilocality) and inheritance through the female matriline (matrilineal inheritance) are also associated with greater prestige for elder women ($r = 0.379$, $p = 0.011$, $n = 44$; $r = 0.415$, $p = 0.015$, $n = 34$, respectively).

19. Though called theories, these ideas were more like cultural observations than scientific theories. Yet they still hold some historical importance today. From the early 1960s, *disengagement theory* contends that it's natural (nay, inevitable!) that older adults disengage from their usual activities and relationships, thereby withdrawing from society. On the other hand, *activity theory* argues more prescriptively that staying active and engaged leads to more successful aging. Here, maintaining personal relationships and important social roles are seen as key for elder well-being. Thus, if roles shift with changing physical capacities and deteriorating health, the idea is that older adults will adjust accordingly to remain active. The last of the three yawn-inducing psychosocial theories of aging is called *continuity theory*. This builds on activity theory, by arguing that older adults with "normal aging" will maintain some continuity in their lifestyle, in terms of activities and relationships, but may have to adapt circumstances in order to make that happen.

20. VanStone 1963, p. 53.

21. Silverman and Maxwell coded data among 102 societies randomly sampled from Murdock and White's Standard Cross-Cultural Sample of 186 societies, but they only presented analyses based on 34. It doesn't appear that they published the full analysis, despite their intent to do so. Note to junior scholars: never make false promises about forthcoming findings if they're not in press. If nothing else, you'll be shamed in a footnote forty-four years later.

22. Fowler 1987, p. 186.

23. Joe Henrich and Francisco Gil-White give a nice treatment of the question, "How did prestige evolve?" and, along the lines discussed in chapter 7, they focus on why it might pay to defer to elders. See Henrich and Gil-White 2001.

24. Cipriani 1961, p. 490.

25. McArdle and Yeracaris 1981; Holmberg 1969.

26. Amoss 1981, p. 231.

27. Ellis 1996, see Ch. 8.

28. Amoss 1981.

29. See Simmons 1945, p. 69.

30. Simmons 1960, p. 87.

31. Gusinde 1937, p. 941.

32. Gusinde 1931, p. 761.

33. In Leo Simmons' less systematic survey of seventy-one societies, he found abandonment mentioned in thirty-eight groups and fairly common in seven. Killing of the aged was common in eleven groups, occasional in ten and presumed absent in twenty-two. No information was given on the remaining twenty-eight groups. Simmons 1945, p. 228.

34. Maxwell and Maxwell 1980.

35. Simmons 1945, p. 83.

36. One Ache man cried as he recounted to Kim Hill how he buried his old mother alive so that the vultures would not peck out her eyes. See also Hill and Hurtado 1996, pp. 157, 164.

37. Simmons 1945, p. 238.

38. Leighton and Hughes 1955.

39. From Edward Weyer's 1932 book, *The Eskimos* (p. 248). Quoted in Leighton and Hughes, pp. 328–329.

40. Hart, Pilling, and Goodale 1988, p. 154.

41. Simmons 1960, p. 48. Simmons quotes from Hawkes 1916, p. 117.

42. Quote from Rasmussen 1931, p. 144.

43. Minois 1989, pp. 277–280.

44. Holmberg 1969, pp. 224–225. Allan Holmberg spent two years living with the Sirionó of eastern Bolivia in the 1940s and described very harsh conditions of life. Prior to Holmberg's visit, the Sirionó suffered from massive epidemics in the 1920s, and a large reduction in their population. I visited the Sirionó community of Ibiato in 2018 and was happy to speak with seasoned elders there.

45. Guemple 1969.

46. Geronticide was found to be more common where the food supply was irregular: correlation between food constancy and killing is $r = -0.310$, $p = 0.048$ (men); $r = -0.330$, $p = 0.038$ (women). Geronticide was found to be less common in societies with more permanent residence in Simmons's sample (correlation is $r = -0.361$, $p = 0.019$ [for men] and $r = -0.377$, $p = 0.015$ [for women]). Though it's unclear how Simmons came up with his codes for food constancy and permanency of residence, they were positively correlated, as you might expect ($r = 0.519$, $p < 0.001$).

47. Simmons 1945, p. 61.

48. Ethnographer Richard Lee said geronticide was rare among Ju/'hoansi, and George Silberbauer said the same for the G/wi. The Austrian anthropologist Viktor Lebzelter was told in the 1930s that it did occur but only during periods of "extreme thirst" (cited in Simmons 1945). Harriet Rosenberg also confirmed that during very extreme conditions an elder might be left behind in the bush (*na a tsi*). Pat Draper also argued that those without redeeming qualities or kin nearby might not fare so well. And, under extreme circumstances, elders who couldn't keep up with group movements might be abandoned (Draper and Buchanan 1992, p. 135).

49. Shalinsky and Glascock 1988. The observation about *mate* in Melanesia was made over a century ago by the English anthropologist W.H.R. Rivers.

50. Tooby and Cosmides 1996.

51. Nason 1981, pp. 168–170.

52. For further discussion of the tension between generations when elders live a long life, see Amoss and Harrell 1981. A nice case study among the Tallensi is given in Chapter 9 of Fortes 1949.

53. Fortes 1949, p. 173.

54. Tsimane now may say *yoshoropaij* to express gratitude, but this word was invented by New Tribes missionaries. As the evangelical missionary/linguist Wayne Gill told me when I visited him in La Cruz back in 2004, he adapted it from the Trinitario language, where the word was a muddling of the Spanish *Dios te lo paga* (May God return the favor). Cross-culturally, many languages have no verbal expression for "thank you." Not to say there's no concept or expression of gratitude, but daily reciprocity may instead rely on "tacit understandings of rights and duties surrounding mutual assistance and collaboration." See Floyd et al. 2018.

55. Quote from Rosenberg 2020.

56. Rosenberg 2020.

57. Shor, Roelfs, Bugyi, et al. 2012; Shor, Roelfs, Curreli, et al. 2012.

58. Hart, Pilling, and Goodale 1988.

59. Other features of marriage affect future well-being in the latter years. A potential spouse's family network may not be something contemporary urbanites think a lot about, but can be critical to helping make ends meet in small-scale societies. For example, among horticulturalists in the Trobriand Islands, having adult brothers improved a woman's chances on the marriage market since it was customary for brothers to help make gardens for their sisters and to support them throughout their lives. Like many groups around the world, Trobrianders also practice cross-cousin marriage, where preferred partners would be your mother's brother's child or your father's sister's child. One perk of cousin marriages is that they intensify shared kinship networks. An old Trobriander man explained how this helped with old age security: "I wanted when I got old to have someone of my family to look after me, to cook my food, to bring me my lime-pot and lime-stick, to pull out my grey hairs. It is bad to have a stranger do that; when it is someone of my own people, I am not afraid." In the norms of their kinship system, a son would likely ignore his father unless the son's wife was a cross-cousin (Simmons 1945, pp. 188–189).

60. Rasmussen 1908, p. 127.

61. For an overview of cases among different Inuit groups, see Leighton and Hughes 1955.

62. Kim Hill and Brian Wood, personal communication.

63. Cukrowicz et al. 2011. The "interpersonal theory" of suicide posits that perceiving yourself as a burden to others, combined with a feeling that you don't belong or are disconnected from the group (in psych jargon, "thwarted belongingness") motivates suicidal ideation. A meta-analysis shows robust associations between perceiving yourself as a burden and not belonging on suicidal thoughts and attempts (Chu et al. 2017).

64. Joiner et al. 2016.

65. Nason 1981, p. 171.

Chapter 8. Saving the One-Hoss Shay

1. Established in 1974, the National Institute on Aging (NIA), a division of the US National Institutes of Health, is devoted to the study of aging. Its origins owe much to the efforts of Nathan Shock. Consider me a big fan. NIA has been the generous funder of almost all of my health- and aging-related work in Bolivia.

2. Belsky et al. 2015.

3. Also called the Barker hypothesis, named after the late epidemiologist and physician David Barker, who first wrote about the connections between early life deficits and later life disease in the 1980s and 1990s. He had sought to understand why heart disease and diabetes were more common in poorer regions of England. His work established that maternal nutrition and early life health are important determinants of adult disease. These ideas helped launch the field of study called the Developmental Origins of Health and Disease. For a more recent evaluation of the hypothesis, see Skogen and Øverland 2012.

4. For background on safety factors applied to organismal design, see Diamond 2002; Weibel 2000.

5. Pedersen et al. 2019; Osterbur et al. 2014.

6. See Stieglitz et al. 2020; Stieglitz et al. 2019.

7. González et al. 2020.

8. For reviews on oxidative stress and aging, see Liguori et al. 2018; Bardaweel et al. 2018.

9. D_M stands for "Mahalanobis distance," a statistical measure that enables one to estimate the physiological state of an individual relative to a healthy baseline. This involves the calculation of the distance between sets of biomarker values in multidimensional space. For an introduction to measures of physiological dysregulation, see Cohen et al. 2013. For comparison of how the "D_M" measure compares against other typical biomarkers of aging, see Belsky et al. 2018.

10. Kraft et al. 2020.

11. See Belsky et al. 2018; Jansen et al. 2021.

12. Silberbauer 1965, pp. 112–113.

13. See Cohen, Legault, and Fülöp 2020.

14. For great reviews of how to think about aging as a complex system, see Cohen et al. 2022.

15. This body of work falls under the "reliability theory of aging," pioneered by the Russian demographers Leonid Gavrilov and Natalia Gavrilova. For a reader-friendly overview, see Gavrilov and Gavrilova 2005. The sweet Dobzhansky take on Hamlet comes from there, borrowed from his 1962 book, *Mankind Evolving: The Evolution of the Human Species*. Reliability theory also provides a clever demonstration of the compensation law of mortality mentioned in chapter 2, whereby mortality rates seem to converge at later ages. And for the bathtub-shaped mortality curve more generally—major defects lead to a "working-in" period early in life (high infant mortality), followed by a "normal working period" of low mortality, and the "aging period" of increasing mortality with age.

16. Blackwell et al. 2016.

17. For a view on dysregulation across systems in industrialized populations, see Li et al. 2015.

18. For nice examples of how this might work, see Laird and Sherratt 2010.

19. For a great textbook that thinks broadly about the design features of the "flexible phenotype" across animals (but especially in shore birds), see Piersma and Van Gils 2011.

20. For a nice discussion of these trade-offs in the dairy industry, see Oltenacu and Broom 2010.

21. Irimia et al. 2021.

22. For details on cognitive testing among the Tsimane, and especially the effects of schooling, see Gurven et al. 2017. For usual patterns of cognitive aging observed in industrialized populations, see Salthouse 2019.

23. Goldberg 2006, p. 21.

24. For a more detailed description of these cognitive and functional changes, see Spreng and Turner 2019.

25. Schwarz et al. 2016; Saha et al. 2022.

26. See Finch and Stanford 2004.

27. See Ledberg 2020. Ledberg's model of damage accumulation is based on the mathematical theory of "queues." It includes just four parameters: rate of damage accumulation (λ), an initial rate of repair (μ_0), the rate at which repair declines (β), and the threshold (θ) beyond which damage kills the organism. Ledberg shows how λ and θ can be estimated once you know the slopes and intercepts from the log mortality hazard for each cohort. From the Human Mortality Database, I chose two Swedish cohorts to compare: those born in 1770 versus those born in 1910. I modeled the logged mortality hazards from ages 40 to 100, which is almost a perfectly straight line (meaning that mortality over this period grows exponentially). Ledberg arbitrarily assigns $\mu_0 = 550$ and $\beta = 0.485$. For 1770-ers, $\ln(\text{mortality}) = -7.2531 + 0.0673^*\text{age}$ $(R^2 = 0.982)$, which leads to $\theta = 71.59$ and $\lambda = 497.85$. For 1910-ers, $\ln(\text{mortality}) = -10.029 + 0.0925^*\text{age}$ $(R^2 = 0.999)$, which leads to $\theta = 98.40$ and $\lambda = 497.55$. Almost all the differences in mortality rates over adulthood are thus explained by differences in the tolerance to damage (θ). Note, we can ask whether part of our modern biology helps slow the rate of decline in repair capacity (β). Even if we decreased β by 25%, most of the mortality differences are best captured by changes in threshold θ rather than rate of damage accumulation λ (θ now increases by 86%, λ increases by 2.6%).

28. See, for example Mitnitski et al. 2017.

29. Tomasetti and Vogelstein 2015. See also Wu et al. 2016; Thomas et al. 2016.

30. Very short telomeres are also associated with increased risk of certain cancers. When telomeres are critically short, they usually stop dividing. Some cells manage to reactivate telomerase to create conditions where they can persist and potentially become cancerous. See Wentzensen et al. 2011.

31. Olshansky 2020, p. 37.

32. Ehrenreich 2018.

33. For an entertaining and evolutionary take on why physical activity is so important for human health, while simultaneously busting many common myths about exercise, see Lieberman 2021.

Chapter 9. A Mismatch Made in Heaven

1. Yeah, it's a thing, but "repoopulate" first referred to synthetic fecal matter made by a robogut that could help cure nasty gastrointestinal infections but without the "ick" factor of using other people's feces. See "Fake fecal transplants for gut rePOOPulation." *SciAm*, January 10, 2013, https://www.scientificamerican.com/podcast/episode/fake-fecal-transplants-for-gut-repo-13 -01-10/.

2. My youngest academic sibling, Andrew Bishop, survived three weeks confronting sizzling hot temperatures, unpredictable storms, armies of bats, and swarms of bees in central Colombia. Season 14, episode 12: "The Labyrinth" episode of *Naked and Afraid*.

3. Lee et al. 1994.

4. Rowley et al. 2001.

5. Gardiner et al. 2021.

6. For an overview of what we know health-wise about Ötzi based on imaging, see Murphy et al. 2003. On a review of genetics of atherosclerosis in mummies, see Zink et al. 2014.

7. Reviewed in Thomas et al. 2014.

8. Thompson et al. 2013.

9. Enos, Holmes, and Beyer 1953.

10. Tanaka et al. 1988.

11. "Unwrapping Health Secrets: Mummy CT Scans Show Preindustrial Hunter Gatherers Had Clogged Arteries," *USC Today*, March 8, 2013, https://today.usc.edu/unwrapping-health -secrets-mummy-ct-scans-show-preindustrial-hunter-gatherers-had-clogged-arteries.

12. Mann et al. 1972; Mann et al. 1964.

13. 14.5% of US adults from a national registry covering over 2 million patients between 1994 and 2006 had a heart attack without having any of the five traditional risk factors (hypertension, family history of heart disease, dyslipidemia, smoking history, diabetes); see Canto et al. 2011. For this reason, some have argued that traditional risk factors don't go far enough. Instead of the typical "healthy" LDL threshold being < 100 mg/dL, some believe that avoiding heart attacks altogether requires your LDL be lower than 70 mg/dL (O'Keefe et al. 2004)—a mean feat possible for most people today only if they're eating very carefully or taking statins. Even in hunter-gatherer groups where average LDL< 70, many individuals will still be above this threshold. Back in 2005, when we first looked at blood cholesterol among Tsimane, only a third of adults over age 40 had LDL < 70.

14. Relevant papers on Tsimane cardiovascular health: Gurven, Kaplan, et al. 2009; Gurven, Blackwell, et al. 2012; Pisor et al. 2013; Kaplan et al. 2017; Rowan et al. 2021.

15. Diabetes was absent in the 1930s, rising to 10% by 1980, then its prevalence doubled to 20% by 1990 among First Nations (mostly of Sioux and Saulteaux ancestry) in Saskatchewan. See Pioro, Dyck, and Gillis 1996.

16. Minimally invasive blood-based biomarkers of Alzheimer's disease risk are nearing prime time for clinical practice. These include amyloid beta-42/40 ratio, phosphorylated tau protein (p-tau), and neurofilament light change (NfL). See Hansson, 2023.

17. Gatz et al. 2022.

18. Irimia et al. 2021.

19. See Catania and Panegyres 2017.

20. Zheng et al. 2023.

21. For an overview of recent trends in cancer incidence and mortality, see Siegel, Giaquinto, and Jemal 2024.

22. For a recent overview of Peto's Paradox rich with the largest dataset of mammals to date, see Vincze et al. 2022. On speed of mutational clocks in short- versus long-lived species, see Cagan et al. 2022.

23. Hochberg and Noble 2017.

24. Marques et al. 2018.

25. Marques, Compton, and Boddy 2022.

26. My longtime colleague and fellow codirector of the Tsimane Health and Life History Project, Hillard Kaplan, gave a talk (unpublished) showing preliminary prevalence estimates among Tsimane: H. Kaplan, B. Trumble, J. Stieglitz, D. Eid Rodriguez, and M. Gurven, Cancer in an Indigenous South American Population, Center for Evolution and Medicine, Arizona State University.

27. In their traditional pharmacy, Tsimane make their own homespun Viagra called *chu'si dyuj*, which translates to "coatimundi penis." Shavings from the raccoon-like coati's penis bone are believed to stimulate male virility.

28. See Blackwell et al. 2016.

29. The neologism *exposome* reflects the full set of exposures an individual experiences over their lifetime, with an eye toward how these might impact health. For an excellent (but not pithy) overview of the exposome throughout human evolutionary history, see Trumble and Finch 2019.

30. Rindos 2013.

31. Hawks et al. 2007; Cochran and Harpending 2009.

32. Lea et al. 2022.

33. Ritchey et al. 2020.

34. O'Hearn et al. 2022.

35. For example, selection on genes affecting coronary artery disease suggests support for the antagonistic pleiotropy hypothesis we discussed in chapter 2. See Byars et al. 2017.

36. See Pennington et al. 2009; Brinkworth and Barreiro 2014; Yao et al. 2018.

37. Perry et al. 2007.

38. Fumagalli et al. 2019.

39. Fortea et al. 2024.

40. Trumble et al. 2023.

41. Medegan Fagla, 2024 #5505.

42. Trumble et al. 2017; Garcia et al. 2021.

43. Hendrie et al. 2014.

44. McMurry et al. 1991. No indication was given how Tarahumara participants were treated once the study was over.

45. Schulz and Chaudhari 2015.

46. Robinson, Gebre, and Pickering 1995.

47. Headey and Alderman 2019.

48. This has been called the protein leverage hypothesis, which has been proposed as one explanation for the obesity epidemic. Experimental and empirical support is mixed, but suggests that an appetite for protein may at least be partly responsible for our overeating when the proportion of protein in the diet is relatively low. For an updated popular view from the original authors of the idea, see Raubenheimer and Simpson 2020.

49. To read about the ad libitum feeding study comparing diets of ultraprocessed versus unprocessed foods, see Hall et al. 2019.

50. The percent of adult diets coming from ultraprocessed foods in the US went from an already high 53.5% in 2001 to 57% in 2018; see Juul et al. 2021. For Europe, see Mertens, 2022 #5504.

51. For a nice overview of the relationship between Westernized diets and chronic disease risk, see Carrera-Bastos et al. 2011.

52. Goncalves et al. 2022.

53. See Raichlen et al. 2017; Gurven et al. 2013; Althoff et al. 2017.

54. Raichlen et al. 2020.

55. For a recent fun book on how exercise and physical activity affects many aspects of health, see Lieberman 2021.

56. Kivimäki et al. 2019.

57. Lear et al. 2017.

58. Guthold et al. 2018; Lee et al. 2012.

59. For an entertaining account about why exercise alone may not reliably shed those pounds off our middle, see Pontzer 2021.

60. For far more detail, see Fiuza-Luces et al. 2018.

61. Forbes et al. 2015.

62. Wang and Holsinger 2018.

63. Figure taken from this study: Wroblewski et al. 2011.

64. Bach, 2002 #4594.

65. Vatanen et al. 2016.

66. For an entertaining perspective on how being helminth-free may be affecting our risks of autoimmunity (and whose lead author came to visit me in Bolivia to see "what it would be like to be surrounded by parasites"), see Velasquez-Manoff 2012.

67. Brestoff and Artis 2015.

68. For reviews on the role of helminths in diabetes and cardiovascular disease, see Wiria et al. 2014; Gurven et al. 2016; De Ruiter et al. 2017. Examples of experimental studies are included in these papers. Newer studies in humans cited in the text include Sanya et al. 2020; Wiria et al. 2015; Tahapary et al. 2017.

69. See Ricciarelli and Fedele 2017.

70. Kumar et al. 2016; Osorio et al. 2019.

71. Wu et al. 2022.

72. UNDP 2019.

73. Crimmins et al. 2019.

74. For a summary of this view and the research behind it, see Natri et al. 2019.

75. Roberts et al. 2018; Pryce and Fontana 2016.

76. Yassin et al. 2019.

77. For a study showing the separate and combined effects of dietary quality and physical activity on cardiovascular and cancer mortality in the UK, see Ding et al. 2022.

Chapter 10. Foraging Alone?

1. See Rosling, Rosling, and Rönnlund 2018. For a user-friendly, data-filled introduction to global living conditions over time, see also the website Our World in Data, https://ourworldindata .org/a-history-of-global-living-conditions.

2. For a systematic overview of the evidence linking social isolation and loneliness to health, see NASEM 2020. For excellent meta-analyses providing some of the best support linking these to mortality, see Holt-Lunstad, Smith, and Layton 2010; Holt-Lunstad et al. 2015; Wang et al. 2023.

3. For example, see Santini et al. 2020.

4. See Li et al. 2021. Also, "The History of Loneliness," *New Yorker*, March 30, 2020, https:// www.newyorker.com/magazine/2020/04/06/the-history-of-loneliness.

5. Diener and Diener 1996.

6. See Biswas-Diener, Vittersø, and Diener 2009 on Maasai and Greenlandic Inuit, and Miñarro et al. 2021 on Solomon Islanders. See also Galbraith et al. 2024 for a study on self-reported happiness in nineteen small-scale societies. When asked to choose your level of life satisfaction on a 3-, 5-, or even 10-point scale, do people respond with pleasantries just to please the interviewer? Probably not. Similar levels of positive well-being about a person are found whether you ask the person directly or you ask their spouse(s), friends, or neighbors. Or gauging what people say against what they do, like smile or laugh. Another good check is to avoid holistic self-evaluation altogether. When people are asked in-the-moment questions like "how are you feeling right now?" the answer is more often than not "reasonably okay."

7. For recent treatments of the evidence for the U-shaped relationship between subjective well-being and age, see Blanchflower and Oswald 2019; Blanchflower and Graham 2022. For some critiques, see Bartram, 2022 #5367; Galambos, 2020 #5373.

8. Becker and Trautmann 2022.

9. See Stone et al. 2010.

10. No, chimpanzees did not pantomime their feelings about life satisfaction. They were rated by humans familiar with the apes on a few items (positive versus negative mood, how much pleasure they get from social situations, how successful they are at achieving their goals, and, my favorite, how happy the rater would be if they were the ape for a week). To my knowledge, this is the only study of satisfaction in apes, and it's worth noting that the apes had been living in captivity. See Weiss et al. 2012.

11. Carstensen 2006.

12. For a popular and uplifting treatment of this topic, see Rauch 2018.

13. The questions include (1) In general, I consider myself [7 point scale between not very happy to very happy person]; (2) Compared to most of my peers, I consider myself [7 point scale from less happy to more happy]; (3) Some people are generally very happy. They enjoy life regardless of what is going on, getting the most out of everything. To what extent does this describe you? [7 point scale from not at all to a great deal]; (4) Some people are generally not very happy. Although they are depressed, they never seem as happy as they might be. To what extent does this describe you? [7 point scale between not at all to a great deal]. Responses

among these questions are reasonably correlated (ranging from 0.51 to 0.79). For more details, see Frankowiak et al. 2020.

14. Djankov, Nikolova, and Zilinsky 2016.

15. The paper comparing Baka, Punan, and Tsimane quality of life is Reyes-García et al. 2021.

16. My own recently collected, but unpublished, Tsimane data on life satisfaction and a "meaningful, purposeful life" both consistently show lower well-being in later adulthood. See also Gurven et al. 2024.

17. As might be expected, older Punan and Tsimane are less likely to report "high" well-being (that is, life is good or very good).

18. Keith et al. 1994.

19. Goode 2000. "Viewing Depression as Tool for Survival," *New York Times*, February 1, 200, https://archive.nytimes.com/www.nytimes.com/library/national/science/health/020100hth -behavior-depression.html.

20. Mann et al. 1964, p. 309.

21. See Sebastian Junger's book on how returning war veterans miss the comraderie, belonging, and community bonding while in the trenches, compared to the postwar isolation and loneliness during peacetime back home. See Junger 2016.

22. Izquierdo 2005.

23. Martin and Cooper 2017.

24. For details about the method and findings, see Stieglitz et al. 2014.

25. Kohler et al. 2017.

26. For an enjoyable and accessible account of the emotions and mental disorders that combines evolutionary insights with clinical expertise, see Nesse 2019.

27. See Stieglitz et al. 2015.

28. Sharp 1981.

29. For a study comparing giving to receiving support on mortality risk, see Brown et al. 2003.

30. Gruenewald, Liao, and Seeman 2012.

31. For a meta-analysis showing how unemployment relates (and causes) mental distress, see Paul and Moser 2009.

32. These and other hypotheses proposing that some depression symptoms are specific functional responses to adversity are easier to propose than to test. Many symptoms certainly don't feel like they're good for us. Severe depression is indeed likely to be maladaptive, but nonetheless some insight can be gained about the triggers and consequences of mental disorders. See Nesse 2019.

33. McPherson, Smith-Lovin, and Brashears 2006.

34. The Hispanic Mortality Paradox (also called the Latino Mortality Advantage) was first identified in epidemiological studies in the 1980s. Recent meta-analyses confirm it is a robust phenomenon in the United States. See Ruiz, Steffen, and Smith 2013. As more data become available, it's evident that mortality advantages may also apply to Latinos living in other countries, like Argentina, Chile, and Belize. See Chen et al. 2020.

35. Cortes-Bergoderi et al. 2013.

36. Arias and Branch 2019.

37. For the view that stereotypes can become "embodied" and affect health and well-being, see Levy 2009.

38. Levy et al. 2009.

39. Lamont, Swift, and Abrams 2015.

40. Ackerman and Chopik 2021.

41. Löckenhoff et al. 2009.

42. Sorokowski et al. 2017.

43. Sorokowski et al. 2022.

44. Reyes-García 2012.

45. "Time Use," Our World in Data, November 2020, https://ourworldindata.org/time-with
-others-lifetime.

46. "New York State Is Giving Out Hundreds of Robots as Companions for the Elderly," *The
Verge*, May 25, 2022, https://www.theverge.com/2022/5/25/23140936/ny-state-distribute
-home-robot-companions-nysofa-elliq.

A preliminary analysis of eight hundred older New Yorkers showed that 96% thought ElliQ
helped reduce loneliness, and 95% said it improved their "wellness." The average user engaged
with EllieQ 37 times per day and 6 days per week. See https://aging.ny.gov/nysofa-and-elliq
-engagement-report.

47. As science writer Jonah Lehrer once remarked, "It's so comforting to press play and enter
into a familiar social network, even if that social network involves the New Jersey mob." For an
interesting discussion of the role that faux, "parasocial" relationships from television might have
on loneliness, see, "Imaginary Friends," *Scientific American*, July 28, 2009, https://www
.scientificamerican.com/article/imaginary-friends/.

Chapter 11. Bringing It Home

1. De Beauvoir 1972.

2. Landry 2014, p. 30.

3. Thomas 2011. For a popular and invigorating introduction to a reimagining of elderhood,
see Thomas 2004 and Thomas and Thomas 2015.

4. "Age No Barrier: How Jo Schoonbroodt Smashed the 70+ Marathon Record," *The Guardian*, May 12, 2022, https://www.theguardian.com/sport/2022/may/12/age-no-barrier-how-jo
-schoonbroodt-smashed-the-70-marathon-record#:~:text=On%20Sunday%20Jo%20
Schoonbroodt%2C%20a,achievement%20is%20still%20sinking%20in.

5. From a preliminary trial, see Fahy et al. 2019. For some promising thoughts about metformin, see Kulkarni, Gubbi, and Barzilai 2020.

6. To hear about the antiaging efforts of a self-proclaimed "rejuvenation athlete," see, "Why
I Spend $2 Million a Year to Look 18 Years Old Again," *Fortune OnDemand*, June 21, 2023, https://
fortune.com/videos/watch/why-i-spend-%242-million-a-year-to-look-18-years-old-again
/7ef735d1-4f4f-433d-9428-60abd7f923e5.

7. See Zuk 2013.

8. For example, see "An Urban Hunter Gatherer," *Weekend Edition Saturday*, NPR, January 10, 2004, https://www.npr.org/templates/story/story.php?storyId=1592235; "Meet the
Contemporary Hunter-Gatherers Who Live Off the Land in the American West," *Business Insider*, December 14, 2017, https://www.businessinsider.com/adrain-chessers-photos-of-hunter
-gatherer-americans-2014-5; "The Noble Scavenger on the Living-Room Couch," *Newsweek*,
September 28, 2007, https://www.newsweek.com/noble-scavenger-living-room-couch-100663.

9. Booth, Roberts, and Laye 2012. On relationships between physical activity and survival
in seventeen countries, see Lear et al. 2017.

10. Not much vigorous activity, especially among non-exercisers, may be needed to reduce
mortality. See Stamatakis et al. 2022.

11. See *Walking for Health, A Harvard Medical School Special Health Report*, available from
Harvard Health Publishing at https://www.health.harvard.edu/exercise-and-fitness/walking
-for-health.

12. Stenner et al. 2020.

13. Ringeval et al. 2020.

14. I measured going downstairs as 56 steps, and upstairs as 48 steps. I go to my office an average
of four days per week, 48 weeks per year. I count seventeen years resident at UCSB, after discarding

time spent in Bolivia, sabbaticals, and pandemic. General estimates suggest 30–40 kcals burned per 1,000 steps. That range suggests 2.9 to 3.9 pounds kept off my middle over my time at UCSB. During my time at UCSB, I've gained 11.5 pounds (good thing I measured my initial weight during one of the early training sessions in Bolivia!). Forgive the crudeness of this exercise, but if these numbers are at all reasonable, I would've gained 30% more weight if not for using the stairs!

15. Among overweight and obese women eating either a typical US diet, or one based on dietary guidelines, for eight weeks, there was no difference in their cholesterol or blood sugar regulation, though blood pressure was lower among those eating the healthier diet. Total calories consumed were identical between diets. To a large extent, it is the high consumption of calories, rather than type of calories, that seems to make the biggest difference for people's health. See Krishnan et al. 2018. For minimal differences in weight loss across various diets, see Johnston et al. 2014.

16. Pontzer 2021.

17. Everyone loves (writing) lists of sure ways to beat aging. Many of these I've covered in this book. The Power 9 stems from observations in the longevity Blue Zones: (1) Move naturally, (2) Cultivate a sense of purpose, (3) Downshift, (4) Eat until 80% full, (5) Eat a largely plant-based diet, (6) Drink in moderation, (7) Foster a sense of belonging, (8) Put family first, (9) Choose "right tribe." The geriatrician Roger Landry gives ten tips to living a long healthy life: (1) Use it or lose it, (2) Keep moving, (3) Challenge your brain, (4) Stay connected, (5) Lower your risks, (6) Don't act your age, (7) Wherever you are, be there, (8) Find your purpose, (9) Have children, (10) Laugh to a better life. These also jibe with the three protective principles leading to "successful aging" according to the groundbreaking MacArthur Study of Aging: (1) Avoid disease and disability, (2) Maintain mental and physical function, (3) Stay engaged with life.

18. Not just with cigarettes, but men reportedly are more likely to use almost every illicit substance compared to women. A typical explanation for men's greater drug use is that men's neurobiological reward pathways are more easily activated than women's. In other words, men smoke more for the nicotine hit, whereas women smoke more to regulate their mood. From an evolutionary perspective, men are more likely than women to engage in all types of risky behavior, especially during adolescence and early adulthood—that is, periods of high competition for mates, status, and resources. The extent of male-biased cigarette smoking across countries is also related to gender inequality and fertility. In high-fertility regions, women's aversion to nicotine may protect their offspring from harmful cancer-causing substances. For further details, see Hagen, Garfield, and Sullivan 2016.

19. See Our World in Data, "Smoking," https://ourworldindata.org/smoking#share-who -smoke.

20. Lin et al. 2020.

21. Yusuf et al. 2020.

22. "Blue zones" are geographical hotspots for extreme longevity. The term was coined by author and entrepreneur Dan Buettner. He and a team of demographers identified five places where people live very long healthy lives (Ikaria, Greece; Loma Linda, California; Nicoya, Costa Rica; Okinawa, Japan; Sardinia, Italy). Though some of the evidence (or lack thereof) for Blue Zones is controversial (see "Do People in Blue Zones Really Live Longer?" *New York Times*, October 24, 2024, https://www.nytimes.com/2024/10/24/well/live/blue-zones -longevity-aging.html), the notion that healthy diets, an active lifestyle, and tight social community can extend our lives is not.

23. On Seventh-Day Adventist longevity and health, see Fraser and Shavlik 2001. Mormons, whose "Word of Wisdom" doctrine advises against the harms of tobacco, alcohol, and other drugs, and also shows health and survival advantages compared to their non-Mormon peers. See Enstrom, 2008 #5506.

24. The Commonwealth Funds, Issue Briefs, "U.S. Health Care from a Global Perspective, 2022: Accelerating Spending, Worsening Outcomes," January 3, 2023, https://www .commonwealthfund.org/publications/issue-briefs/2023/jan/us-health-care-global -perspective-2022.

25. French et al. 2017; Aldridge and Kelley 2015.

26. Barendregt, Bonneux, and van der Maas 1997.

27. For a similar view, see Bayer and Galea 2015. For a lengthy report exploring how and why US health lags behind that of other high-income countries, see National Research Council 2013.

28. Aburto et al. 2020.

29. Taylor et al. 2009.

30. Diamond 2013, p. 191.

31. Sallis et al. 2006; Buettner and Skemp 2016.

32. See Fitzgerald and Caro 2014 for examples of age-friendly cities around the world.

33. Mead 1971.

34. De Beauvoir 1972, p. 474.

35. For perspectives of disappointed parents whose desire to be grandparents has been thwarted, see "The Unspoken Grief of Never Becoming a Grandparent," *New York Times*, November 11, 2024, https://www.nytimes.com/2024/11/11/well/family/grandparent-grandchild -childfree.html.

36. Biesele and Howell 1981, p. 87.

37. These self-reports are convincing, though empirical studies of the effects of grandparenting on mental and physical health are more equivocal. When under financial strain, with low levels of schooling, or under other difficult conditions, grandparental caregiving may lead to poorer health. Positive relationships among active grandparenting, health, and well-being may be due to causality going in the opposite direction: those in better shape are eager to take on a more active role in the lives of their grandchildren. For an example of a study examining the causal relationship between grandparental caregiving and mental health, see Komonpaisarn and Loichinger 2019. When all caretaking resides with grandparents (that is, in the absence of parents), such custodial grandparenting tends to bring lower health and well-being, whereas positive effects are observed most in non-co-residing grandparents. For a systematic review, see Danielsbacka, Křenková, and Tanskanen 2022. Sometimes positive effects are observed more in grandmothers than grandfathers: Di Gessa, Bordone, and Arpino 2020.

38. Pipher 1999, p. 226.

39. Moorman and Stokes 2016.

40. "V.R. 'Reminiscence Therapy' Lets Seniors Relive the Past," *New York Times*, May 6, 2022, https://www.nytimes.com/2022/05/06/well/mind/virtual-reality-therapy -seniors.html.

41. See "Singapore's Seniors Pick Up Smartphones to Cope with Covid-19," *GovInsider*, August 31, 2020, https://govinsider.asia/digital-gov/singapore-trains-seniors-to-adapt-routines -with-smartphones/.

42. Pew Research Center 2022.

43. For reviews and meta-analyses on benefits of volunteering and intergenerational cooperation for elders, see Anderson et al. 2014; Su 2017.

44. A great example is Project SHINE, which pairs college students with elderly immigrants and refugees. The students tutor and coach elders in community centers, in English language and US history to prepare for their citizenship exam. A mix of learning and connection. Started in Philadelphia, SHINE has spread across eighteen US cities. To date, almost 40,000 immigrants have been matched with over 9,000 students.

45. For more information on intergenerational schooling in Ohio, see https://igschools.org/.

46. "College Students and Senior Citizens Living Together? It's More Common Than You Think," *Business Insider*, February 15, 2020, https://www.insider.com/intergenerational-living -senior-citizens-college-students-2020-1.

47. For information on AmeriCorps Seniors programs and evaluation of these programs on elder health, see Frazier et al. 2019.

48. Frazier et al. 2019. See also Richman, Bennett, and Gleason 2023.

49. Here are a few online platforms designed to match elders with younger generations: Eldera: https://www.eldera.ai/; Big & Mini: https://bigandmini.org/; Papa: https://www.papa.com/.

50. When asked about his secret to having a long life, Roke replied, "Less meat and more vegetables. Keep your body moving. Forget about politics. And above all, stay single." His life is fascinating. To read an interview with him, see "Santa Barbara's Century Man," *Santa Barbara Independent*, December 22, 2022, https://www.independent.com/2022/12/21/santa-barbara -century-man-roke-fukumura/. For his biography, see https://sbgen.org/fukumura-family/.

51. "The 50 Jobs Where People Work the Longest," *Time*, April 5, 2017, https://time.com /4726657/retirement-age-jobs/.

52. Here are two lists of elder-friendly employers: https://www.monster.com/career-advice /article/companies-friendly-toward-older-workers-1217; https://www.flexjobs.com/blog/post /8-best-employers-with-flexible-jobs-for-workers-over-50/.

53. Certified Age Friendly Employer Program, Age-Friendly Institute, https://institute .agefriendly.org/initiatives/certified-age-friendly-employer-program/.

54. These workshops cost about $5,000, though discounts are available. At least the price includes a massage.

55. See "How One Entrepreneur Is Turning Intergenerational Relationships into Organizational Success," *Forbes*, February 9, 2021, https://www.forbes.com/sites/sheilacallaham /2021/02/09/how-one-entrepreneur-is-turning-intergenerational-relationships-into -organizational-success/?sh=1989560a482a.

56. For information on universities offering free courses, see *Best Colleges* (blog), https:// www.bestcolleges.com/blog/free-college-tuition-senior-citizens/; and https://www .thepennyhoarder.com/save-money/free-college-courses-for-senior-citizens/.

57. According to the 2020 Profile of Older Americans, 4% of US elders age 65+ live in nursing homes (1% aged 65–74, 2% aged 75–84 and 8% aged 85+), and 2% live in assisted living facilities.

58. It's estimated that 70% of Americans reaching age 65 today will develop a disability serious enough to need long-term support and services for activities of daily living, though the average duration of support needed will be less than two years. Roughly one in seven adults will need care for more than five years. For more details and how costs of care may be covered, see Johnson et al. 2021.

59. Poo and Conrad 2015, p. 143

60. Hayashi 2012.

61. See Choosing Wisely: An Initiative of the ABIM Foundation, https://choosingwisely.org/.

62. Kavalieratos et al. 2016.

63. In the US, hospice care is covered by Medicare, and therefore has stricter eligibility requirements than palliative care. Hospice patients usually need to have a terminal life expectancy of six months or less and declining functional capacity, with curative treatment no longer deemed viable. Other disease-specific criteria may also apply.

64. Wettstein and Zulkarnain 2017.

65. Butler 2014, pp. 195, 255.

66. Source: 2019 Federal Reserve Board Survey of Consumer Finances, as cited in https:// www.federalreserve.gov/econres/notes/feds-notes/wealth-and-income-concentration-in-the -scf-20200928.html.

67. "Tribal Elders Are Dying from the Pandemic, Causing a Cultural Crisis for American Indians," *New York Times*, January 12, 2021, https://www.nytimes.com/2021/01/12/us/tribal-elders-native-americans-coronavirus.html.

68. "Grand-Mates: Generations Sharing a Special Bond (and Sometimes the Rent)," *New York Times*, September 30, 2022, https://www.nytimes.com/2022/09/30/realestate/grandparents-grandchildren-living-together.html.

69. Tsimane now represent themselves at the municipal and regional government levels. Coordination on a variety of health services currently involves partnerships with multiple institutions, including with the Tsimane Health and Life History Project. A nonprofit started by my former student Helen Davis, One Pencil Project provides school supplies for communities in Bolivia and Namibia. It also aims to provide scholarships for higher education in medicine, nursing, agronomy, and law. See https://www.onepencilproject.org/. The first cohort is soon to graduate. Tsimane graduates will be well poised to help resolve ongoing problems and conflicts with respect to territory, economy, and health.

70. Simmons 1960, p. 50.

71. Nuland 2008, p. 283.

REFERENCES

AARP. 2019. *2018 Grandparents Today National Survey.* https://www.aarp.org/content/dam /aarp/research/surveys_statistics/life-leisure/2019/aarp-grandparenting-study.doi.10 .26419-2Fres.00289.001.pdf.

Abrams, Peter. 1993. "Does increased mortality favor the evolution of more rapid senescence?" *Evolution* 47 (3): 877–887.

Aburto, José Manuel, Jonas Schöley, Ilya Kashnitsky, Luyin Zhang, Charles Rahal, Trifon I. Missov, Melinda C. Mills, Jennifer B. Dowd, and Ridhi Kashyap. 2021. "Quantifying impacts of the COVID-19 pandemic through life expectancy losses: A population-level study of 29 countries." *International Journal of Epidemiology* 21:1–12.

Aburto, José Manuel, Francisco Villavicencio, Ugofilippo Basellini, Søren Kjærgaard, and James W. Vaupel. 2020. "Dynamics of life expectancy and life span equality." *Proceedings of the National Academy of Sciences* 117 (10): 5250–5259.

Acierno, Ron, Kenneth Ruggiero, Dean Kilpatrick, Heidi Resnick, and Sandro Galea. 2006. "Risk and protective factors for psychopathology among older versus younger adults after the 2004 Florida hurricanes." *American Journal of Geriatric Psychiatry* 14 (12): 1051–1059.

Ackerman, Lindsay, and William Chopik. 2021. "Cross-cultural comparisons in implicit and explicit age bias." *Personality and Social Psychology Bulletin* 47 (6): 953–968.

Acsádi, György, János Nemeskéri, and Kornél Balás. 1970. *History of Human Life Span and Mortality.* Budapest: Akadémiai Kiadó Budapest.

Alami, Sarah, Christopher Von Rueden, Edmond Seabright, Thomas S. Kraft, Aaron D. Blackwell, Jonathan Stieglitz, Hillard Kaplan, and Michael Gurven. 2020. "Mother's social status is associated with child health in a horticulturalist population." *Proceedings of the Royal Society B* 287 (1922): 20192783.

Aldridge, Melissa, and Amy Kelley. 2015. "The myth regarding the high cost of end-of-life care." *American Journal of Public Health* 105 (12): 2411–2415.

Allen, Sian, and Will Hopkins. 2015. "Age of peak competitive performance of elite athletes: A systematic review." *Sports Medicine* 45 (10): 1431–1441.

Allman, J., A. Rosin, R. Kumar, and A. Hasenstaub. 1998. "Parenting and survival in anthropoid primates: Caretakers live longer." *Proceedings of the National Academy of Sciences* 95 (12) 6866–6869.

Alstott, Anne L. 2016. *A New Deal for Old Age: Toward a Progressive Retirement.* Cambridge, MA: Harvard University Press.

Althoff, Tim, Jennifer L. Hicks, Abby C. King, Scott L. Delp, and Jure Leskovec. 2017. "Large-scale physical activity data reveal worldwide activity inequality." *Nature* 547 (7663): 336.

Amoss, Pamela. 1981. "Coast Salish Elders." In *Other Ways of Growing Old*, edited by Pamela Amoss and Stevan Harrell. Stanford, CA: Stanford University Press.

Amoss, Pamela, and Stevan Harrell. 1981. "Introduction: An anthropological perspective on aging." In *Other Ways of Growing Old*, edited by Pamela Amoss and Stevan Harrell, 1–24. Stanford, CA: Stanford University Press.

Anderson, Nicole, Thecla Damianakis, Edeltraut Kröger, Laura M. Wagner, Deirdre R. Dawson, Malcolm A. Binns, Syrelle Bernstein, Eilon Caspi, and Suzanne L. Cook. 2014. "The benefits associated with volunteering among seniors: A critical review and recommendations for future research." *Psychological Bulletin* 140 (6): 1505.

Antón, Susan, Richard Potts, and Leslie Aiello. 2014. "Evolution of early Homo: An integrated biological perspective." *Science* 345 (6192).

Ardesch, Dirk Jan, Lianne H. Scholtens, Longchuan Li, Todd M. Preuss, James K. Rilling, and Martijn P. van den Heuvel. 2019. "Evolutionary expansion of connectivity between multimodal association areas in the human brain compared with chimpanzees." *Proceedings of the National Academy of Sciences* 116 (14): 7101–7106.

Arias, Elizabeth, and Mortality Statistics Branch. 2019. "Race crossover in longevity." In *Encyclopedia of Gerontology and Population Aging*, edited by Danan Gu and Matthew E. Dupre, 1–10. New York: Springer.

Ashida, Sato, Erin L. Robinson, Jane Gay, and Marizen Ramirez. 2016. "Motivating rural older residents to prepare for disasters: Moving beyond personal benefits." *Ageing and Society* 36 (10): 2117–2140.

Austad, Steven N. 1993. "Retarded senescence in an insular population of Virginia opossums (Didelphis virginiana)." *Journal of Zoology* 229 (4): 695–708.

Austad, Steven N., and Jessica M. Hoffman. 2018. "Is antagonistic pleiotropy ubiquitous in aging biology?" *Evolution, Medicine, and Public Health* 2018 (1): 287–294.

Babkin, Igor V., and Irina N. Babkina. 2015. "The origin of the variola virus." *Viruses* 7 (3): 1100–1112. https://doi.org/10.3390/v7031100.

Bach, Jean-François. 2002. "The effect of infections on susceptibility to autoimmune and allergic diseases." *New England Journal of Medicine* 347 (12): 911–920.

Bagnall, Roger S., and Bruce W. Frier. 1994. *The Demography of Roman Egypt*. Cambridge, UK: Cambridge University Press.

Bailey, Robert C. 1985. *The Socioecology of Efe Pygmy Men in the Ituri Forest*. Anthropological Papers Series. Ann Arbor: University of Michigan, Museum of Anthropology.

Bardaweel, Sanaa K., Mustafa Gul, Muhammad Alzweiri, Aman Ishaqat, Husam A. ALSalamat, and Rasha M. Bashatwah. 2018. "Reactive oxygen species: The dual role in physiological and pathological conditions of the human body." *Eurasian Journal of Medicine* 50 (3): 193.

Barendregt, Jan J., Luc Bonneux, and Paul J. van der Maas. 1997. "The health care costs of smoking." *New England Journal of Medicine* 337 (15): 1052–1057.

Barton, Robert A., and Chris Venditti. 2014. "Rapid evolution of the cerebellum in humans and other great apes." *Current Biology* 24 (20): 2440–2444.

Bayer, Ronald, and Sandro Galea. 2015. "Public health in the precision-medicine era." *New England Journal of Medicine* 373 (6): 499–501.

Becker, Christoph K., and Stefan T. Trautmann. 2022. "Does happiness increase in old age? Longitudinal evidence from 20 European countries." *Journal of Happiness Studies* 23:3625–3654.

Becker, Noémie S. A., Paul Verdu, Barry Hewlett, and Samuel Pavard. 2010. "Can life history trade-offs explain the evolution of short stature in human pygmies? A response to Migliano et al. (2007)." *Human Biology* 82 (1): 17–27.

Bellhouse, David R. 2011. "A new look at Halley's life table." *Journal of the Royal Statistical Society: Series A (Statistics in Society)* 174 (3): 823–832.

Belsky, Daniel W., Avshalom Caspi, Renate Houts, Harvey J. Cohen, David L. Corcoran, Andrea Danese, HonaLee Harrington, et al. 2015. "Quantification of biological aging in young adults." *Proceedings of the National Academy of Sciences* 112 (30): E4104–E4110.

Belsky, Daniel W., Terrie E. Moffitt, Alan A. Cohen, David L. Corcoran, Morgan E. Levine, Joseph A. Prinz, Jonathan Schaefer, et al. 2018. "Eleven telomere, epigenetic clock, and biomarker-composite quantifications of biological aging: Do they measure the same thing?" *American Journal of Epidemiology* 187 (6): 1220–1230.

Bering, J. M., and D. J. Povinelli. 2003. "Comparing cognitive development." In *Primate Psychology*, edited by D. Maestripieri, 205–233. Cambridge, MA: Harvard University Press.

Bertoni, M. 1941. "Los Guayakies. Caracteres antropológicos, razas etnológicas y reseña cultural." *Revista de la Sociedad Científica del Paraguay* 5 (2): 1–62.

Bertrand, Benoit, Eugenia Cunha, Anne Bécart, Didier Gosset, and Valery Hédouin. 2019. "Age at death estimation by cementochronology: Too precise to be true or too precise to be accurate?" *American Journal of Physical Anthropology* 169 (3): 464–481.

Bhui, Rahul, Maciej Chudek, and Joseph Henrich. 2019. "Work time and market integration in the original affluent society." *Proceedings of the National Academy of Sciences* 116 (44): 22100–22105.

Biesele, Megan. 1993. *Women Like Meat: The Folklore and Foraging Ideology of the Kalahari Ju/'hoan*. Bloomington: Indiana University Press.

Biesele, Megan, and Nancy Howell. 1981. "'The old people give you life': Aging among !Kung hunter-gatherers." In *Other Ways of Growing Old*, edited by Pamela Amoss and Stevan Harrell, 77–98. Stanford, CA: Stanford University Press.

Binford, Lewis Roberts, and W. J. Chasko. 1976. "Nunamiut demographic history: A provocative case." In *Demographic Anthropology: Quantitative Approaches*, edited by Ezra B. W. Zubrow, 63–143. Albuquerque: University of New Mexico Press.

Birks, Jacqueline, and John Grimley Evans. 2009. "Ginkgo biloba for cognitive impairment and dementia." *Cochrane Database of Systematic Reviews* (1). https://doi.org/10.1002/14651858 .CD003120.pub3.

Biswas-Diener, Robert, Joar Vittersø, and Ed Diener. 2009. "Most people are pretty happy, but there is cultural variation: The Inughuit, the Amish, and the Maasai." In *Culture and Well-Being: The Collected Works of Ed Diener*, edited by Ed Diener, 245–260. Dordrecht: Springer Netherlands.

Black, Francis L. 1975. "Infectious disease in primitive societies." *Science* 187 (4176): 515–518.

Blackwell, Aaron, Benjamin Trumble, Ivan Maldonado Suarez, Jonathan Stieglitz, Bret Beheim, J. Josh Snodgrass, Hillard Kaplan, and Michael Gurven. 2016. "Immune function in Amazonian horticulturalists." *Annals of Human Biology* 43 (4): 382–396. https://doi.org/10.1080 /03014460.2016.1189963.

Blanchflower, David, and Carol Graham. 2022. "The mid-life dip in well-being: A critique." *Social Indicators Research* 161 (1): 287–344.

Blanchflower, David, and Andrew Oswald. 2019. "Do humans suffer a psychological low in midlife? Two approaches (with and without controls) in seven data sets." In *The Economics of Happiness*, edited by Mariano Rojas, 439–453. Cham, Switzerland: Springer.

Blurton Jones, Nicholas 1991. "Review: Land filled with flies: A political economy of the Kalahari." *Journal of Anthropological Research* 47 (1): 95–97.

Blurton Jones, Nicholas. 2016. *Demography and Evolutionary Ecology of Hadza Hunter-Gatherers*. Cambridge Studies in Biological and Evolutionary Anthropology, Series Number 71: Cambridge, UK: Cambridge University Press.

Blurton Jones, Nicholas, Kristen Hawkes, and James O'Connell. 2005. "Older Hadza men and women as helpers: Residence data." In *Hunter-Gatherer Childhoods: Evolutionary, Developmental, and Cultural Perspectives*, edited by B. S. Hewlett and M. E. Lamb, 214–236. Piscataway: Transactions.

Blurton Jones, Nicholas, L. Smith, James O'Connell, Kristen Hawkes, and C. L. Samuzora. 1992. "Demography of the Hadza, an increasing and high density population of savanna foragers." *American Journal of Physical Anthropology* 89:159–181.

Bock, John. 2002. "Learning, life history, and productivity: Children's lives in the Okavango Delta, Botswana." *Human Nature* 13 (2): 161–198.

Bocquet-Appel, Jean-Pierre, and Claude Masset. 1982. "Farewell to paleodemography." *Journal of Human Evolution* 11 (4): 321–333.

Bocquet-Appel, Jean Pierre, and Claude Masset. 1996. "Paleodemography: Expectancy and false hope." *American Journal of Physical Anthropology* 99 (4): 571–583.

Boldsen, Jesper L., George R. Milner, Lyle W. Konigsberg, James W. Wood, Robert D. Hoppa, and James W. Vaupel. 2002. "Transition analysis: A new method for estimating age from skeletons." In *Paleodemography: Age Distributions from Skeletal Samples*, edited by R. D. Hoppa and J. W. Vaupel, 73–106. Cambridge, UK: Cambridge University Press.

Bolland, Mark J., Alison Avenell, John A. Baron, Andrew Grey, Graeme S. MacLennan, Greg D. Gamble, and Ian R. Reid. 2010. "Effect of calcium supplements on risk of myocardial infarction and cardiovascular events: Meta-analysis." *BMJ* 341:c3691.

Boone, James L. 2002. "Subsistence strategies and early human population history: An evolutionary ecological perspective." *World Archaeology* 34 (1): 6–25.

Booth, Frank W., Christian K. Roberts, and Matthew J. Laye. 2012. "Lack of exercise is a major cause of chronic diseases." *Comprehensive Physiology* 2 (2): 1143.

Borgerhoff Mulder, Monique, Lameck Msalu, Tim Caro, and Jonathan Salerno. 2012. "Remarkable rates of lightning strike mortality in Malawi." *PLoS One* 7 (1): e29281.

Boyd, Brian. 2010. *On the Origin of Stories: Evolution, Cognition, and Fiction*. Cambridge, MA: Harvard University Press.

Bradford, Phillips Verner, and Harvey Blume. 1992. *Ota Benga: The Pygmy in the Zoo*. New York: St. Martin's Press.

Brainard, Jean. 1981. *Herders to Farmers: The Effects of Settlement on the Demography of the Turkana Population of Kenya*. Binghamton, NY: SUNY Press.

Brainard, Jean. 1986. "Differential mortality in Turkana agriculturalists and pastoralists." *American Journal of Physical Anthropology* 70 (4): 525–536.

Brent, Lauren J. N., Daniel W. Franks, Emma A. Foster, Kenneth C. Balcomb, Michael A. Cant, and Darren P. Croft. 2015. "Ecological knowledge, leadership, and the evolution of menopause in killer whales." *Current Biology* 25 (6): 746–750.

Brestoff, Jonathan R., and David Artis. 2015. "Immune regulation of metabolic homeostasis in health and disease." *Cell* 161 (1): 146–160. http://dx.doi.org/10.1016/j.cell.2015.02.022.

Brinkworth, Jessica F., and Luis B. Barreiro. 2014. "The contribution of natural selection to present-day susceptibility to chronic inflammatory and autoimmune disease." *Current Opinion in Immunology* 31:66–78. https://doi.org/10.1016/j.coi.2014.09.008.

Brown, A. Radcliffe. 1922. *The Andaman Islanders: A Study in Social Anthropology (Anthony Wilkin Studentship Research, 1906)*. Cambridge, UK: Cambridge University Press.

Brown, Stephanie L., Randolph M. Nesse, Amiram D. Vinokur, and Dylan M. Smith. 2003. "Providing social support may be more beneficial than receiving it: Results from a prospective study of mortality." *Psychological Science* 14 (4): 320–327. https://doi.org/10.1111/1467-9280.1.

Buettner, Dan, and Sam Skemp. 2016. "Blue zones: Lessons from the world's longest lived." *American Journal of Lifestyle Medicine* 10 (5): 318–321.

Buffenstein, Rochelle. 2008. "Negligible senescence in the longest living rodent, the naked mole-rat: Insights from a successfully aging species." *Journal of Comparative Physiology B* 178 (4): 439–445.

Bugos, P., and L. McCarthy. 1983. "Ayoreo infanticide: A case study." In *Infanticide: Comparative and Evolutionary Perspectives*, edited by G. Hausfater and S. Hrdy, 503–570. New York: Aldine.

Bullock, Meggan, Lourdes Márquez, Patricia Hernández, and Fernando Ruíz. 2013. "Paleodemographic age-at-death distributions of two Mexican skeletal collections: A comparison of transition analysis and traditional aging methods." *American Journal of Physical Anthropology* 152 (1): 67–78.

Bulterijs, Sven, Raphaella S. Hull, Victor C. E. Björk, and Avi G. Roy. 2015. "It is time to classify biological aging as a disease." *Frontiers in Genetics* 6:205.

Burgess, Adriane. 2014. "An evolutionary concept analysis of labor support." *International Journal of Childbirth Education* 29 (2).

Burstein, Stanley M., and Caleb E. Finch. 2018. "Longevity examined: An ancient Greek's very modern views on ageing." *Nature* 560 (7718): 430–431.

Butler, Katy. 2014. *Knocking on Heaven's Door: The Path to a Better Way of Death.* New York: Simon and Schuster.

Byars, Sean G., Qin Qin Huang, Lesley-Ann Gray, Andrew Bakshi, Samuli Ripatti, Gad Abraham, Stephen C. Stearns, and Michael Inouye. 2017. "Genetic loci associated with coronary artery disease harbor evidence of selection and antagonistic pleiotropy." *PLoS Genetics* 13 (6): e1006328.

Byars, Sean G., and Konstantinos Voskarides. 2019. "Genes that improved fitness also cost modern humans: Evidence for genes with antagonistic effects on longevity and disease." *Evolution, Medicine, and Public Health* 2019 (1): 4–6.

Byars, Sean G., and Konstantinos Voskarides. 2020. "Antagonistic pleiotropy in human disease." *Journal of Molecular Evolution* 88 (1): 12–25.

Byers, D. A., and A. Ugan. 2005. "Should we expect large game specialization in the late Pleistocene? An optimal foraging perspective on early Paleoindian prey choice." *Journal of Archaeological Science* 32:1624–1640.

Cagan, Alex, Adrian Baez-Ortega, Natalia Brzozowska, Federico Abascal, Tim H. H. Coorens, Mathijs A. Sanders, Andrew R. J. Lawson, et al. 2022. "Somatic mutation rates scale with lifespan across mammals." *Nature* 604 (7906): 517–524.

Cannata-Andia, Jorge B., Pablo Roman-Garcia, and Keith Hruska. 2011. "The connections between vascular calcification and bone health." *Nephrology Dialysis Transplantation* 26 (11): 3429–3436.

Canto, John G., Catarina I. Kiefe, William J. Rogers, Eric D. Peterson, Paul D. Frederick, William J. French, C. Michael Gibson, et al. 2011. "Number of coronary heart disease risk factors and mortality in patients with first myocardial infarction." *JAMA* 306 (19): 2120–2127.

Canudas-Romo, Vladimir. 2008. "The modal age at death and the shifting mortality hypothesis." *Demographic Research* 19:1179–1204.

Carey, James R. 1989. "The multiple decrement life table: A unifying framework for cause-of-death analysis in ecology." *Oecologia* 78 (1): 131–137.

Carmichael, Sarah. 2011. "Marriage and power: Age at first marriage and spousal age gap in lesser developed countries." *History of the Family* 16 (4): 416–436. https://doi.org/10.1016/j.hisfam .2011.08.002.

Carnes, B. A., S. J. Olshansky, and D. Grahn. 1996. "Continuing the search for a law of mortality." *Population and Development Review* 22:231–264.

Carnes, Bruce A., David O. Staats, and William E. Sonntag. 2008. "Does senescence give rise to disease?" *Mechanisms of Ageing and Development* 129 (12): 693–699.

Carrera-Bastos, Pedro, Maelán Fontes-Villalba, Michael Gurven, Frits A. J. Muskiet, Torbjörn Åkerfeldt, Ulf Lindblad, Lennart Råstam, et al. 2020. "C-reactive protein in traditional Melanesians on Kitava." *BMC Cardiovascular Disorders* 20 (1): 1–8.

Carrera-Bastos, Pedro, Maelan Fontes-Villalba, James H. O'Keefe, Staffan Lindeberg, and Loren Cordain. 2011. "The western diet and lifestyle and diseases of civilization." *Research Reports in Clinical Cardiology* 2 (1): 15–35.

Carstensen, Laura L. 2006. "The influence of a sense of time on human development." *Science* 312 (5782): 1913–1915.

Caspari, Rachel, and Sang-Hee Lee. 2004. "Older age becomes common late in human evolution." *Proceedings of the National Academy of Science* 101 (30): 10895–10900.

Caswell, Hal. 2007. "Extrinsic mortality and the evolution of senescence." *Trends in Ecology and Evolution* 22 (4): 173–174.

Catania, K., and P. K. Panegyres. 2017. "Dementia in indigenous populations." *Journal of Neurological Disorders* 5 (362): 2.

Chagnon, N. 1968. *Yanomamo: The Fierce People.* New York: Holt, Rinehart and Winston.

Chapais, Bernard. 2009. *Primeval Kinship: How Pair-Bonding Gave Birth to Human Society.* Cambridge, MA: Harvard University Press.

Chen, Brian H., Riccardo E. Marioni, Elena Colicino, Marjolein J. Peters, Cavin K. Ward-Caviness, Pei-Chien Tsai, Nicholas S. Roetker, et al. 2016. "DNA methylation-based measures of biological age: Meta-analysis predicting time to death." *Aging (Albany NY)* 8 (9): 1844.

Chen, Yingxi, Neal D. Freedman, Erik J. Rodriquez, Meredith S. Shiels, Anna M. Napoles, Diana R. Withrow, Susan Spillane, Byron Sigel, Eliseo J. Perez-Stable, and Amy Berrington de González. 2020. "Trends in premature deaths among adults in the United States and Latin America." *JAMA Network Open* 3 (2): e1921085–e1921085.

Cheung, Siu Lan Karen, Jean-Marie Robine, Edward Jow-Ching, and Graziella Caselli. 2005. "Three dimensions of the survival curve: Horizontalization, verticalization and longevity extension." *Demography* 42 (2): 243–258.

Chisena, Simone, and Christophe Delage. 2018. "On the attribution of palaeolithic artworks: The case of La Marche (Lussac-les-Chateaux, Vienne)." *Open Archaeology* 4 (1): 239–261.

Chu, Carol, Jennifer M. Buchman-Schmitt, Ian H. Stanley, Melanie A. Hom, Raymond P. Tucker, Christopher R. Hagan, Megan L. Rogers, et al. 2017. "The interpersonal theory of suicide: A systematic review and meta-analysis of a decade of cross-national research." *Psychological Bulletin* 143 (12): 1313.

Ciecka, James E. 2008. "Edmond Halley's life table and its uses." *Journal of Legal Economics* 15:65.

Cipriani, Lidio. 1961. "Hygiene and medical practices among the Onge (Little Andaman)." *Anthropos* 56 (3/4): 481–500.

Clark, Gregory. 2007. *A Farewell to Alms.* Princeton, NJ: Princeton University Press.

Coale, A. J., and P. Demeny. 1966. *Regional Model Life Tables and Stable Populations.* Princeton, NJ: Rutgers University Press.

Coall, David A., and Ralph Hertwig. 2010. "Grandparental investment: Past, present, and future." *Behavioral and Brain Sciences* 33 (1): 1–19.

Cochran, G., and H. Harpending. 2009. *The 10,000 Year Explosion: How Civilization Accelerated Human Evolution.* New York: Basic Books.

Cohen, Alan A. 2004. "Female post-reproductive lifespan: A general mammalian trait." *Biological Reviews* 79 (4): 733–750.

Cohen, Alan A., Luigi Ferrucci, Tamàs Fülöp, Dominique Gravel, Nan Hao, Andres Kriete, Morgan E. Levine, et al. 2022. "A complex systems approach to aging biology." *Nature Aging* 2:580–591. https://doi.org/10.1038/s43587-022-00252-6.

Cohen, Alan A., Véronique Legault, and Tamàs Fülöp. 2020. "What if there's no such thing as 'aging'?" *Mechanisms of Ageing and Development* 192:111344.

Cohen, Alan A., Emmanuel Milot, Jian Yong, Christopher L. Seplaki, Tamàs Fülöp, Karen Bandeen-Roche, and Linda P. Fried. 2013. "A novel statistical approach shows evidence for multi-system physiological dysregulation during aging." *Mechanisms of Ageing and Development* 134 (3/4): 110–117.

Cohen, Mark Nathan, and George J. Armelagos. 1984. *Paleopathology at the Origins of Agriculture.* Orlando, FL: Academic Press.

Colchero, Fernando, Roland Rau, Owen R. Jones, Julia A. Barthold, Dalia A. Conde, Adam Lenart, Laszlo Nemeth, et al. 2016. "The emergence of longevous populations." *Proceedings of the National Academy of Sciences* 113 (48): E7681–E7690.

Conrad, Donald F., Jonathan E. M. Keebler, Mark A. DePristo, Sarah J. Lindsay, Yujun Zhang, Ferran Casals, Youssef Idaghdour, et al. 2011. "Variation in genome-wide mutation rates within and between human families." *Nature Genetics* 43 (7): 712.

Cordain, L., J. Brand Miller, S. B. Eaton, N. Mann, S.H.A. Holt, and J. D. Speth. 2000. "Plant-animal subsistence ratios and macronutrient energy estimations in hunter-gatherer diets." *American Journal of Clinical Nutrition* 71:682–692.

Cortes-Bergoderi, Mery, Kashish Goel, Mohammad Hassan Murad, Thomas Allison, Virend K. Somers, Patricia J. Erwin, Ondrej Sochor, and Francisco Lopez-Jimenez. 2013. "Cardiovascular mortality in Hispanics compared to non-Hispanic whites: A systematic review and meta-analysis of the Hispanic paradox." *European Journal of Internal Medicine* 24 (8): 791–799.

Crimmins, Eileen M., Hyunju Shim, Yuan S. Zhang, and Jung Ki Kim. 2019. "Differences between men and women in mortality and the health dimensions of the morbidity process." *Clinical Chemistry* 65 (1): 135–145.

Crimmins, Eileen M., Yuan Zhang, and Yasuhiko Saito. 2016. "Trends over 4 decades in disability-free life expectancy in the United States." *American Journal of Public Health* 106 (7): 1287–1293.

Croft, Darren P., Lauren J. N. Brent, Daniel W. Franks, and Michael A. Cant. 2015. "The evolution of prolonged life after reproduction." *Trends in Ecology and Evolution* 30 (7): 407–416.

Cukrowicz, Kelly C., Jennifer S. Cheavens, Kimberly A. Van Orden, R. Michael Ragain, and Ronald L. Cook. 2011. "Perceived burdensomeness and suicide ideation in older adults." *Psychology and Aging* 26 (2): 331.

Cumming, Elaine, and William E. Henry. 1961. *Growing Old, the Process of Disengagement*. New York: Basic Books.

Cummins, Neil. 2017. "Lifespans of the European elite, 800–1800." *Journal of Economic History* 77 (2): 406–439.

Cunningham, Andrew J., Steven Worthington, Vivek V. Venkataraman, and Richard W. Wrangham. 2019. "Do modern hunter-gatherers live in marginal habitats?" *Journal of Archaeological Science: Reports* 25:584–599.

Danielsbacka, Mirkka, Lenka Křenková, and Antti O. Tanskanen. 2022. "Grandparenting, health, and well-being: A systematic literature review." *European Journal of Ageing* 19:1–28.

Davison, Raziel, and Michael Gurven. 2022. "The importance of elders: Extending Hamilton's force of selection to include intergenerational transfers." *Proceedings of the National Academy of Sciences* 119 (28): e2200073119.

De Beauvoir, Simone. 1972. *The Coming of Age [Old Age]. Translated by Patrick O'Brian*. London: André Deutsch / George Weidenfeld and Nicolson.

DeKosky, Steven T., Jeff D. Williamson, Annette L. Fitzpatrick, Richard A. Kronmal, Diane G. Ives, Judith A. Saxton, Oscar L. Lopez, et al. 2008. "Ginkgo biloba for prevention of dementia: A randomized controlled trial." *JAMA* 300 (19): 2253–2262.

De la Croix, David, and Omar Licandro. 2015. "The longevity of famous people from Hammurabi to Einstein." *Journal of Economic Growth* 20 (3): 263–303.

Delage, Christophe. 2016. "Comments on a recent challenge to the authenticity of the La Marche engravings." *Fornvännen* 111 (3): 192–197.

De Magalhaes, J. P., and J. Costa. 2009. "A database of vertebrate longevity records and their relation to other life-history traits." *Journal of Evolutionary Biology* 22 (8): 1770–1774.

De Ruiter, K., Dicky Levenus Tahapary, E. Sartono, Pradana Soewondo, Taniawati Supali, J.W.A. Smit, and M. Yazdanbakhsh. 2017. "Helminths, hygiene hypothesis and type 2 diabetes." *Parasite Immunology* 39 (5): e12404.

DeWitte, Sharon N., Jessica C. Boulware, and Rebecca C. Redfern. 2013. "Medieval monastic mortality: Hazard analysis of mortality differences between monastic and nonmonastic cemeteries in England." *American Journal of Physical Anthropology* 152 (3): 322–332.

Diamond, Jared. 2002. "Quantitative evolutionary design." *Journal of Physiology* 542 (2): 337–345.

Diamond, Jared. 2013. *The World until Yesterday: What Can We Learn from Traditional Societies?* New York: Penguin.

Diekmann, Yoan, Daniel Smith, Pascale Gerbault, Mark Dyble, Abigail E. Page, Nikhil Chaudhary, Andrea Bamberg Migliano, and Mark G. Thomas. 2017. "Accurate age estimation in small-scale societies." *Proceedings of the National Academy of Sciences* 114 (31): 8205–8210.

Diener, Ed, and Carol Diener. 1996. "Most people are happy." *Psychological Science* 7 (3): 181–185.

Di Gessa, Giorgio, Valeria Bordone, and Bruno Arpino. 2020. "Becoming a grandparent and its effect on well-being: The role of order of transitions, time, and gender." *Journals of Gerontology: Series B* 75 (10): 2250–2262.

Ding, Ding, Joe Van Buskirk, Binh Nguyen, Emmanuel Stamatakis, Mona Elbarbary, Nicola Veronese, Philip J. Clare, et al. 2022. "Physical activity, diet quality and all-cause cardiovascular disease and cancer mortality: A prospective study of 346 627 UK Biobank participants." *British Journal of Sports Medicine.* https://doi.org/10.1136/bjsports-2021-105195.

Djankov, Simeon, Elena Nikolova, and Jan Zilinsky. 2016. "The happiness gap in Eastern Europe." *Journal of Comparative Economics* 44 (1): 108–124.

Dobzhansky, Theodosius. 1973. "Nothing in biology makes sense except in the light of evolution." *American Biology Teacher* 35 (3): 125–129.

Domínguez-Rodrigo, Manuel, Enrique Baquedano, Elia Organista, Lucía Cobo-Sánchez, Audax Mabulla, Vivek Maskara, Agness Gidna, et al. 2021. "Early Pleistocene faunivorous hominins were not kleptoparasitic, and this impacted the evolution of human anatomy and socio-ecology." *Scientific Reports* 11(1): 16135.

Draper, Patricia, and Anne Buchanan. 1992. "If you have a child you have a life: Demographic and cultural perspectives on fathering in old age in !Kung society." In *Father-Child Relations: Cultural and Biosocial Contexts*, edited by Barry S. Hewlett, 131–152. New York: Routledge.

Drefahl, Sven, Anders Ahlbom, and Karin Modig. 2014. "Losing ground—Swedish life expectancy in a comparative perspective." *PLoS One* 9 (2): e88357.

Dyson, Tim. 1977. "The demography of the Hadza in historical perspective." *African Historical Demography.* Edinburgh: Centre for African Studies, University of Edinburgh.

Early, John. 1990. *The Population Dynamics of the Mucajai Yanomama.* San Diego, CA: Academic Press.

Early, John D., and Thomas N. Headland. 1998. *Population Dynamics of a Philippine Rain Forest People: The San Ildefonso Agta.* Gainesville: University Press of Florida.

Early, John D., and John F. Peters. 2000. *The Xilixana Yanomami of the Amazon: History, Social Structure, and Population Dynamics.* Gainsville: University Press of Florida.

Eder, J. F. 1987. *On the Road to Tribal Extinction: Depopulation, Deculturation, and Adaptive Well-Being among the Batak of the Philippines.* Berkeley: University of California Press.

Ehrenreich, Barbara. 2018. *Natural Causes: An Epidemic of Wellness, the Certainty of Dying, and Killing Ourselves to Live Longer.* London: Hachette UK.

Ellis, Rebecca. 1996. A taste for movement: An exploration of the social ethics of the Tsimane of lowland Bolivia. PhD thesis, University of St. Andrews, Scotland.

Ellis, Samuel, Daniel W. Franks, Stuart Nattrass, Michael A. Cant, Destiny L. Bradley, Deborah Giles, Kenneth C. Balcomb, and Darren P. Croft. 2018. "Postreproductive lifespans are rare in mammals." *Ecology and Evolution* 8 (5): 2482–2494.

Endicott, Kirk. 1988. "Property, power and conflict among the Batek of Malaysia." In *Hunters and Gatherers: Property, Power and Ideology*, edited by T. Ingold, D. Riches, and J. Woodburn, 110–128. New York: St. Martin's Press.

Enos, William F., Robert H. Holmes, and James Beyer. 1953. "Coronary disease among United States soldiers killed in action in Korea: Preliminary report." *Journal of the American Medical Association* 152 (12): 1090–1093.

Estioko-Griffin, Agnes A. 1985. "Women as hunters: The case of an Eastern Cagayan Agta group." In *The Agta of Northeastern Luzon: Recent Studies*, edited by P. B. Griffin and A. Estioko-Griffin, 18–32. Cebu City, Philippines: San Carlos Publications.

Evans, Alistair R., E. Susanne Daly, Kierstin K. Catlett, Kathleen S. Paul, Stephen J. King, Matthew M. Skinner, Hans P. Nesse, et al. 2016. "A simple rule governs the evolution and development of hominin tooth size." *Nature* 530 (7591): 477–480.

Fabian, Daniel, and Thomas Flatt. 2011. "The evolution of aging." *Nature Education Knowledge* 3 (3): 1–10.

Fahy, Gregory M., Robert T. Brooke, James P. Watson, Zinaida Good, Shreyas S. Vasanawala, Holden Maecker, Michael D. Leipold, David T. S. Lin, Michael S. Kobor, and Steve Horvath. 2019. "Reversal of epigenetic aging and immunosenescent trends in humans." *Aging Cell* 18 (6): e13028.

Feng, Ying, and Jie Ren. 2022. "Within marriage age gap across countries." *Economics Letters* 210: 110190.

Fernández-Llamazares, Álvaro, Isabel Díaz-Reviriego, and Victoria Reyes-García. 2017. "Defaunation through the eyes of the Tsimane." In *Hunter-Gatherers in a Changing World*, edited by Victoria Reyes-García and Aili Pyhälä, 77–90. New York: Springer.

Finch, Caleb E. 1990. *Longevity, Senescence and the Genome*. Chicago: University of Chicago Press.

Finch, Caleb E. 2007. *The Biology of Human Longevity*. San Diego, CA: Academic Press.

Finch, Caleb E. 2009. "Update on slow aging and negligible senescence—a mini-review." *Gerontology* 55 (3): 307–313.

Finch, Caleb E., Hiram Beltrán-Sánchez, and Eileen M. Crimmins. 2014. "Uneven futures of human lifespans: Reckonings from Gompertz mortality rates, climate change, and air pollution." *Gerontology* 60 (2): 183–188.

Finch, Caleb E., Malcolm C. Pike, and Matthew Whitten. 1990. "Slow mortality rate accelerations during aging in animals approximate that of humans." *Science* 249:902–905.

Finch, Caleb E., and Craig B. Stanford. 2004. "Meat-adaptive genes and the evolution of slower aging in humans." *Quarterly Review of Biology* 79 (1): 3–50.

Firth, R. 1929. *Primitive Economics of the New Zealand Maori*. New York: Routledge.

Fisher, R. A. 1930. *The Genetical Theory of Natural Selection*. London: Oxford University Press.

Fitzgerald, Kelly G., and Francis G. Caro. 2014. "An overview of age-friendly cities and communities around the world." *Journal of Aging and Social Policy* 26 (1/2): 1–18.

Fiuza-Luces, Carmen, Alejandro Santos-Lozano, Michael Joyner, Pedro Carrera-Bastos, Oscar Picazo, José L. Zugaza, Mikel Izquierdo, Luis M. Ruilope, and Alejandro Lucia. 2018. "Exercise benefits in cardiovascular disease: Beyond attenuation of traditional risk factors." *Nature Reviews Cardiology* 15 (12): 731–743.

Fix, Alan G. 1977. *The Demography of the Semai Senoi*. Anthropological Papers Series, vol. 62. Ann Arbor: University of Michigan Museum of Anthropology.

Flannery, Regina. 1953. *The Gros Ventres of Montana—Part I: Social life*. Washington, DC: Catholic University of America Press.

Floyd, Simeon, Giovanni Rossi, Julija Baranova, Joe Blythe, Mark Dingemanse, Kobin H. Kendrick, Jörg Zinken, and N. J. Enfield. 2018. "Universals and cultural diversity in the expression of gratitude." *Royal Society Open Science* 5 (5): 180391.

Forbes, Dorothy, Scott C. Forbes, Catherine M. Blake, Emily J. Thiessen, and Sean Forbes. 2015. "Exercise programs for people with dementia." *Cochrane Database of Systematic Reviews*, no. 4, art. CD006489. https://doi.org/10.1002/14651858.CD006489.pub4.

Fortea, Juan, Jordi Pegueroles, Daniel Alcolea, Olivia Belbin, Oriol Dols-Icardo, Lídia Vaqué-Alcázar, Laura Videla, et al. 2024. "APOE4 homozygozity represents a distinct genetic form of Alzheimer's disease." *Nature Medicine* 30 (5): 1284–1291.

Fortes, Meyer. 1949. *The Web of Kinship among the Tallensi*. African Ethnographic Studies of the 20th Century. Oxford: Oxford University Press.

Fortier, Jana. 2001. "Sharing, hoarding, and theft: Exchange and resistance in forager-farmer relations." *Ethnology* 40 (3): 193–211.

Fowler, Loretta. 1987. *Shared Symbols, Contested Meanings: Gros Ventre Culture and History, 1778–1984*. Ithaca, NY: Cornell University Press.

Frackowiak, Tomasz, Anna Oleszkiewicz, Marina Butovskaya, Agata Groyecka, Maciej Karwowski, Marta Kowal, and Piotr Sorokowski. 2020. Subjective happiness among Polish and Hadza people." *Frontiers in Psychology* 11: 1173.

Franceschi, Claudio, Paolo Garagnani, Cristina Morsiani, Maria Conte, Aurelia Santoro, Andrea Grignolio, Daniela Monti, et al. 2018. "The continuum of aging and age-related diseases: common mechanisms but different rates." *Frontiers in Medicine* 5:61.

Fraser, Gary E., and David J. Shavlik. 2001. "Ten years of life: Is it a matter of choice?" *Archives of Internal Medicine* 161 (13): 1645–1652. https://doi.org/10.1001/archinte.161.13.1645.

Frazier, Rebecca S., Claudia Birmingham, Victoria Wheat, and Annie Georges. 2019. "A systematic review of Senior Corps' impact on volunteers and program beneficiaries." Report prepared for the Corporation for National and Community Service. Bethesda, MD: JBS International.

French, Eric B., Jeremy McCauley, Maria Aragon, Pieter Bakx, Martin Chalkley, Stacey H. Chen, Bent J. Christensen, et al. 2017. "End-of-life medical spending in last twelve months of life is lower than previously reported." *Health Affairs* 36 (7): 1211–1217.

Fries, J. F. 1980. "Ageing, natural death and the compression of morbidity." *New England Journal of Medicine* 303:130–136.

Fries, J. F. 1989. "The compression of morbidity: Near or far?" *Milbank Quarterly* 67:208–232.

Fumagalli, Matteo, Stephane M. Camus, Yoan Diekmann, Alice Burke, Marine D. Camus, Paul J. Norman, Agnel Joseph, et al. 2019. "Genetic diversity of CHC22 clathrin impacts its function in glucose metabolism." *eLife* 8:e41517.

Furuse, Yuki, Akira Suzuki, and Hitoshi Oshitani. 2010. "Origin of measles virus: Divergence from rinderpest virus between the 11th and 12th centuries." *Virology Journal* 7:52–52. https://doi.org/10.1186/1743-422X-7-52.

Gage, Timothy B. 1998. "The comparative demography of primates: With some comments on the evolution of life histories." *Annual Review of Anthropology* 27:197–221.

Gaillard, Jean-Michel, Michael Garratt, and Jean-François Lemaître. 2017. "Senescence in mammalian life history traits." In *The Evolution of Senescence in the Tree of Life*, edited by Richard P. Shefferson, Owen R. Jones, and Roberto Salguero-Gómez, 126–155. Cambridge, UK: Cambridge University Press.

Galbraith, Eric D., Christopher Barrington-Leigh, Sara Miñarro, Santiago Álvarez-Fernández, Emmanuel M.N.A.N. Attoh, Petra Benyei, Laura Calvet-Mir, et al. 2024. "High life satisfaction reported among small-scale societies with low incomes." *Proceedings of the National Academy of Sciences* 121 (7): e2311703121.

Garcia, Angela R., Caleb Finch, Margaret Gatz, Thomas Kraft, Daniel Eid Rodriguez, Daniel Cummings, Mia Charifson, et al. 2021. "APOE4 is associated with elevated blood lipids and lower levels of innate immune biomarkers in a tropical Amerindian subsistence population." *eLife* 10:e68231.

Gardiner, Fergus W., Kristopher Rallah-Baker, Angela Dos Santos, Pritish Sharma, Leonid Churilov, Geoffrey A. Donnan, Stephen M. Davis, et al. 2021. "Indigenous Australians have a greater prevalence of heart, stroke, and vascular disease, are younger at death, with higher hospitalisation and more aeromedical retrievals from remote regions." *eClinicalMedicine* 42:101181.

Garfield, Zachary H., Kristen L. Syme, and Edward H. Hagen. 2020. "Universal and variable leadership dimensions across human societies." *Evolution and Human Behavior* 41 (5): 397–414.

Garmany, Armin, Satsuki Yamada, and Andre Terzic. 2021. "Longevity leap: Mind the healthspan gap." *NPJ Regenerative Medicine* 6 (1): 1–7.

Gat, Azar. 2008. *War in Human Civilization*. Oxford: Oxford University Press.

Gatz, Margaret, Wendy J. Mack, Helena C. Chui, E. Meng Law, Giuseppe Barisano, M. Linda Sutherland, James D. Sutherland, et al. 2022. "Prevalence of dementia and mild cognitive impairment in indigenous Bolivian forager-horticulturalists." *Alzheimer's and Dementia*. https://doi.org/10.1002/alz.12626.

Gavrilov, Leonid A., and Natalia S. Gavrilova. 1991. *The Biology of the Lifespan: A Quantitative Approach*. New York: Harwood Academic.

Gavrilov, Leonid A., and Natalia S. Gavrilova. 2005. "Reliability theory of aging and longevity." In *Handbook of the Biology of Aging*, edited by Nicolas Musi and Peter Hornsby, 3–42. New York: Academic Press.

Gavrilova, Natalia S., and Leonid A. Gavrilov. 2020. "Are we approaching a biological limit to human longevity?" *Journals of Gerontology: Series A* 75 (6): 1061–1067.

Getz, Sara M. 2020. "The use of transition analysis in skeletal age estimation." *Wiley Interdisciplinary Reviews: Forensic Science* 2 (6): e1378.

Glascock, Anthony P. 1991. "Nothing is without cost: The effects of development on the health of older people in South Central Somalia." *Journal of Cross-Cultural Gerontology* 6 (3): 287–299.

Glascock, Anthony P., and Susan L. Feinman. 1980. "A holocultural analysis of old age." *Comparative Social Research* 3 (3): 1.

Glascock, Anthony P., and Susan L. Feinman. 1981. "Social asset or social burden: Treatment of the aged in non-industrial societies." In *Dimensions: Aging, Culture, and Health*, edited by Christine L. Fry, 13–31. New York: Praeger.

Goldberg, Elkhonon. 2006. *The Wisdom Paradox: How Your Mind Can Grow Stronger as Your Brain Grows Older*. New York: Penguin.

Gompertz, Benjamin. 1825. "On the nature of the function expressive of the law of human mortality, and on a new mode of determining the value of life contingencies." *Philosophical Transactions of the Royal Society of London B* 115:513–585.

Gompertz, Benjamin. 1871. "On one uniform law of mortality from birth to extreme old age, and on the law of sickness." *Journal of the Institute of Actuaries* 16 (5): 329–344.

Goncalves, Natalia, Naomi Vidal Ferreira, Neha Khandpur, Euridice Martinez Steele, Renata Bertazzi Levy, Paulo Lotufo, Isabela Bensenor, Dirce Lobo Marchioni, and Claudia Suemoto. 2022. "Consumption of ultra-processed foods and cognitive decline in the ELSA-Brasil study: A prospective study." In *Alzheimer's Association International Conference*. https://alz.confex.com/alz/2022/meetingapp.cgi/Paper/63301.

González, Nicole Thompson, Emily Otali, Zarin Machanda, Martin N. Muller, Richard Wrangham, and Melissa Emery Thompson. 2020. "Urinary markers of oxidative stress respond to infection and late-life in wild chimpanzees." *PLoS One* 15 (9): e0238066.

Goodall, J. 1986. *The Chimpanzees of the Gombe: Patterns of Behavior*. Cambridge, UK: Cambridge University Press.

Gruber, Kenneth J., Susan H. Cupito, and Christina F. Dobson. 2013. "Impact of doulas on healthy birth outcomes." *Journal of Perinatal Education* 22 (1): 49–58. https://doi.org/10.1891/1058-1243.22.1.49.

Gruenewald, Tara L., Diana H. Liao, and Teresa E. Seeman. 2012. "Contributing to others, contributing to oneself: Perceptions of generativity and health in later life." *Journals of Gerontology: Series B* 67 (6): 660–665. https://doi.org/10.1093/geronb/gbs034.

Guemple, D. Lee. 1969. "Human resource management: The dilemma of the aging Eskimo." Paper read at Sociological Symposium. https://www.historymuseum.ca/collections /archive/3193679.

Gurven, Michael. 2012. "Infant and fetal mortality among a high fertility and mortality population in the Bolivian Amazon." *Social Science and Medicine* 75 (12): 2493–2502. http://dx.doi .org/10.1016/j.socscimed.2012.09.030.

Gurven, Michael, Aaron Blackwell, Daniel Eid Rodríguez, Jonathan Stieglitz, and Hillard Kaplan. 2012. "Does blood pressure inevitably rise with age? Longitudinal evidence among forager-horticulturalists." *Hypertension* 60 (1): 25–33.

Gurven, Michael, Yoann Buoro, Daniel Eid Rodriguez, Katherine Sayre, Benjamin Trumble, Aili Pyhälä, Hillard Kaplan, et al. 2024. "Subjective well-being across the life course among non-industrialized populations." *Science Advances* 10 (43): eadoo952.

Gurven, Michael, and Raziel Davison. 2019. "Periodic catastrophes over human evolutionary history are necessary to explain the forager population paradox." *Proceedings of the National Academy of Sciences* 116 (26): 12758–12766.

Gurven, Michael, Eric Fuerstenberg, Benjamin Trumble, Jonathan Stieglitz, Bret Beheim, Helen Davis, and Hillard Kaplan. 2017. "Cognitive performance across the life course of Bolivian forager-farmers with limited schooling." *Developmental Psychology* 53 (1): 160–176. http://dx .doi.org/10.1037/dev0000175.

Gurven, Michael, and Kim Hill. 2009. "Why do men hunt? A re-evaluation of "Man the Hunter" and the sexual division of labor." *Current Anthropology* 50 (1): 51–74.

Gurven, Michael, Adrian Jaeggi, Hillard Kaplan, and Daniel Cummings. 2013. "Physical activity and modernization among Bolivian Amerindians." *PLoS One* 8 (1): e55679.

Gurven, Michael, and Hillard Kaplan. 2006. "Determinants of time allocation to production across the lifespan among the Machiguenga and Piro Indians of Peru." *Human Nature* 17 (1): 1–49.

Gurven, Michael, and Hillard Kaplan. 2007. "Longevity among hunter-gatherers: A cross-cultural comparison." *Population and Development Review* 33 (2): 321–365.

Gurven, Michael, and Hillard Kaplan. 2008. "Beyond the grandmother hypothesis: Evolutionary models of longevity." In *Cultural Context of Aging: Worldwide Perspectives*, 3rd ed., edited by Jay Sokolovosky, 53–60. Westport, CT: Greenwood Press

Gurven, Michael, Hillard Kaplan, and Maguin Gutierrez. 2006. "How long does it take to become a proficient hunter? Implications for the evolution of delayed growth." *Journal of Human Evolution* 51:454–470.

Gurven, Michael, Hillard Kaplan, and Alfredo Zelada Supa. 2007. "Mortality experience of Tsimane Amerindians: Regional variation and temporal trends." *American Journal of Human Biology* 19:376–398.

Gurven, Michael, Hillard Kaplan, Jeffrey Winking, Daniel Eid Rodriguez, Sarinnapha Vasunilashorn, Jung Ki Kim, Caleb Finch, and Eileen Crimmins. 2009. "Inflammation and infection do not promote arterial aging and cardiovascular disease among lean Tsimane forager-horticulturalists." *PLoS One* 4 (8): e6590. https://doi.org/10.1371/journal.pone.0006590.

Gurven, Michael, Jonathan Stieglitz, Paul Hooper, Cristina Gomes, and Hillard Kaplan. 2012. "From the womb to the tomb: The role of transfers in shaping the evolved human life history." *Experimental Gerontology* 47:807–813.

Gurven, Michael, Benjamin Trumble, Jonathan Stieglitz, Aaron Blackwell, David Michalik, Caleb Finch, and Hillard Kaplan. 2016. "Cardiovascular disease and type 2 diabetes in evolutionary perspective: A critical role for helminths?" *Evolution, Medicine, and Public Health* 2016 (1): 338–357. https://doi.org/10.1093/emph/eow028.

Gurven, Michael, and Robert Walker. 2006. "Energetic demand of multiple dependents and the evolution of slow human growth." *Proceedings of the Royal Society of London, Series B: Biological Sciences* 273:835–841.

Gurven, Michael, Jeffrey Winking, Hillard Kaplan, Christopher von Rueden, and Lisa McAllister. 2009. "A bioeconomic approach to marriage and the sexual division of labor." *Human
Nature* 20 (2):151–183.

Gusinde, Martin. 1937. "The Yahgan: The life and thought of the water nomads of Cape
Horn." Translated by Frieda Schütze. Chapter from *Die Feuerland-Indianer* [The Fuegian
Indians]. New Haven, CT: electronic Human Relations Area Files, Yahgan, Doc no. 1.
https://ehrafworldcultures.yale.edu/cultures/sh06/documents/001.

Guthold, Regina, Gretchen A. Stevens, Leanne M. Riley, and Fiona C. Bull. 2018. "Worldwide
trends in insufficient physical activity from 2001 to 2016: A pooled analysis of 358 population-
based surveys with 1.9 million participants." *Lancet Global Health* 6 (10): e1077–e1086.

Hagen, Edward H., Melissa J. Garfield, and Roger J. Sullivan. 2016. "The low prevalence of female
smoking in the developing world: Gender inequality or maternal adaptations for fetal protection?" *Evolution, Medicine, and Public Health* 2016 (1): 195–211. https://doi.org/10.1093
/emph/eow013.

Haile-Selassie, Yohannes, Stephanie M. Melillo, Antonino Vazzana, Stefano Benazzi, and Timothy M. Ryan. 2019. "A 3.8-million-year-old hominin cranium from Woranso-Mille, Ethiopia."
Nature 573 (7773): 214–219.

Haldane, J.B.S. 1942. *New Paths in Genetics*. London: Harper and Brothers.

Hall, Kevin D., Alexis Ayuketah, Robert Brychta, Hongyi Cai, Thomas Cassimatis, Kong Y.
Chen, Stephanie T. Chung, Elise Costa, Amber Courville, and Valerie Darcey. 2019.
"Ultra-processed diets cause excess calorie intake and weight gain: An inpatient randomized
controlled trial of ad libitum food intake." *Cell Metabolism* 30 (1): 67–77, e3.

Hamilton, James B., and Gordon E. Mestler. 1969. "Mortality and survival: Comparison of eunuchs with intact men and women in a mentally retarded population." *Journal of Gerontology*
24 (4): 395–411.

Hamilton, W. D. 1966. "The molding of senescence by natural selection." *Journal of Theoretical
Biology* 12:12–45.

Hamilton W. D. 1996. *Narrow Roads of Gene Land: Volume 1: Evolution of Social Behaviour*. New
York: Oxford University Press

Hammer, M., and R. Foley. 1996. "Longevity, life history and allometry: How long did hominids
live?" *Journal of Human Evolution* 11:61–66.

Harpending, Henry C., and LuAnn Wandsnider. 1982. "Population structures of Ghanzi and
Ngamiland !Kung." *Current Developments in Anthropological Genetics* 2:29–50.

Hart, Charles William Merton, Arnold R. Pilling, and Jane Carter Goodale. 1988. *The Tiwi of
North Australia*. San Diego, CA: Harcourt College.

Hartz, Arthur J., Evelyn M. Kuhn, and Jose Pulido. 1999. "Prestige of training programs and
experience of bypass surgeons as factors in adjusted patient mortality rates." *Medical Care*
37 (1): 93–103.

Harvey, Barbara. 1993. *Living and Dying in England 1100–1540: The Monastic Experience*. Oxford,
UK: Clarendon Press.

Hatcher, John. 1986. "Mortality in the fifteenth century: Some new evidence." *Economic History
Review*. 39 (1): 19–38.

Hatcher, John, Alan John Piper, and David Stone. 2006. "Monastic mortality: Durham Priory,
1395–1529 1." *Economic History Review* 59 (4): 667–687.

Hawkes, Ernest William. 1916. *The Labrador Eskimo*. Anthropological Series, Memoir 91. Ottawa: Government Printing Bureau.

Hawkes, Kristen. 2003. "Grandmothers and the evolution of human longevity." *American Journal
of Human Biology* 15 (3): 380–400.

Hawkes, Kristen, and James O'Connell. 2005. "How old is human longevity." *Journal of Human
Evolution* 49 (5): 650–653.

Hawkes, Kristen, James O'Connell, and Nicholas Blurton Jones. 1991. "Hunting income patterns among the Hadza: Big game, common goods, foraging goals and the evolution of the human diet." *Phil. Trans. R. Soc. London (B)* 334: 243–251.

Hawkes, Kristen, James O'Connell, and Nicholas Blurton Jones. 1997. "Hadza women's time allocation, offspring provisioning, and the evolution of long postmenopausal life spans." *Current Anthropology* 38 (4): 551–577.

Hawks, John, Eric T. Wang, Gregory M. Cochran, Henry C. Harpending, and Robert K. Moyzis. 2007. "Recent acceleration of human adaptive evolution." *Proceedings of the National Academy of Sciences, USA* 104 (52): 20753–20758. https://doi.org/10.1073/pnas.0707650104.

Hayashi, Mayumi. 2012. "Japan's Fureai Kippu time-banking in elderly care: Origins, development, challenges and impact." *International Journal of Community Currency Research* 16 (A): 30–44.

Hayflick, Leonard. 1994. *How and Why We Age.* New York: Ballantine Books.

Hayflick, Leonard, and Paul S. Moorhead. 1961. "The serial cultivation of human diploid cell strains." *Experimental Cell Research* 25 (3): 585–621.

Headey, Derek, and Harold Alderman. 2019. "The relative caloric prices of healthy and unhealthy foods differ systematically across income levels and continents." *Journal of Nutrition.* https://doi.org/10.1093/jn/nxz158.

Headland, Thomas. 1986. Why foragers do not become farmers: A historical study of a changing ecosystem and its effect on a Negrito hunter-gatherer group in the Philippines. PhD dissertation, University of Hawaii at Manoa.

Headland, Thomas. 1990. "Paradise revised." *The Sciences* 30 (5): 45–50.

Headland, Thomas, and Robert Bailey. 1991. "Have hunter-gatherers ever lived in tropical rain forest independently of agriculture?" *Human Ecology* 19 (2): 115–122.

Hearne, Samuel. 1958. *Narrative of a Journey from Prince of Wales's Fort in Hudson Bay to the Northern Ocean in the Years 1769, 1770, 1771, and 1772.* Toronto: Macmillan.

Heinen, Heinz Dieter. 1972. *Adaptive Changes in the Tribal Economy: A Case Study of the Winikina-Warao.* University of California at Los Angeles, PhD diss. https://ehrafworldcultures.yale.edu/cultures/ss18/documents/009.

Heinen, Heinz Dieter, and Kenneth Ruddle. 1974. "Ecology, ritual, and economic organization in the distribution of palm starch among the Warao of the Orinoco Delta." *Journal of Anthropological Research* 30 (2): 116–138.

Heldstab, Sandra, Karin Isler, Judith Burkart, and Carel van Schaik. 2019. "Allomaternal care, brains and fertility in mammals: Who cares matters." *Behavioral Ecology and Sociobiology* 73 (6): 71.

Heligman, L., and J. H. Pollard. 1980. "The age pattern of mortality." *Journal of the Institute of Actuaries* 107 (1): 49–80. http://doi.org/10.1017/S0020268100040257.

Hendrie, Hugh, Jill Murrell, Olusegun Baiyewu, Kathleen Lane, Christianna Purnell, Adesola Ogunniyi, Frederick W. Unverzagt, et al. 2014. "APOE ε4 and the risk for Alzheimer disease and cognitive decline in African Americans and Yoruba." *International Psychogeriatrics* 26 (6): 977–985.

Henrich, Joseph. 2017. *The Secret of Our Success: How Culture Is Driving Human Evolution, Domesticating Our Species, and Making Us Smarter.* Princeton, NJ: Princeton University Press.

Henrich, Joseph, and Francisco Gil-White. 2001. "The evolution of prestige: Freely conferred status as a mechanism for enhancing the benefits of cultural transmission." *Evolution and Human Behavior* 22:1–32.

Hewlett, Barry S. 1991. "Demography and childcare in preindustrial societies." *Journal of Anthropological Research* 47:1–37.

Hewlett, Barry, J.M.H. van de Koppel, and M. van de Koppel. 1986. "Causes of death among Aka pygmies of the Central African Republic." In *African Pygmies,* edited by L. L. Cavalli Sforza, 45–63. New York: Academic Press.

Hill, Kim, Michael Barton, and Ana Magdalena Hurtado. 2009. "The emergence of human uniqueness: Characters underlying behavioral modernity." *Evolutionary Anthropology* 18 (5): 187–200.

Hill, Kim, and Ana Magdalena Hurtado. 1996. *Ache Life History: The Ecology and Demography of a Foraging People*. New York: Aldine de Gruyter.

Hill, Kim, Ana Magdalena Hurtado, and Robert Walker. 2007. "High adult mortality among Hiwi hunter-gatherers: Implications for human evolution." *Journal of Human Evolution* 52:443–454.

Hochberg, Michael E., and Robert J. Noble. 2017. "A framework for how environment contributes to cancer risk." *Ecology Letters* 20 (2): 117–134.

Hoebel, Edward Adamson, and Ernest Wallace. 1952. *The Comanches: Lords of the South Plains*. Norman: University of Oklahoma Press.

Hoffman, Jordie, Kyle Farquharson, and Vivek Venkataraman. 2023. "The ecological and social context of women's hunting in small-scale societies." *Hunter-Gatherer Research* 7 (5): 1–32

Hoffmann, Dirk L., Christopher D. Standish, Marcos García-Diez, Paul B. Pettitt, James A. Milton, João Zilhão, Javier J. Alcolea-González, et al. 2018. "U-Th dating of carbonate crusts reveals Neandertal origin of Iberian cave art." *Science* 359 (6378): 912–915.

Hollingsworth, Thomas H. 1977. "Mortality in the British peerage families since 1600." *Population (French Edition)* 32:323–352.

Holmberg, Allan R. 1969. *Nomads of the Long Bow: The Sirionó of Eastern Bolivia*. Garden City, NY: Natural History Press (Reprint, 1968).

Holt-Lunstad, Julianne, Timothy B. Smith, Mark Baker, Tyler Harris, and David Stephenson. 2015. "Loneliness and social isolation as risk factors for mortality: A meta-analytic review." *Perspectives on Psychological Science* 10 (2): 227–237.

Holt-Lunstad, Julianne, Timothy B. Smith, and J. Bradley Layton. 2010. "Social relationships and mortality risk: A meta-analytic review." *PLoS Medicine* 7 (7): e1000316.

Horowitz, Sheryl, George Armelagos, and Ken Wachter. 1988. "On generating birth rates from skeletal populations." *American Journal of Physical Anthropology* 76 (2): 189–196.

Horstman, Connie, and Donald V. Kurtz. 1979. "Compadrazgo and adaptation in sixteenth century central Mexico." *Journal of Anthropological Research* 35 (3): 361–372.

Horvath, Steve. 2013. "DNA methylation age of human tissues and cell types." *Genome Biology* 14 (10): 1–20.

Horvath, Steve, Michael Gurven, Morgan Levine, Benjamin Trumble, Hillard Kaplan, Hooman Allayee, B. R. Ritz, et al. 2016. "An epigenetic age analysis of race/ethnicity, gender and coronary heart disease addresses several paradoxes surrounding mortality." *Genome Biology* 17:171.

Howell, Nancy. 1976. "Toward a uniformitarian theory of human paleodemography." *Journal of Human Evolution* 5 (1): 25–40.

Howell, Nancy. 1979. *Demography of the Dobe !Kung*. New York: Academic Press.

Howell, Nancy. 2010. *Life Histories of the Dobe !Kung: Food, Fatness, and Well-Being over the Life Span*. Origins of Human Behavior and Culture, vol. 4. Berkeley: University of California Press.

Hrdy, Sarah. 1981. "'Nepotists' and 'altruists': The behavior of old females among macaques and langur monkeys." In *Other Ways of Growing Old*, edited by Pamela Amoss and Stevan Harrell, 59–76. Stanford, CA: Stanford University Press.

Hrdy, Sarah. 2005a. "Comes the child before the man: How cooperative breeding and prolonged post-weaning dependence shaped human potentials." In *Hunter-Gatherer Childhoods*, edited by B. S. Hewlett and M. E. Lamb, 65–91. Piscataway, NJ: Transactions.

Hrdy, Sarah. 2005b. "Cooperative breeders with an ace in the hole." In *Grandmotherhood: The Evolutionary Significance of the Second Half of Female Life*, edited by E. Voland, A. Chasiotis, and W. Schiefenhovel. New Brunswick, NJ: Rutgers University Press.

Huanca, Tomás. 2008. *Tsimane' Oral Tradition, Landscape, and Identity in Tropical Forest*. La Paz, BO: SEPHIS.

Hublin, Jean-Jacques, Abdelouahed Ben-Ncer, Shara E. Bailey, Sarah E. Freidline, Simon Neubauer, Matthew M. Skinner, Inga Bergmann, et al. 2017. "New fossils from Jebel Irhoud, Morocco and the pan-African origin of Homo sapiens." *Nature* 546 (7657): 289–292.

Huntingford, George Wynn Brereton. 1951. "The social institutions of the Dorobo." *Anthropos* 46 (1/2): 1–48.

Huntingford, George Wynn Brereton. 1954. "The political organization of the Dorobo." *Anthropos* 49 (1/2): 123–148.

Irimia, Andrei, Nikhil N. Chaudhari, David J. Robles, Kenneth A. Rostowsky, Alexander S. Maher, Nahian F. Chowdhury, Maria Calvillo, et al. 2021. "The indigenous South American Tsimane exhibit relatively modest decrease in brain volume with age despite high systemic inflammation." *Journals of Gerontology: Series A* 76 (12): 2147–2155.

Isendahl, Christian. 2011. "The domestication and early spread of manioc (Manihot esculenta Crantz): A brief synthesis." *Latin American Antiquity* 22 (4): 452–468.

Izquierdo, Carolina. 2005. "When 'health' is not enough: Societal, individual and biomedical assessments of well-being among the Matsigenka of the Peruvian Amazon." *Social Science and Medicine* 61 (4): 767–783.

Jagger, Carol, Clare Gillies, Francesco Moscone, Emmanuelle Cambois, Herman Van Oyen, Wilma Nusselder, and Jean-Marie Robine. 2008. "Inequalities in healthy life years in the 25 countries of the European Union in 2005: A cross-national meta-regression analysis." *The Lancet* 372 (9656): 2124–2131.

James, Spencer L., Degu Abate, Kalkidan Hassen Abate, Solomon M. Abay, Cristiana Abbafati, Nooshin Abbasi, Hedayat Abbastabar, et al. 2018. "Global, regional, and national incidence, prevalence, and years lived with disability for 354 diseases and injuries for 195 countries and territories, 1990–2017: A systematic analysis for the Global Burden of Disease Study 2017." *The Lancet* 392 (10159): 1789–1858.

Jamison, C., L. Cornell, P. Jamison, and H. Nakazato. 2002. "Are all grandmothers equal? A review and a preliminary test of the grandmother hypothesis in Tokugawa Japan." *American Journal of Physical Anthropology* 119 (1): 67–76.

Jansen, Rick, Laura K. M. Han, Josine E. Verhoeven, Karolina A. Aberg, Edwin C. G. J. van den Oord, Yuri Milaneschi, and Brenda W. J. H. Penninx. 2021. "An integrative study of five biological clocks in somatic and mental health." *eLife* 10:e59479.

Jenkins, Nicole L., Gawain McColl, and Gordon J. Lithgow. 2004. "Fitness cost of extended lifespan in Caenorhabditis elegans." *Proceedings of the Royal Society of London. Series B: Biological Sciences* 271 (1556): 2523–2526.

Johnson, Patricia Lyons. 1981. "When dying is better than living: Female suicide among the Gainj of Papua New Guinea." *Ethnology* 20 (4): 325–334.

Johnson, Richard W., Melissa M. Favreault, Judith Dey, William Marton, and Lauren Anderson. 2021. "Most older adults are likely to need and use long-term services and supports issue brief." US Department of Health and Human Services. https://aspe.hhs.gov/reports/most-older-adults-are-likely-need-use-long-term-services-supports-issue-brief-0.

Johnson, Steven. 2021. *Extra Life: A Short History of Living Longer*. New York: Riverhead Books.

Johnston, Bradley C., Steve Kanters, Kristofer Bandayrel, Ping Wu, Faysal Naji, Reed A. Siemieniuk, Geoff D. C. Ball, et al. 2014. "Comparison of weight loss among named diet programs in overweight and obese adults: A meta-analysis." *JAMA* 312 (9): 923–933. https://doi.org/10.1001/jama.2014.10397.

Joiner, Thomas E., Melanie A. Hom, Christopher R. Hagan, and Caroline Silva. 2016. "Suicide as a derangement of the self-sacrificial aspect of eusociality." *Psychological Review* 123 (3): 235.

Jones, Owen R., Alexander Scheuerlein, Roberto Salguero-Gomez, Carlo Giovanni Camarda, Ralf Schaible, Brenda B. Casper, Johan P. Dahlgren, et al. 2014. "Diversity of ageing across the tree of life." *Nature* 505 (7482): 169–173.

Juckett, David A., and Barnett Rosenberg. 1993. "Comparison of the Gompertz and Weibull functions as descriptors for human mortality distributions and their intersections." *Mechanisms of Ageing and Development* 69 (1/2): 1–31.

Judge, D. S., and J. R. Carey. 2000. "Postreproductive life predicted by primate patterns." *Journal of Gerontology: Biological Sciences* 55A (4): B201–209.

Junger, Sebastian. 2016. *Tribe: On Homecoming and Belonging.* New York: HarperCollins.

Juul, Filippa, Niyati Parekh, Euridice Martinez-Steele, Carlos Augusto Monteiro, and Virginia W. Chang. 2021. "Ultra-processed food consumption among US adults from 2001 to 2018." *American Journal of Clinical Nutrition* 115 (1): 211–221. https://doi.org/10.1093/ajcn/nqab305.

Kannisto, V. 2001. "Mode and dispersion of the length of life." *Population: An English Selection* 13:159–171.

Kaplan, David. 2000. "The darker side of the 'original affluent society.'" *Journal of Anthropological Research* 56 (3): 301–324.

Kaplan, H., K. Hill, J. B. Lancaster, and A. M. Hurtado. 2000. "A theory of human life history evolution: Diet, intelligence, and longevity." *Evolutionary Anthropology* 9 (4): 156–185.

Kaplan, H. S., and A. J. Robson. 2009. "We age because we grow." *Proceedings of the Royal Society: Biological Sciences* 276 (1663): 1837–1844.

Kaplan, Hillard S., Eric Schniter, Vernon L. Smith, and Bart J. Wilson. 2012. "Risk and the evolution of human exchange." *Proceedings of the Royal Society B: Biological Sciences* 279 (1740): 2930–2935.

Kaplan, Hillard, Randall C. Thompson, Benjamin C. Trumble, L. Samuel Wann, Adel H. Allam, Bret Beheim, Bruno Frohlich, et al. 2017. "Coronary atherosclerosis in indigenous South American Tsimane: A cross-sectional cohort study." *Lancet* 389 (10080): 1730–1739. https://doi.org/10.1016/s0140-6736(17)30752-3.

Kaushik, Susmita, and Ana Maria Cuervo. 2015. "Proteostasis and aging." *Nature Medicine* 21 (12): 1406–1415.

Kavalieratos, Dio, Jennifer Corbelli, D. I. Zhang, J. Nicholas Dionne-Odom, Natalie C. Ernecoff, Janel Hanmer, Zachariah P. Hoydich, et al. 2016. "Association between palliative care and patient and caregiver outcomes: A systematic review and meta-analysis." *JAMA* 316 (20): 2104–2114.

Keeley, Lawrence H. 1996. *War before Civilization.* New York: Oxford University Press.

Keith, Jennie, Christine L. Fry, Anthony P. Glascock, Charlotte Ikels, Jeanette Dickerson-Putman, Henry C. Harpending, and Patricia Draper. 1994. *The Aging Experience: Diversity and Commonality across Cultures.* London: Sage Publications.

Kelly, Robert L. 2013. *The Lifeways of Hunter-Gatherers: The Foraging Spectrum.* Cambridge, UK: Cambridge University Press.

Kennedy, Brian K., Shelley L. Berger, Anne Brunet, Judith Campisi, Ana Maria Cuervo, Elissa S. Epel, Claudio Franceschi, et al. 2014. "Geroscience: Linking aging to chronic disease." *Cell* 159 (4): 709–713.

Kennedy, G. E. 2003. "Palaeolithic grandmothers? Life history theory and early *Homo*." *Journal of the Royal Anthropological Institute* 9:549–572.

Kirkwood, Thomas B. L. 1977. "Evolution of ageing." *Nature* 270 (5635): 301–304.

Kirkwood, Thomas B. L. 2011. "Systems biology of ageing and longevity." *Philosophical Transactions of the Royal Society B: Biological Sciences* 366 (1561): 64–70.

Kirkwood, Thomas B. L. 2015. "Deciphering death: A commentary on Gompertz (1825) 'On the nature of the function expressive of the law of human mortality, and on a new mode of determining the value of life contingencies.'" *Philosophical Transactions of the Royal Society B: Biological Sciences* 370 (1666): 20140379.

Kirkwood, Thomas B. L., and S. N. Austad. 2000. "Why do we age?" *Nature* 408: 233–238.

Kirkwood, Thomas B. L., and M. R. Rose. 1991. "Evolution of senescence: Late survival sacrificed for reproduction." *Philosophical Transactions of the Royal Society of London B: Biological Sciences* 332 (1262): 15–24. https://doi.org/10.1098/rstb.1991.0028.

Kivimäki, Mika, Archana Singh-Manoux, Jaana Pentti, Séverine Sabia, Solja T. Nyberg, Lars Alfredsson, Marcel Goldberg, et al. 2019. "Physical inactivity, cardiometabolic disease, and risk of dementia: An individual-participant meta-analysis." *BMJ* 365. https://doi.org/10.1136/bmj.l1495.

Kohler, Iliana V., Collin F. Payne, Chiwoza Bandawe, and Hans-Peter Kohler. 2017. "The demography of mental health among mature adults in a low-income, high-HIV-prevalence context." *Demography* 54 (4): 1529–1558.

Komonpaisarn, Touchanun, and Elke Loichinger. 2019. "Providing regular care for grandchildren in Thailand: An analysis of the impact on grandparents' health." *Social Science and Medicine* 229:117–125.

Konigsberg, L. W., and N. P. Herrmann. 2006. "The osteological evidence for human longevity in the recent past." In *The Evolution of Human Life History*, edited by K. Hawkes and R. R. Paine, 267–306. Santa Fe, NM: School of American Research Press.

Konigsberg, Lyle W., and Susan R. Frankenberg. 1994. "Paleodemography: 'Not quite dead.'" *Evolutionary Anthropology* 3 (3): 92–105.

Koster, Jeremy, Richard McElreath, Kim Hill, Douglas Yu, Glenn Shepard, Nathalie van Vliet, Michael Gurven, et al. 2020. "The life history of human foraging: Cross-cultural and individual variation." *Science Advances* 6 (26): eaax9070. https://doi.org.10.1126/sciadv.aax9070.

Kraft, Thomas, Jonathan Stieglitz, Benjamin Trumble, Angela Garcia, Hillard Kaplan, and Michael Gurven. 2020. "Multi-system physiological dysregulation and ageing in a subsistence population." *Philosophical Transactions of the Royal Society B* 375 (1811): 20190610.

Kraft, Thomas, Vivek Venkataraman, and Nathaniel Dominy. 2014. "A natural history of human tree climbing." *Journal of Human Evolution* 71:105–118.

Kraft, Thomas, Vivek Venkataraman, Ian Wallace, Nicholas Holowka, David Raichlen, Alyssa Crittenden, Brian Wood, Michael Gurven, and Herman Pontzer. 2021. "The energetics of uniquely human subsistence strategies." *Science* 374 (6575): eabf0130.

Kramer, Karen L. 2005. "Children's help and the pace of reproduction: Cooperative breeding in humans." *Evolutionary Anthropology* 14 (6): 224–237.

Kramer, Karen L., and Peter T. Ellison. 2010. "Pooled energy budgets: Resituating human energy-allocation trade-offs." *Evolutionary Anthropology: Issues, News, and Reviews* 19 (4): 136–147.

Krishnan, Sridevi, Sean H. Adams, Lindsay H. Allen, Kevin D. Laugero, John W. Newman, Charles B. Stephensen, Dustin J. Burnett, et al. 2018. "A randomized controlled-feeding trial based on the Dietary Guidelines for Americans on cardiometabolic health indexes." *American Journal of Clinical Nutrition* 108 (2): 266–278.

Kulkarni, Ameya S., Sriram Gubbi, and Nir Barzilai. 2020. "Benefits of metformin in attenuating the hallmarks of aging." *Cell Metabolism* 32 (1): 15–30.

Kumar, Deepak Kumar Vijaya, Se Hoon Choi, Kevin J. Washicosky, William A. Eimer, Stephanie Tucker, Jessica Ghofrani, Aaron Lefkowitz, et al. 2016. "Amyloid-β peptide protects against microbial infection in mouse and worm models of Alzheimer's disease." *Science Translational Medicine* 8 (340): 340ra72–340ra72. https://doi.org/10.1126/scitranslmed.aaf1059.

Laird, Robert A., and Thomas N. Sherratt. 2010. "The economics of evolution: Henry Ford and the Model T." *Oikos* 119 (1): 3–9.

Lamont, Ruth A., Hannah J. Swift, and Dominic Abrams. 2015. "A review and meta-analysis of age-based stereotype threat: Negative stereotypes, not facts, do the damage." *Psychology and Aging* 30 (1): 180.

Lancaster Jones, Frank. 1963. *A Demographic Survey of the Aboriginal Population of the Northern Territory, with Special Reference to Bathurst Island Mission*. Canberra: Australian Institute of Aboriginal Studies.

Lancaster Jones, Frank. 1965. "The Demography of the Australian Aborigines." *International Social Science Journal* 17 (2): 232–245.

Landry, Roger. 2014. *Live Long, Die Short: A Guide to Authentic Health and Successful Aging*. Austin, TX: Greenleaf Book Group.

Lanikova, Lucie, N. Scott Reading, Hao Hu, Tsewang Tashi, Tatiana Burjanivova, Anna Shestakova, Bhola Siwakoti, et al. 2017. "Evolutionary selected Tibetan variants of HIF pathway and risk of lung cancer." *Oncotarget* 8 (7): 11739.

Larsen, Clark Spencer. 2003. "Equality for the sexes in human evolution? Early hominid sexual dimorphism and implications for mating systems and social behavior." *Proceedings of the National Academy of Sciences* 100 (16): 9103–9104.

Lea, Amanda, Angela Garcia, J. Arevalo, J. Ayroles, K. Buetow, S.W. Cole, D. Eid Rodriguez, M. Gutierrez Cayuba, et al. 2022. "Natural selection of immune and metabolic genes associated with health in South American Amerindians." *Proceedings of the National Academy of Sciences* 120 (1): e2207544120.

Lea, Amanda, Dino Martins, Joseph Kamau, Michael Gurven, and Julien F. Ayroles. 2020. "Urbanization and market integration have strong, nonlinear effects on cardiometabolic health in the Turkana." *Science Advances* 6 (43): eabb1430.

Lear, Scott A., Weihong Hu, Sumathy Rangarajan, Danijela Gasevic, Darryl Leong, Romaina Iqbal, Amparo Casanova, et al. 2017. "The effect of physical activity on mortality and cardiovascular disease in 130 000 people from 17 high-income, middle-income, and low-income countries: The PURE study." *The Lancet* 390 (10113): 2643–2654.

Ledberg, Anders. 2020. "Exponential increase in mortality with age is a generic property of a simple model system of damage accumulation and death." *PLoS One* 15 (6): e0233384.

Lee, Amanda, Anne Bailey, Daisy Yarmirr, Kerin O'Dea, and John Mathews. 1994. "Survival tucker: Improved diet and health indicators in an Aboriginal community." *Australian Journal of Public Health* 18 (3): 277–285.

Lee, I-Min, Eric J. Shiroma, Felipe Lobelo, Pekka Puska, Steven N. Blair, Peter T. Katzmarzyk, and Lancet Physical Activity Series Working Group. 2012. "Effect of physical inactivity on major non-communicable diseases worldwide: An analysis of burden of disease and life expectancy." *The Lancet* 380 (9838): 219–229.

Lee, Richard B. 1979. *The !Kung San: Men, Women, and Work in a Foraging Society*. Cambridge, UK: Cambridge University Press.

Lee, Ronald D. 2003. "Rethinking the evolutionary theory of aging: Transfers, not births, shape senescence in social species." *Proceedings of the National Academy of Sciences* 100 (16): 9637–9642.

Lehman, Harvey C. 1943. "The longevity of the eminent." *Science* 98 (2543): 270–273.

Leighton, Alexander H., and Charles C. Hughes. 1955. "Notes on Eskimo patterns of suicide." *Southwestern Journal of Anthropology* 11 (4): 327–338.

Leng, Sean X., and Brian K. Kennedy. 2019. "International investment in geroscience." *Public Policy and Aging Report* 29 (4): 134–138.

Levine, Morgan, Ake T. Lu, David A. Bennett, and Steve Horvath. 2015. "Epigenetic age of the pre-frontal cortex is associated with neuritic plaques, amyloid load, and Alzheimer's disease related cognitive functioning." *Aging* 7 (12): 1198–1211. https://doi.org/10.18632/aging.100864.

Levitin, Daniel J. 2020. *Successful Aging: A Neuroscientist Explores the Power and Potential of Our Lives*. New York: Penguin.

Levitis, Daniel, Oskar Burger, and Laurie Bingaman Lackey. 2013. "The human post-fertile lifespan in comparative evolutionary context." *Evolutionary Anthropology: Issues, News, and Reviews* 22 (2): 66–79.

Levy, Becca. 2009. "Stereotype embodiment: A psychosocial approach to aging." *Current Directions in Psychological Science* 18 (6): 332–336.

Levy, Becca, Alan B. Zonderman, Martin D. Slade, and Luigi Ferrucci. 2009. "Age stereotypes held earlier in life predict cardiovascular events in later life." *Psychological Science* 20 (3): 296–298.

Lexis, W. 1878. "Sur la Durée Normale de la Vie Humaine et sur la Théorie de la Stabilité des Rapports Statistiques." *Annales de Démographie Internationale* 2:447–460.

Li, Qing, Shengrui Wang, Emmanuel Milot, Patrick Bergeron, Luigi Ferrucci, Linda P. Fried, and Alan A. Cohen. 2015. "Homeostatic dysregulation proceeds in parallel in multiple physiological systems." *Aging Cell* 14 (6): 1103–1112.

Li, Ting, and James J. Anderson. 2015. "The Strehler-Mildvan correlation from the perspective of a two-process vitality model." *Population Studies* 69 (1): 91–104.

Li, Zeyang, Anna Wei, Vishanth Palanivel, and Joshua Conrad Jackson. 2021. "A data-driven analysis of sociocultural, ecological, and economic correlates of depression across nations." *Journal of Cross-Cultural Psychology* 52 (8/9): 822–843.

Liebenberg, Louis. 1990. *The Art of Tracking: The Origin of Science*. Cape Town, SA: David Phillip.

Lieberman, Daniel. 2021. *Exercised: Why Something We Never Evolved to Do Is Healthy and Rewarding*. New York: Vintage.

Liebert, Melissa, J. Josh Snodgrass, Felicia Madimenos, Tara Cepon, Aaron Blackwell, and Lawrence S. Sugiyama. 2013. "Implications of market integration for cardiovascular and metabolic health among an indigenous Amazonian Ecuadorian population." *Annals of Human Biology* 40 (3): 228–242.

Liguori, Ilaria, Gennaro Russo, Francesco Curcio, Giulia Bulli, Luisa Aran, David Della-Morte, Gaetano Gargiulo, et al. 2018. "Oxidative stress, aging, and diseases." *Clinical Interventions in Aging* 13:757–772.

Lin, Xiling, Yufeng Xu, Xiaowen Pan, Jingya Xu, Yue Ding, Xue Sun, Xiaoxiao Song, Yuezhong Ren, and Peng-Fei Shan. 2020. "Global, regional, and national burden and trend of diabetes in 195 countries and territories: An analysis from 1990 to 2025." *Scientific Reports* 10 (1): 1–11.

Lindeberg, S., P. Nilsson-Ehle, A. Terent, B. Vessby, and B. Scherstén. 1994. "Cardiovascular risk factors in a Melanesian population apparently free from stroke and ischaemic heart disease: The Kitava study." *Journal of Internal Medicine* 236 (3): 331–340.

Littwin, Susan. 1986. *The Postponed Generation: Why American Youth Are Growing Up Later*. New York: William Morrow.

Löckenhoff, Corinna E., Filip De Fruyt, Antonio Terracciano, Robert R. McCrae, Marleen De Bolle, Paul T. Costa, Maria E. Aguilar-Vafaie, et al. 2009. "Perceptions of aging across 26 cultures and their culture-level associates." *Psychology and Aging* 24 (4): 941.

Logue, Barbara J. 1990. "Modernization and the status of the frail elderly: Perspectives on continuity and change." *Journal of Cross-Cultural Gerontology* 5 (4): 345–374.

López-Otín, Carlos, Maria A. Blasco, Linda Partridge, Manuel Serrano, and Guido Kroemer. 2013. "The hallmarks of aging." *Cell* 153 (6): 1194–1217.

López-Otín, Carlos, Lorenzo Galluzzi, José M. P. Freije, Frank Madeo, and Guido Kroemer. 2016. "Metabolic control of longevity." *Cell* 166 (4): 802–821.

Lovejoy, C. O., R. S. Meindl, T. R. Pryzbeck, T. S. Barton, K. G. Heiple, and D. Kotting. 1977. "Paleodemography of the Libben site, Ottawa County, Ohio." *Science* 198:291–293.

Macdonell, W. R. 1913. "On the expectation of life in ancient Rome, and in the provinces of Hispania and Lusitania, and Africa." *Biometrika* 9 (3/4): 366–380.

Malthus, Thomas. 2013. *An Essay on the Principle of Population (1798)*. New Haven, CT: Yale University Press.

Mancilha-Carvalho, J. J., and Douglas E. Crews. 1990. "Lipid profiles of Yanomamo Indians of Brazil." *Preventive Medicine* 19 (1): 66–75.

Mann, George V., Oswald A. Roels, Donald L. Price, and Joseph M. Merrill. 1962. "Cardiovascular disease in African Pygmies: A survey of the health status, serum lipids and diet of Pygmies in Congo." *Journal of Chronic Diseases* 15 (4): 341–371.

Mann, George V., R. D. Shaffer, R. S. Anderson, H. H. Sandstead, H. Prendergast, J. C. Mann, S. Rose, et al. 1964. "Cardiovascular disease in the Masai." *Journal of Atherosclerosis Research* 4 (4): 289–312.

Mann, George V., Anne Spoerry, Margarete Gary, and Debra Jarashow. 1972. "Atherosclerosis in the Masai." *American Journal of Epidemiology* 95 (1): 26–37.

Marlowe, Frank. 2010. *The Hadza: Hunter-Gatherers of Tanzania*. Origins of Human Behavior and Culture, vol. 3. Berkeley: University of California Press.

Marlowe, Frank, J. Colette Berbesque, Brian Wood, Alyssa Crittenden, Claire Porter, and Audax Mabulla. 2014. "Honey, Hadza, hunter-gatherers, and human evolution." *Journal of Human Evolution* 71:119–128.

Marques, Carina, Zachary Compton, and Amy M. Boddy. 2022. "Connecting palaeopathology and evolutionary medicine to cancer research: Past and present." *Palaeopathology and Evolutionary Medicine: An Integrated Approach*. https://doi.org/10.1093/oso/9780198849711.003.0013.

Marques, Carina, Vítor Matos, Tiago Costa, Albert Zink, and Eugénia Cunha. 2018. "Absence of evidence or evidence of absence? A discussion on paleoepidemiology of neoplasms with contributions from two Portuguese human skeletal reference collections (19th–20th century)." *International Journal of Paleopathology* 21:83–95.

Marshall, Lorna J. 1976. *The !Kung of Nyae Nyae*. Cambridge, MA: Harvard University Press.

Marshall Thomas, Elizabeth. 1959. *The Harmless People*. London: Secker and Warburg.

Martin, Robert William, and Andrew J. Cooper. 2017. "Subjective well-being in a remote culture: The Himba." *Personality and Individual Differences* 115:19–22.

Maxwell, Eleanor Krassen, and Robert J. Maxwell. 1980. *Contempt for the Elderly: A Cross-Cultural Analysis*. Chicago: University of Chicago Press.

Maxwell, Robert J., and Philip Silverman. 1970. "Information and esteem: Cultural considerations in the treatment of the aged." *Aging and Human Development* 1 (4): 361–392.

Maxwell, Robert J., and Philip Silverman. 1989. "Gerontocide." In *The Content of Culture: Constants and Variants*, edited by Ralph Bolton, 511–523. New Haven, CT: Human Relations Area Files.

Maxwell, Robert J., Philip Silverman, and Eleanor Maxwell. 1982. "The motive for geronticide." *Studies in Third World Societies* 22 (special issue): 67–84.

Mbalilaki, Julia Aneth, Zablon Masesa, Sigmund Bjarne Strømme, Arne Torbjørn Høstmark, Jan Sundquist, Per Wändell, Annika Rosengren, and Mai-Lis Hellenius. 2010. "Daily energy expenditure and cardiovascular risk in Masai, rural and urban Bantu Tanzanians." *British Journal of Sports Medicine* 44 (2): 121–126.

McArdle, Joan L., and Constantine Yeracaris. 1981. "Respect for the elderly in preindustrial societies as related to their activity." *Behavior Science Research* 16 (3/4): 307–339.

McDonald Pavelka, Mary S., and Linda Marie Fedigan. 1999. "Reproductive termination in female Japanese monkeys: A comparative life history perspective." *American Journal of Physical Anthropology* 109 (4): 455–464.

McMurry, Martha P., Maria Teresa Cerqueira, Sonja L. Connor, and William E. Connor. 1991. "Changes in lipid and lipoprotein levels and body weight in Tarahumara Indians after consumption of an affluent diet." *New England Journal of Medicine* 325 (24): 1704–1708.

McPherson, Miller, Lynn Smith-Lovin, and Matthew E. Brashears. 2006. "Social isolation in America: Changes in core discussion networks over two decades." *American Sociological Review* 71 (3): 353–375.

Mead, Margaret. 1928. *Coming of Age in Samoa: A Psychological Study of Primitive Youth for Western Civilisation.* New York: W. Morrow.

Mead, Margaret. 1971. "A new style of aging." *Christianity and Crisis* 31.

Medawar, Peter B. 1952. *An Unsolved Problem in Biology.* London: Lewis.

Medvedev, Zhores A. 1990. "An attempt at a rational classification of theories of ageing." *Biological Reviews* 65 (3): 375–398.

Meehan, Betty. 1982. *Shell Bed to Shell Midden.* Canberra: Australian Institute of Aboriginal Studies.

Meehan, Courtney L., Courtney Helfrecht, and Robert J. Quinlan. 2014. "Cooperative breeding and Aka children's nutritional status: Is flexibility key?" *American Journal of Physical Anthropology* 153 (4): 513–525.

Meindl, Richard S., Robert P. Mensforth, and C. Owen Lovejoy. 2008. "The Libben site: A hunting, fishing, and gathering village from the eastern late woodlands of North America. Analysis and implications for palaeodemography and human origins." In *Recent Advances in Palaeodemography*, edited by Jean-Pierre Bocquet-Appel, 259–275. Dordrecht: Springer.

Migliano, Andrea Bamberg, Lucio Vinicius, and Marta Mirazón Lahr. 2007. "Life history trade-offs explain the evolution of human pygmies." *Proceedings of the National Academy of Sciences* 104 (51): 20216–20219.

Min, Kyung-Jin, Cheol-Koo Lee, and Han-Nam Park. 2012. "The lifespan of Korean eunuchs." *Current Biology* 22 (18): R792–R793.

Miñarro, Sara, Victoria Reyes-García, Shankar Aswani, Samiya Selim, Christopher P. Barrington-Leigh, and Eric D. Galbraith. 2021. "Happy without money: Minimally monetized societies can exhibit high subjective well-being." *PLoS One* 16 (1): e0244569.

Minge-Klevana, Wanda, Kwame Arhin, P.T.W. Baxter, T. Carlstein, Charles J. Erasmus, Michael P. Freedman, Allen Johnson, et al. 1980. "Does labor time decrease with industrialization? A survey of time-allocation studies [and comments and reply]." *Current Anthropology* 21 (3): 279–298.

Minois, Georges. 1989. *History of Old Age: From Antiquity to the Renaissance.* Chicago: University of Chicago Press.

Mitnitski, A. B., A. D. Rutenberg, S. Farrell, and K. Rockwood. 2017. "Aging, frailty and complex networks." *Biogerontology* 18 (4): 433–446.

Moorad, Jacob A., and Craig A. Walling. 2017. "Measuring selection for genes that promote long life in a historical human population." *Nature Ecology and Evolution* 1 (11): 1773.

Moorman, Sara M., and Jeffrey E. Stokes. 2016. "Solidarity in the grandparent–adult grandchild relationship and trajectories of depressive symptoms." *The Gerontologist* 56 (3): 408–420.

Moussouami, Simplice Innocent, Eddie Janvier Bouhika, Florent Nsompi, Jean Michel Bazaba Kayilou, François Mbemba, and Alphonse Massamba. 2016. "Prevalence and risk factors of cardiovascular diseases in the Congo-Brazzaville Pygmies." *World Journal of Cardiovascular Diseases* 6 (07): 211.

Mueller, Noel T., Oscar Noya-Alarcon, Monica Contreras, Lawrence J. Appel, and Maria Gloria Dominguez-Bello. 2018. "Association of age with blood pressure across the lifespan in isolated Yanomami and Yekwana villages." *JAMA Cardiology* 3 (12): 1247–1249.

Mulder, Monique Borgerhoff. 2007. "Hamilton's rule and kin competition: The Kipsigis case." *Evolution and Human Behavior* 28 (5): 299–312.

Muller, Martin N., Nicholas G. Blurton Jones, Fernando Colchero, Melissa Emery Thompson, Drew K. Enigk, Joseph T. Feldblum, Beatrice H. Hahn, et al. 2020. "Sexual dimorphism in chimpanzee (Pan troglodytes schweinfurthii) and human age-specific fertility." *Journal of Human Evolution* 144:102795.

Murphy, W. A. Jr., D. zur Nedden, P. Gostner, R. Knapp, W. Recheis, and H. Seidler. 2003. "The iceman: Discovery and imaging." *Radiology* 226:614–629.

Murphy, Yolanda, and Robert F. Murphy. 1974. *Women of the Forest*. New York: Columbia University Press.

Muthukrishna, Michael, Michael Doebeli, Maciej Chudek, and Joseph Henrich. 2018. "The Cultural Brain Hypothesis: How culture drives brain expansion, sociality, and life history." *PLoS Computational Biology* 14 (11): e1006504.

NASEM (National Academies of Sciences Engineering and Medicine). 2020. *Social Isolation and Loneliness in Older Adults: Opportunities for the Health Care System*. Washington, DC: National Academies Press. https://nap.nationalacademies.org/catalog/25663/social -isolation-and-loneliness-in-older-adults-opportunities-for-the.

Nason, J. 1981. "Respected elder or old person: Aging in a Micronesian community." In *Other Ways of Growing Old*, edited by Pamela Amoss and Stevan Harrell,155–173. Stanford, CA: Stanford University Press.

National Resource Council. 2013. US Health in International Perspective: Shorter lives, poorer health. Washington, DC: The National Academies Press.

Natri, Heini, Angela R. Garcia, Kenneth H. Buetow, Benjamin C. Trumble, and Melissa A. Wilson. 2019. "The pregnancy pickle: Evolved immune compensation due to pregnancy underlies sex differences in human diseases." *Trends in Genetics* 35 (7): 478–488.

Neel, J. V., and K. M. Weiss. 1975. "The genetic structure of a tribal population, the Yanomama Indian." *American Journal of Physical Anthropology* 42:25–52.

Nesse, Randolph M. 2019. *Good Reasons for Bad Feelings: Insights from the Frontier of Evolutionary Psychiatry*. New York: Penguin.

Nieschlag, Eberhard, Susan Nieschlag, and Hermann M. Behre. 1993. "Lifespan and testosterone." *Nature* 366 (6452): 215–215.

Nove, Andrea, Ingrid K. Friberg, Luc de Bernis, Fran McConville, Allisyn C. Moran, Maria Najjemba, Petra ten Hoope-Bender, Sally Tracy, and Caroline S. E. Homer. 2021. "Potential impact of midwives in preventing and reducing maternal and neonatal mortality and stillbirths: A Lives Saved Tool modelling study." *Lancet Global Health* 9 (1): e24–e32.

Nuland, Sherwin B. 2008. *The Art of Aging: A Doctor's Prescription for Well-Being*. New York: Random House Trade Paperbacks.

Nussey, Daniel H., Tim Coulson, Michel Festa-Bianchet, and Jean-Michel Gaillard. 2008. "Measuring senescence in wild animal populations: Towards a longitudinal approach." *Functional Ecology* 22:393–406.

Nussey, Daniel H., Hannah Froy, Jean-François Lemaitre, Jean-Michel Gaillard, and Steve N. Austad. 2013. "Senescence in natural populations of animals: Widespread evidence and its implications for bio-gerontology." *Ageing Research Reviews* 12 (1): 214–225.

O'Connell, J. F., K. Hawkes, and N. G. Blurton Jones. 1999. "Grandmothering and the evolution of *Homo erectus*." *Journal of Human Evolution* 36:461–485.

O'Hearn, Meghan, Brianna N. Lauren, John B. Wong, David D. Kim, and Dariush Mozaffarian. 2022. "Trends and disparities in cardiometabolic health among US adults, 1999–2018." *Journal of the American College of Cardiology* 80 (2): 138–151.

Ohnuki-Tierney, Emiko. 1974. *The Ainu of the Northwest Coast of Southern Sakhalin*. New York: Holt, Rinehart and Winston.

O'Keefe, J. H., L. Cordain, W. H. Harris, R. M. Moe, and R. Vogel. 2004. "Optimal low-density lipoprotein is 50 to 70 mg/dl: Lower is better and physiologically normal." *Journal of the American College of Cardiology* 43 (11): 2142–2146.

Olshansky, S. Jay. 2020. "What is a healthy body? A biodemographer's view." In *Explaining Health across the Sciences*, edited by Jonathan Sholl and Suresh I. S. Rattan, 35–41. Cham: Springer.

Olshansky, S. Jay, and Bruce A. Carnes. 1997. "Ever since Gompertz." *Demography* 34 (1): 1–15.

Olshansky, S. Jay, Bruce A. Carnes, and Douglas Grahn. 1998. "Confronting the boundaries of human longevity." *American Scientist* 86 (1): 52–61.

Oltenacu, Pascal A., and Donald M. Broom. 2010. "The impact of genetic selection for increased milk yield on the welfare of dairy cows." *Animal Welfare* 19 (1): 39–49.

Osorio, Carolina, Tulasi Kanukuntla, Eddie Diaz, Nyla Jafri, Michael Cummings, and Adonis Sfera. 2019. "The post-amyloid era in Alzheimer's disease: Trust your gut feeling." *Frontiers in Aging Neuroscience* 11 (143). https://doi.org/10.3389/fnagi.2019.00143.

Osterbur, K., F. A. Mann, K. Kuroki, and A. DeClue. 2014. "Multiple organ dysfunction syndrome in humans and animals." *Journal of Veterinary Internal Medicine* 28 (4): 1141–1151.

Page, Abigail E., and Jennifer C. French. 2020. "Reconstructing prehistoric demography: What role for extant hunter-gatherers?" *Evolutionary Anthropology: Issues, News, and Reviews* 29 (6): 332–345.

Paul, Karsten I., and Klaus Moser. 2009. "Unemployment impairs mental health: Meta-analyses." *Journal of Vocational Behavior* 74 (3): 264–282.

Pavard, Samuel, and Christophe F. D. Coste. 2021. "Evolutionary demographic models reveal the strength of purifying selection on susceptibility alleles to late-onset diseases." *Nature Ecology and Evolution* 5 (3): 392–400.

Pearl, Raymond, and John R. Miner. 1935. "Experimental studies on the duration of life. XIV. The comparative mortality of certain lower organisms." *Quarterly Review of Biology* 10:60–79.

Pearl, Raymond, and Sylvia L. Parker. 1922. "Experimental studies on the duration of life. V. On the influence of certain environmental factors on duration of life in Drosophila." *American Naturalist* 56 (646): 385–405.

Pearson, Karl. 1902. "On the change in expectation of life in man during a period of circa 2000 years." *Biometrika* 1 (2): 261–264.

Peccei, J. S. 2001. "Menopause: Adaptation or epiphenomenon?" *Evolutionary Anthropology* 10:43–57.

Pedersen, Peter Bank, Daniel Pilsgaard Henriksen, Mikkel Brabrand, and Annmarie Touborg Lassen. 2019. "Prevalence of organ failure and mortality among patients in the emergency department: A population-based cohort study." *BMJ Open* 9 (10): e032692.

Pennington, Renee, Chandler Gatenbee, Brett Kennedy, Henry Harpending, and Gregory Cochran. 2009. "Group differences in proneness to inflammation." *Infection, Genetics and Evolution* 9 (6): 1371–1380. https://doi.org/10.1016/j.meegid.2009.09.017.

Pennington, Renee, and Henry Harpending. 1991. "Infertility in Herero pastoralists of southern Africa." *American Journal of Human Biology* 3 (2):135–153. https://doi.org/10.1002/ajhb.1310030209.

Pennington, Renee, and Henry Harpending. 1993. *The Structure of an African Pastoralist Community: Demography, History, and Ecology of the Ngamiland Herero.* Oxford: Clarendon Press.

Peoples, Hervey C., Pavel Duda, and Frank W. Marlowe. 2016. "Hunter-gatherers and the origins of religion." *Human Nature* 27 (3): 261–282.

Perlman, Robert M. 1954. "The aging syndrome." *Journal of the American Geriatrics Society* 2 (2): 123–129.

Perry, George H., Nathaniel J. Dominy, Katrina G. Claw, Arthur S. Lee, Heike Fiegler, Richard Redon, John Werner, et al. 2007. "Diet and the evolution of human amylase gene copy number variation." *Nature Genetics* 39 (10): 1256.

Pew Research Center. 2022. "Financial issues top the list of reasons US adults live in multigenerational homes." https://www.pewresearch.org/social-trends/2022/03/24/financial-issues-top-the-list-of-reasons-u-s-adults-live-in-multigenerational-homes/.

Picchi, Debra. 1994. "Observations about a Central Brazilian indigenous population: The Bakairi." *South American Indian Studies* 4:37–46.

Piersma, Theunis, and Jan A. Van Gils. 2011. *The Flexible Phenotype: A Body-Centered Integration of Ecology, Physiology, and Behaviour.* Oxford: Oxford University Press.

Pinker, Steven. 2012. *The Better Angels of Our Nature: Why Violence Has Declined.* New York: Penguin Books.

Pioro, Maggie P., Roland F. Dyck, and Debra C. Gillis. 1996. "Diabetes prevalence rates among First Nations adults on Saskatchewan reserves in 1990: Comparison by tribal grouping, geography and with non-First Nations people." *Canadian Journal of Public Health = Revue Canadienne de Sante Publique* 87 (5): 325–328.

Pipher, Mary Bray. 1999. *Another Country: Navigating the Emotional Terrain of Our Elders.* New York: Riverhead Books.

Pisor, Anne C., Michael Gurven, Aaron D. Blackwell, Hillard Kaplan, and Gandhi Yetish. 2013. "Patterns of senescence in human cardiovascular fitness: VO2max in subsistence and industrialized populations." *American Journal of Human Biology* 25 (6): 756–769.

Pontzer, Herman. 2021. *Burn: New research Blows the Lid Off How We Really Burn Calories, Lose Weight, and Stay Healthy.* Garden City Park: Avery.

Poo, Ai-jen, and Ariane Conrad. 2015. *The Age of Dignity: Preparing for the Elder Boom in a Changing America.* New York: New Press.

Porter, Claire C., and Frank W. Marlowe. 2007. "How marginal are forager habitats?" *Journal of Archaeological Science* 34 (1): 59–68.

Potts, Richard. 1998. "Environmental hypotheses of hominin evolution." *American Journal of Physical Anthropology* 107 (S27): 93–106.

Preston, Samuel, Patrick Heuveline, and Michael Guillot. 2001. *Demography: Measuring and Modeling Population Processes.* Malden, MA: Blackwell.

Pryce, Christopher R., and Adriano Fontana. 2016. "Depression in autoimmune diseases." In *Inflammation-Associated Depression: Evidence, Mechanisms and Implications*, edited by Robert Dantzer and Lucile Capuron, 139–154. Cambridge, MA: Harvard University Press.

Raichlen, David A., Herman Pontzer, Jacob A. Harris, Audax Z. P. Mabulla, Frank W. Marlowe, J. Josh Snodgrass, Geeta Eick, J. Colette Berbesque, Amelia Sancilio, and Brian M. Wood. 2017. "Physical activity patterns and biomarkers of cardiovascular disease risk in hunter-gatherers." *American Journal of Human Biology* 29 (2): e22919.

Raichlen, David A., Herman Pontzer, Theodore W. Zderic, Jacob A. Harris, Audax Z. P. Mabulla, Marc T. Hamilton, and Brian M. Wood. 2020. "Sitting, squatting, and the evolutionary biology of human inactivity." *Proceedings of the National Academy of Sciences* 117 (13): 7115–7121.

Rasmussen, Knud. 1908. *The People of the Polar North; A Record.* London: Kegan Paul, Trench, Trubner.

Rasmussen, Knud. 1931. *The Netsilik Eskimo: Social Life and Spiritual Culture.* Report of the Fifth Thule Expedition, vol. 8. Copenhagen: Gyldendalske Boghandel.

Raubenheimer, David, and Stephen Simpson. 2020. *Eat Like the Animals: What Nature Teaches Us about the Science of Healthy Eating.* Boston: Houghton Mifflin.

Rauch, Jonathan. 2018. *The Happiness Curve: Why Life Gets Better after 50.* New York: Thomas Dunne Books.

Reed, Kaye E. 1997. "Early hominid evolution and ecological change through the African Plio-Pleistocene." *Journal of Human Evolution* 32 (2/3): 289–322.

Remolina, Silvia C., and Kimberly A. Hughes. 2008. "Evolution and mechanisms of long life and high fertility in queen honey bees." *Age* 30 (2): 177–185.

Reyes-García, Victoria. 2012. "Happiness in the Amazon: Folk explanations of happiness in a hunter-horticulturalist society in the Bolivian Amazon." In *Happiness across Cultures*, edited by H. Selin and G. Davey, 209–225. Dordrecht: Springer.

Reyes-García, Victoria, James Broesch, Laura Calvet-Mir, Nuria Fuentes-Peláez, Thomas W. McDade, Sorush Parsa, Susan Tanner, Tomás Huanca, William R. Leonard, and Maria R. Martínez-Rodríguez. 2009. "Cultural transmission of ethnobotanical knowledge and skills: An empirical analysis from an Amerindian society." *Evolution and Human Behavior* 30 (4): 274–285.

Reyes-García, Victoria, Sandrine Gallois, Aili Pyhälä, Isabel Díaz-Reviriego, Álvaro Fernández-Llamazares, Eric Galbraith, Sara Miñarro, and Lucentezza Napitupulu. 2021. "Happy just because. A cross-cultural study on subjective wellbeing in three Indigenous societies." *PLoS One* 16 (5): e0251551.

Reznick, D. N., M. J. Bryant, D. Roff, C. K. Ghalambor, and D. E. Ghalambor. 2004. "Effect of extrinsic mortality on the evolution of senescence in guppies." *Nature* 431:1095–1099.

Ricciarelli, Roberta, and Ernesto Fedele. 2017. "The amyloid cascade hypothesis in Alzheimer's disease: It's time to change our mind." *Current Neuropharmacology* 15 (6): 926–935. https://doi.org/10.2174/1570159X15666170116143743.

Richerson, Peter J., Robert L. Bettinger, and Robert Boyd. 2005. "Evolution on a restless planet: Were environmental variability and environmental change major drivers of human evolution?" In *Handbook of Evolution, Vol. 2: The Evolution of Living Systems*, edited by Franz M. Wuketits and Francisco J. Ayala, 223–242. Hoboken, NJ: Wiley-Blackwell.

Richman, S., M. Bennett, and K. Gleason. 2023. "AmeriCorps 2023 state of the evidence report." *Mathematica,* September.

Ricklefs, Robert W. 1998. "Evolutionary theories of aging: Confirmation of a fundamental prediction, with implications for the genetic basis and evolution of the life span." *American Naturalist* 152 (1): 24–44.

Rilling, James K. 2006. "Human and non-human primate brains: Are they allometrically scaled versions of the same design?" *Evolutionary Anthropology* 15:65–77.

Rindos, David. 2013. *The Origins of Agriculture: An Evolutionary Perspective.* New York: Academic Press.

Ringeval, Mickael, Gerit Wagner, James Denford, Guy Paré, and Spyros Kitsiou. 2020. "Fitbit-based interventions for healthy lifestyle outcomes: Systematic review and meta-analysis." *Journal of Medical Internet Research* 22 (10): e23954. https://doi.org/10.2196/23954.

Ringman, John M. 2007. "The Huntington disease of Woody Guthrie: Another man done gone." *Cognitive and Behavioral Neurology* 20 (4): 238–243.

Ritchey, Matthew D., Hilary K. Wall, Mary G. George, and Janet S. Wright. 2020. "US trends in premature heart disease mortality over the past 50 years: Where do we go from here?" *Trends in Cardiovascular Medicine* 30 (6): 364–374.

Roberts, Andrea L., Laura D. Kubzansky, Susan Malspeis, Candace H. Feldman, and Karen H. Costenbader. 2018. "Association of depression with risk of incident systemic lupus erythematosus in women assessed across 2 decades." *JAMA Psychiatry* 75 (12): 1225–1233.

Robinson, Elizabeth J., Yitades Gebre, and Joyce L. L. Pickering. 1995. "Effect of bush living on aboriginal Canadians of the Eastern James Bay Region with non-insulin-dependent diabetes mellitus." *Chronic Diseases in Canada* 16 (4): 144–148.

Rosenberg, Harriet G. 2020. "Complaint discourse, aging and caregiving among the Ju/'hoansi of Botswana-with 2018 update." In *The Cultural Context of Aging: Worldwide Perspectives*, edited by J. Sokolovsky. Santa Barbara, CA: Praeger.

Rosling, Hans, O. Rosling, and A. R. Rönnlund. 2018. *Factfulness: Ten Reasons We're Wrong about the World and Why Things Are Better Than You Think*. New York: Flatiron Books.

Roth, Eric Abella. 1981. "Sedentism and changing fertility patterns in a northern Athapascan isolate." *Journal of Human Evolution* 10 (5): 413–425.

Roth, Eric Abella, and Ajit Kisor Ray. 1985. "Demographic patterns of sedentary and nomadic Juang of Orissa." *Human Biology* 57 (3): 319–325.

Rowan, Christopher J., Michael A. Eskander, Edmond Seabright, Daniel Eid Rodriguez, Edhitt Cortez Linares, Raul Quispe Gutierrez, Juan Copajira Adrian, et al. 2021. "Very low prevalence and incidence of atrial fibrillation among Bolivian forager-farmers." *Annals of Global Health* 87 (1).

Rowley, Kevin G., Qing Su, Marion Cincotta, Michelle Skinner, Karen Skinner, Benedicta Pindan, Gwyneth A. White, and Kerin O'Dea. 2001. "Improvements in circulating cholesterol, antioxidants, and homocysteine after dietary intervention in an Australian Aboriginal community." *American Journal of Clinical Nutrition* 74 (4): 442–448.

Ruby, J. Graham, Megan Smith, and Rochelle Buffenstein. 2018. "Naked mole-rat mortality rates defy Gompertzian laws by not increasing with age." *eLife* 7:e31157.

Ruiz, John M., Patrick Steffen, and Timothy B. Smith. 2013. "Hispanic mortality paradox: A systematic review and meta-analysis of the longitudinal literature." *American Journal of Public Health* 103 (3): e52–e60. https://doi.org/10.2105/ajph.2012.301103.

Russell, Josiah Cox. 1948. *British Medieval Population*. Albuquerque: University of New Mexico Press.

Ryan, Calen. 2021. "'Epigenetic clocks': Theory and applications in human biology." *American Journal of Human Biology* 33 (3): e23488.

Sacher, George A. 1975. "Maturation and longevity in relation to cranial capacity in hominid evolution." In *Primate Functional Morphology and Evolution*, edited by R. Tuttle, 417–441. The Hague: Mouton.

Sacher, George A. 1977. "Life table modification and life prolongation." *Handbook of the Biology of Aging*, edited by Nicolas Musi and Peter Hornsby. New York: Academic Press.

Saha, Sudeshna, Naazneen Khan, Troy Comi, Andrea Verhagen, Aniruddha Sasmal, Sandra Diaz, Hai Yu, et al. 2022. "Evolution of human-specific alleles protecting cognitive function of grandmothers." *Molecular Biology and Evolution* 39 (8). https://doi.org/10.1093/molbev/msac151.

Sahlins, Marshall. 1968. "Notes on the original affluent society." *Man the Hunter*, edited by R. Lee and I. Devore, 85–89. London: Routledge.

Sallis, James, Robert Cervero, William Ascher, Karla Henderson, M. Katherine Kraft, and Jacqueline Kerr. 2006. "An ecological approach to creating active living communities." *Annual Review of Public Health* 27:297–322.

Salthouse, Timothy A. 2019. "Trajectories of normal cognitive aging." *Psychology and Aging* 34 (1): 17.

Sanderson, Warren, and Sergei Scherbov. 2008. *Rethinking Age and Aging*. Washington, DC: Population Reference Bureau.

Sanderson, Warren C., and Sergei Scherbov. 2010. "Remeasuring aging." *Science* 329 (5997): 1287–1288.

Santini, Ziggi Ivan, Paul E. Jose, Erin York Cornwell, Ai Koyanagi, Line Nielsen, Carsten Hinrichsen, Charlotte Meilstrup, Katrine R. Madsen, and Vibeke Koushede. 2020. "Social disconnectedness, perceived isolation, and symptoms of depression and anxiety among older Americans (NSHAP): A longitudinal mediation analysis." *Lancet Public Health* 5 (1): e62–e70.

Sanya, Richard E., Emily L. Webb, Christopher Zziwa, Robert Kizindo, Moses Sewankambo, Josephine Tumusiime, Esther Nakazibwe, Gloria Oduru, Emmanuel Niwagaba, et al. 2020. "The effect of helminth infections and their treatment on metabolic outcomes: Results of a cluster-randomised trial." *Clinical Infectious Diseases* 71 (3): 601–613.

Sasaki, Tomohiko, and Osamu Kondo. 2016. "Maximum likelihood estimate of life expectancy in the prehistoric Jomon: Canine pulp volume reduction suggests a longer life expectancy than previously thought." *American Journal of Physical Anthropology* 161 (1): 170–180.

Sasson, Isaac. 2021. "Age and COVID-19 mortality: A comparison of Gompertz doubling time across countries and causes of death." *Demographic Research* 44:379–396.

Sattenspiel, Lisa, and Henry Harpending. 1983. "Stable populations and skeletal age." *American Antiquity* 48 (3): 489–498.

Sattler, David N., Charles F. Kaiser, and James B. Hittner. 2000. "Disaster preparedness: Relationships among prior experience, personal characteristics, and distress." *Journal of Applied Social Psychology* 30 (7): 1396–1420.

Sawyer, Susan M., Peter S. Azzopardi, Dakshitha Wickremarathne, and George C. Patton. 2018. "The age of adolescence." *Lancet Child and Adolescent Health* 2 (3): 223–228.

Scalise Sugiyama, Michelle. 2017. "Oral storytelling as evidence of pedagogy in forager societies." *Frontiers in Psychology* 8:471.

Schebesta, Paul, and Frieda Schütze. 1954. *Negritos of Asia, Vol. 2. Ethnography of the Negritos: Half-Vol. 1, Economy and Sociology*. Studia Instituti Anthropos. https://ehrafworldcultures .yale.edu/cultures/an07/documents/002.

Scheper-Hughes, Nancy. 1993. *Death without Weeping: The Violence of Everyday Life in Brazil*. Berkeley: University of California Press.

Schniter, Eric, Hillard Kaplan, and Michael Gurven. 2023. "Cultural transmission vectors of essential knowledge and skills among Tsimane forager-horticulturalists." *Evolution and Human Behavior* 44:530–540.

Schniter, Eric, Nathaniel T. Wilcox, Bret A. Beheim, Hillard S. Kaplan, and Michael Gurven. 2018. "Information transmission and the oral tradition: Evidence of a late-life service niche for Tsimane Amerindians." *Evolution and Human Behavior* 39 (1): 94–105.

Schrire, Carmel. 1984. *Past and Present in Hunter-Gatherer Studies*. New York: Academic Press.

Schulz, Leslie O., and Lisa S. Chaudhari. 2015. "High-risk populations: The Pimas of Arizona and Mexico." *Current Obesity Reports* 4 (1): 92–98.

Schuppli, Caroline, Sereina M. Graber, Karin Isler, and Carel P. van Schaik. 2016. "Life history, cognition and the evolution of complex foraging niches." *Journal of Human Evolution* 92:91–100.

Schuppli, Caroline, Karin Isler, and Carel P. Van Schaik. 2012. "How to explain the unusually late age at skill competence among humans." *Journal of Human Evolution* 63:843–850.

Schwarz, Flavio, Stevan A. Springer, Tasha K. Altheide, Nissi M. Varki, Pascal Gagneux, and Ajit Varki. 2016. "Human-specific derived alleles of CD33 and other genes protect against postreproductive cognitive decline." *Proceedings of the National Academy of Sciences* 113 (1): 74–79.

Sear, Rebecca. 2016. "Beyond the nuclear family: An evolutionary perspective on parenting." *Current Opinion in Psychology* 7:98–103.

Sear, Rebecca, and Ruth Mace. 2008. "Who keeps children alive? A review of the effects of kin on child survival." *Evolution and Human Behavior* 29 (1): 1–18.

Sear, Rebecca, Ruth Mace, and Ian A. McGregor. 2000. "Maternal grandmothers improve the nutritional status and survival of children in rural Gambia." *Proceedings of the Royal Society of London, Series B* no. 267:1641–1647.

Sear, Rebecca, Ruth Mace, and Ian A. McGregor. 2003. "The effects of kin on female fertility in rural Gambia." *Evolution and Human Behavior* 24:25–42.

Shalinsky, Audrey, and Anthony Glascock. 1988. "Killing infants and the aged in nonindustrial societies: Removing the liminal." *Social Science Journal* 25 (3): 277–287.

Sharp, Henry S. 1977. "The Chipewyan hunting unit." *American Ethnologist* 4 (2): 377–393.

Sharp, Henry S. 1981. "Old age among the Chipewyan." *Other Ways of Growing Old: Anthropological Perspectives*, edited by Pamela Amoss and Stevan Harrell, 99–109. Stanford, CA: Stanford University Press.

Sherratt, T. N., C. Hassall, R. A. Laird, D. J. Thompson, and A. Cordero-Rivera. 2011. "A comparative analysis of senescence in adult damselflies and dragonflies (Odonata)." *Journal of Evolutionary Biology* 24 (4): 810–822.

Shor, Eran, David J. Roelfs, Paul Bugyi, and Joseph E. Schwartz. 2012. "Meta-analysis of marital dissolution and mortality: Reevaluating the intersection of gender and age." *Social Science and Medicine* 75 (1): 46–59.

Shor, Eran, David J. Roelfs, Misty Curreli, Lynn Clemow, Matthew M. Burg, and Joseph E. Schwartz. 2012. "Widowhood and mortality: A meta-analysis and meta-regression." *Demography* 49 (2): 575–606.

Siegel, Rebecca L., Angela N. Giaquinto, and Ahmedin Jemal. 2024. "Cancer statistics, 2024." *CA: A Cancer Journal for Clinicians* 74 (1): 12–49.

Silberbauer, G. 1981. *Hunter and Habitat in the Central Kalahari Desert*. Cambridge, UK: Cambridge University Press.

Silvert, Martin, Lluis Quintana-Murci, and Maxime Rotival. 2019. "Impact and evolutionary determinants of Neanderthal introgression on transcriptional and post-transcriptional regulation." *American Journal of Human Genetics* 104 (6): 1241–1250.

Simmons, Leo W. 1945. *The Role of the Aged in Primitive Society*. New Haven, CT: Yale University Press.

Simmons, Leo W. 1946. "Attitudes toward aging and the aged: Primitive-societies." *Journal of Gerontology* 1 (Part 1): 72–95.

Simmons, Leo W. 1960. "Aging in preindustrial societies." *Handbook of Social Gerontology*, edited by Clark Tibbits, 62–91. Chicago: University of Chicago Press.

Simmons, Leo W., and Harold G. Wolff. 1954. *Social Science in Medicine*. Russell Sage Foundation. https://www.russellsage.org/publications/social-science-medicine.

Simonton, Dean Keith. 1997. "Creative productivity: A predictive and explanatory model of career trajectories and landmarks." *Psychological Review* 104 (1): 66.

Simoons, F. J. 1994. *Eat Not of This Flesh: Food Avoidances from Prehistory to the Present*. Madison: University of Wisconsin Press.

Sinclair, David A., and Matthew D. LaPlante. 2019. *Lifespan: Why We Age—and Why We Don't Have To*. New York: Atria Books.

Singh, Manvir. 2018. "The cultural evolution of shamanism." *Behavioral and Brain Sciences* 41.

Sinnett, P. F., and H. M. Whyte. 1973. "Epidemiological studies in a total highland population, Tukisenta, New Guinea: Cardiovascular disease and relevant clinical, electrocardiographic, radiological and biochemical findings." *Journal of Chronic Diseases* 26 (5): 265–290.

Skirbekk, Vegard. 2004. "Age and individual productivity: A literature survey." *Vienna Yearbook of Population Research* 2:133–153.

Skogen, Jens Christoffer, and Simon Øverland. 2012. "The fetal origins of adult disease: A narrative review of the epidemiological literature." *JRSM Short Reports* 3 (8): 1–7.

Smith, Daniel, Philip Schlaepfer, Katie Major, Mark Dyble, Abigail E. Page, James Thompson, Nikhil Chaudhary, et al. 2017. "Cooperation and the evolution of hunter-gatherer storytelling." *Nature Communications* 8 (1): 1853.

Smith, Eric Alden, Kim Hill, Frank Marlowe, David Nolin, Polly Wiessner, Michael Gurven, Samuel Bowles, et al. 2010. "Wealth transmission and inequality among hunter-gatherers." *Current Anthropology* 51 (1): 19.

Snopkowski, Kristin, Cristina Moya, and Rebecca Sear. 2014. "A test of the intergenerational conflict model in Indonesia shows no evidence of earlier menopause in female-dispersing groups." *Proceedings of the Royal Society of London B: Biological Sciences* 281 (1788): 20140580.

Solway, Jacqueline S., and Richard B. Lee. 1990. "Foragers, genuine or spurious? Situating the Kalahari San in history." *Current Anthropology* 31 (2): 109–146.

Sorokowski, Piotr, Anna Oleszkiewicz, Corinna E. Löckenhoff, Marta Kowal, Tomasz Frackowiak, and W. P. Malecki. 2022. "Literacy and perceptions of aging: Evidence from the Dani in Papua." *Language and Communication* 82:1–7.

Sorokowski, Piotr, Agnieszka Sorokowska, Tomasz Frackowiak, and Corinna E. Löckenhoff. 2017. "Aging perceptions in Tsimane' Amazonian forager-farmers compared with two industrialized societies." *Journals of Gerontology Series B: Psychological Sciences and Social Sciences* 72 (4): 561–570.

Spencer, Paul. 1965. "The Samburu." In *The Samburu*. Berkeley: University of California Press.

Spottiswoode, Claire N., and Brian M. Wood. 2023. "Culturally determined interspecies communication between humans and honeyguides." *Science* 382 (6675): 1155–1158.

Spreng, R. Nathan, and Gary R. Turner. 2019. "The shifting architecture of cognition and brain function in older adulthood." *Perspectives on Psychological Science* 14 (4): 523–542.

Stamatakis, Emmanuel, Matthew N. Ahmadi, Jason M. R. Gill, Cecilie Thøgersen-Ntoumani, Martin J. Gibala, Aiden Doherty, and Mark Hamer. 2022. "Association of wearable device-measured vigorous intermittent lifestyle physical activity with mortality." *Nature Medicine* 28:2521–2529.

Stearns, S. C., M. Ackermann, M. Doebeli, and M. Kaiser. 2000. "Experimental evolution of aging, growth and reproduction in fruit flies." *Proceedings of the National Academy of Sciences, USA* 97:3309–3313.

Stefansson, Vilhjalmur. 1960. *Cancer: Disease of Civilization? An Anthropological and Historical Study*. New York: Hill and Wang.

Stelter, Robert, David De la Croix, and Mikko Myrskylä. 2021. "Leaders and laggards in life expectancy among European scholars from the sixteenth to the early twentieth century." *Demography* 58 (1): 111–135.

Stenner, Hedwig T., Johanna Boyen, Markus Hein, Gudrun Protte, Momme Kück, Armin Finkel, Alexander A. Hanke, and Uwe Tegtbur. 2020. "Everyday Pedelec use and its effect on meeting physical activity guidelines." *International Journal of Environmental Research and Public Health* 17 (13): 4807.

Steyn, H. P. 1994. "Role and position of elderly !Xu in the Schmidtsdrift Bushman community." *South African Journal of Ethnology* 17 (2): 31–37.

Stieglitz, Jonathan, Paul Hooper, Benjamin Trumble, Hillard Kaplan, and Michael Gurven. 2020. "Productivity loss associated with functional disability in a contemporary small-scale subsistence population." *eLife* 9:e62883.

Stieglitz, Jonathan, Eric Schniter, Christopher Von Rueden, Hillard Kaplan, and Michael Gurven. 2014. "Functional disability and social conflict increase risk of depression in older adulthood among Bolivian forager-farmers." *Journals of Gerontology Series B: Psychological Sciences and Social Sciences*. https://doi.org/10.1093/geronb/gbu080.

Stieglitz, Jonathan, Benjamin Trumble, Caleb Finch, Dong Li, Matthew J. Budoff, Hillard Kaplan, Michael Gurven, and HORUS Study Team. 2019. "Computed tomography shows high fracture prevalence among physically active forager-horticulturalists with high fertility." *eLife* 8:e48607.

Stieglitz, Jonathan, Benjamin Trumble, Melissa Emery Thompson, Aaron Blackwell, Hillard Kaplan, and Michael Gurven. 2015. "Depression as sickness behavior? A test of the host defense hypothesis in a high pathogen population." *Brain, Behavior, and Immunity* 49:130–139.

Stiles, Daniel. 1992. "The hunter-gatherer' revisionist' debate." *Anthropology Today* 8 (2): 13–17.

Stiner, Mary C. 2002. "Carnivory, coevolution, and the geographic spread of the genus Homo." *Journal of Archaeological Research* 10 (1): 1–63.

Stiner, Mary C., Natalie D. Munro, Todd A. Surovell, Eitan Tchernov, and Ofer Bar-Yosef. 1999. "Paleolithic population growth pulses evidenced by small animal exploitation." *Science* 283 (5399): 190–194.

Stone, Arthur A., Joseph E. Schwartz, Joan E, Broderick, and Angus Deaton. 2010. "A snapshot of the age distribution of psychological well-being in the United States." *Proceedings of the National Academy of Sciences* 107 (22): 9985–9990.

Strassmann, Beverly, and Wendy Garrard. 2011. "Alternatives to the grandmother hypothesis." *Human Nature* 22 (1): 201–222.

Strehler, B. L., and A. S. Mildvan. 1960. "General theory of mortality and aging." *Science* 132: 14–21.

Su, Yan. 2017. *Impact of Intergenerational Programs on Older Adults' Psychological Well-Being: A Meta-Analysis*. Ames: Iowa State University Press.

Sugrue, Victoria J., Joseph A. Zoller, Pritika Narayan, Ake T. Lu, Oscar J. Ortega-Recalde, Matthew J. Grant, C. Simon Bawden, et al. 2021. "Castration delays epigenetic aging and feminizes DNA methylation at androgen-regulated loci." *eLife* 10: e64932. https://doi.org/10.7554/eLife.64932.

Suwa, Gen, Berhane Asfaw, Reiko T. Kono, Daisuke Kubo, C. Owen Lovejoy, and Tim D. White. 2009. "The Ardipithecus ramidus skull and its implications for hominid origins." *Science* 326 (5949): 68–68e7.

Suzman, Richard M., David P. Willis, and Kenneth G. Manton. 1995. *The Oldest Old*. New York: Oxford University Press.

Swartz, Aimee. 2008. "James Fries: Healthy aging pioneer." *American Journal of Public Health* 98 (7): 1163–1166.

Tahapary, Dicky L., Karin de Ruiter, Ivonne Martin, Eric A. T. Brienen, Lisette van Lieshout, Christa M. Cobbaert, Pradana Soewondo, et al. 2017. "Effect of anthelmintic treatment on insulin resistance: A cluster-randomized, placebo-controlled trial in Indonesia." *Clinical Infectious Diseases* 65 (5): 764–771. https://doi.org/10.1093/cid/cix416.

Tanaka, Kenzo, Junichi Masuda, Tsukasa Imamura, Katsuo Sueishi, Teruyuki Nakashima, Isamu Sakurai, Tsuyoshi Shozawa, et al. 1988. "A nation-wide study of atherosclerosis in infants, children and young adults in Japan." *Atherosclerosis* 72 (2/3): 143–156.

Taylor, Paul, Rich Morin, Kim Parker, D'Vera Cohn, and Wendy Wang. 2009. Growing old in America: Expectations vs. reality. Pew Research Center. June 29. https://www.pewresearch.org/social-trends/2009/06/29/growing-old-in-america-expectations-vs-reality/.

Temple, Stanley A. 1987. "Do predators always capture substandard individuals disproportionately from prey populations?" *Ecology* 68 (3): 669–674.

Terashima, Hideaki, and Barry S. Hewlett. 2016. *Social Learning and Innovation in Contemporary Hunter-Gatherers*. Cham: Springer.

Thomas, Bill, and William H. Thomas. 2015. *Second Wind: Navigating the Passage to a Slower, Deeper, and More Connected Life*. New York: Simon and Schuster.

Thomas, Frédéric, Randolph M. Nesse, Robert Gatenby, Cindy Gidoin, François Renaud, Benjamin Roche, and Beata Ujvari. 2016. "Evolutionary ecology of organs: A missing link in cancer development?" *Trends in Cancer* 2 (8): 409–415.

Thomas, Gregory S., L. Samuel Wann, Adel H. Allam, Randall C. Thompson, David E. Michalik, M. Linda Sutherland, James D. Sutherland, et al. 2014. "Why did ancient people have atherosclerosis? From autopsies to computed tomography to potential causes." *Global Heart* 9 (2): 229–237.

Thomas, William H. 2004. *What Are Old People For? How Elders Will Save the World*. New York: VanderWyk & Burnham.

Thomas, William H. 2011. "Eldertopia." *AARP: The Journal* 8:26–31.

Thompson, Jessica C., Susana Carvalho, Curtis W. Marean, and Zeresenay Alemseged. 2019. "Origins of the human predatory pattern: The transition to large-animal exploitation by early hominins." *Current Anthropology* 60 (1): 1–23.

Thompson, Randall C., Adel H. Allam, Guido P. Lombardi, L. Samuel Wann, M. Linda Sutherland, James D. Sutherland, Muhammad Al-Tohamy Soliman, et al. 2013. "Atherosclerosis across 4000 years of human history: The Horus study of four ancient populations." *The Lancet* 381 (9873): 1211–1222.

Thornton, Arland. 2005. *Reading History Sideways: The Fallacy and Enduring Impact of the Developmental Paradigm on Family Life*. Chicago: University of Chicago Press.

Tilley, Lorna. 2022. "Disability and care in the bioarchaeological record: Meeting the challenges of being human." In *The Routledge Handbook of Paleopathology*, edited by Anne Grauer, 457–481. New York: Routledge.

Tomasetti, Cristian, and Bert Vogelstein. 2015. "Variation in cancer risk among tissues can be explained by the number of stem cell divisions." *Science* 347 (6217): 78–81.

Tomczyk, Szymon, Kathleen Fischer, Steven Austad, and Brigitte Galliot. 2015. "Hydra, a powerful model for aging studies." *Invertebrate Reproduction and Development* 59 (sup1): 11–16.

Tooby, John, and Leda Cosmides. 1996. "Friendship and the banker's paradox: Other pathways to the evolution of adaptations for altruism." Paper read at Proceedings—British Academy. https://www.researchgate.net/publication/232419330_Friendship_and_the_Banker%27s _Paradox_Other_Pathways_to_the_Evolution_of_Adaptations_for_Altruism.

Tooby, John, and Irven DeVore. 1987. "The reconstruction of hominid behavioral evolution through strategic modeling." In *The Evolution of Human Behavior: Primate Models*, edited by W. G. Kinzey, 183–237. Albany: SUNY Press.

Totterdell, John, Rebecca Wellard, Isabella Reeves, Brodie Elsdon, Pia Markovic, Machi Yoshida, Ashleigh Fairchild, Gemma Sharp, and Robert Pitman. "The first three records of killer whales (Orcinus orca) killing and eating blue whales (Balaenoptera musculus)." 2022. *Marine Mammal Science* 38 (3): 1286–1301.

Trinkaus, Erik. 1995. "Neanderthal mortality patterns." *Journal of Archaeological Science* 22 (1): 121–142.

Trinkaus, Erik. 2011. "Late Pleistocene adult mortality patterns and modern human establishment." *Proceedings of the National Academy of Sciences* 108 (4): 1267–1271.

Trumble, Benjamin, Mia Charifson, Tom Kraft, Angela Garcia, Daniel Cummings, Paul Hooper, Amanda Lea, et al. 2023. "Apolipoprotein-ε 4 is associated with higher fecundity in a natural fertility population." *Science Advances* 9 (32): eade9797.

Trumble, Benjamin, and Caleb Finch. 2019. "The exposome in human evolution: From dust to diesel." *Quarterly Review of Biology* 94 (4): 333–394.

Trumble, Benjamin, Jonathan Stieglitz, Aaron Blackwell, Hooman Allayee, Bret Beheim, Caleb Finch, Michael Gurven, and Hillard Kaplan. 2017. "Apolipoprotein E4 is associated with improved cognitive function in Amazonian forager-horticulturalists with a high parasite burden." *FASEB Journal* 31 (4): 1508–1515.

Truswell, A. Stewart, and John D. L. Hansen. 1976. "8 Medical Research among the !Kung." In *Kalahari Hunter-Gatherers: Studies of the !Kung San and Their Neighbors*, edited by Richard B. Lee and Irven Devore, 166–194. Cambridge, MA: Harvard University Press.

Tuljapurkar, Shripad, Cedric Puleston, and Michael Gurven. 2007. "Why men matter: Mating pattern drives evolution of post-reproductive lifespan." *PLoS One* 2 (8): e785.

Turnbull, Colin M. 1965. *The Mbuti Pygmies: An Ethnographic Survey*. New York: American Museum of Natural History.

Turnbull, Colin M. 1983. *The Mbuti Pygmies: Change and Adaptation*. Case Studies in Cultural Anthropology. New York: Holt, Rinehart, and Winston.

UNDESA (United Nations Department of Economic and Social Affairs). 2019. *World Population Ageing 2019 Highlights*. New York: United Nations.

UNDP. 2019. Human Development Report. In *United Nations Development Report*. New York: UN Development Programme.

Ungar, Peter. 2017. *Evolution's Bite*. Princeton, NJ: Princeton University Press.

Vallois, H. V. 1961. "The social life of early man: The evidence of skeletons." In *Social Life of Early Man*, edited by S. L. Washburn, 214–235. Chicago: Aldine de Gruyter.

Van Arsdale, P. 1981. "Disintegration of the ritual support network among aged Asmat hunter-gatherers of New Guinea." In *Dimensions: Aging, Culture and Health*, by C. Fry and contributors, 33–46. New York: Praeger.

Vanarsdale, Peter W. 2021. "The elderly Asmat of New Guinea." In *Other Ways of Growing Old*, edited by Pamela Amoss and Stevan Harrell, 111–124. Stanford, CA: Stanford University Press.

van den Eerenbeemt, Marie-Louise. 1985. "A demographic profile of the Fulani of Central Mali with special emphasis on infant and child mortality." In *Population, Health and Nutrition in the Sahel: Issues in the Welfare of Selected West African Communities*, edited by Allan Hill, 79–103. London: London School of Hygiene and Tropical Medicine.

VanStone, James W. 1963. *The Snowdrift Chipewyan*. Vol. 63. NCRC 63-4. Northern Co-ordination and Research Centre, Department of Northern Affairs, Canada. https://publications.gc.ca/collections/collection_2017/aanc-inac/R42-3-1963-4-eng.pdf.

Van Voorhies, Wayne A., Jacqueline Fuchs, and Stephen Thomas. 2005. "The longevity of Caenorhabditis elegans in soil." *Biology Letters* 1 (2): 247–249.

Vatanen, Tommi, Aleksandar D. Kostic, Eva d'Hennezel, Heli Siljander, Eric A. Franzosa, Moran Yassour, Raivo Kolde, et al. 2016. "Variation in microbiome LPS immunogenicity contributes to autoimmunity in humans." *Cell* 165 (4): 842–853.

Velasquez-Manoff, Moises. 2012. *An Epidemic of Absence: A New Way of Understanding Allergies and Autoimmune Diseases*. New York: Simon and Schuster.

Vilchez, David, Isabel Saez, and Andrew Dillin. 2014. "The role of protein clearance mechanisms in organismal ageing and age-related diseases." *Nature Communications* 5 (1): 1–13.

Vincze, Orsolya, Fernando Colchero, Jean-Francois Lemaître, Dalia A. Conde, Samuel Pavard, Margaux Bieuville, Araxi O. Urrutia, et al. 2022. "Cancer risk across mammals." *Nature* 601 (7892): 263–267.

Voland, E., and J. Beise. 2002. "Opposite effects of maternal and paternal grandmothers on infant survival in historical Krummhorn." *Behavioral Ecology and Sociobiology* 52:435–443.

Volk, Anthony A., and Jeremy A. Atkinson. 2013. "Infant and child death in the human environment of evolutionary adaptation." *Evolution and Human Behavior* 34 (3): 182–192.

von Rueden, Christopher, Sarah Alami, Hillard Kaplan, and Michael Gurven. 2018. "Sex differences in political leadership in an egalitarian society." *Evolution and Human Behavior* 39 (4): 402–411.

von Rueden, Christopher, Michael Gurven, and Hillard Kaplan. 2008. "The multiple dimensions of male social status in an Amazonian society." *Evolution and Human Behavior* 29:402–415.

von Rueden, Christopher, Michael Gurven, and Hillard Kaplan. 2011. "Why do men seek status? Fitness payoffs to dominance and prestige." *Proceedings of the Royal Society B: Biological Sciences* 278 (1715): 2223–2232.

Voskarides, Konstantinos. 2018. "Combination of 247 genome-wide association studies reveals high cancer risk as a result of evolutionary adaptation." *Molecular Biology and Evolution* 35 (2): 473–485.

Walker, David W., Gawain McColl, Nicole L. Jenkins, Jennifer Harris, and Gordon J. Lithgow. 2000. "Evolution of lifespan in C. elegans." *Nature* 405 (6784): 296–297.

Walker, Philip L., John Johnson, and Patricia Lambert. 1988. "Age and sex biases in the preservation of human skeletal remains." *American Journal of Physical Anthropology* 76:183–188.

Walker, Robert, Kim Hill, Mark Flinn, and Ryan Ellsworth. 2011. "Evolutionary history of hunter-gatherer marriage practices." *PLoS One* no. 6 (4): e19066.

Walker, Robert, Lisa Sattenspiel, and Kim Hill. 2015. "Mortality from contact-related epidemics among indigenous populations in Greater Amazonia." *Scientific Reports* 5:14032.

Wang, Fan, Yu Gao, Zhen Han, Yue Yu, Zhiping Long, Xianchen Jiang, Yi Wu, et al. 2023. "A systematic review and meta-analysis of 90 cohort studies of social isolation, loneliness and mortality." *Nature Human Behaviour* 7 (8): 1307–1319.

Wang, Li, Jiawen Cui, Biao Jin, Jianguo Zhao, Huimin Xu, Zhaogeng Lu, Weixing Li, et al. 2020. "Multifeature analyses of vascular cambial cells reveal longevity mechanisms in old Ginkgo biloba trees." *Proceedings of the National Academy of Sciences* 117 (4): 2201–2210.

Wang, Rosy, and R. M. Damian Holsinger. 2018. "Exercise-induced brain-derived neurotrophic factor expression: Therapeutic implications for Alzheimer's dementia." *Ageing Research Reviews* 48:109–121. https://doi.org/10.1016/j.arr.2018.10.002.

Washburn, Sherwood. 1981. "Longevity in primates." In *Aging: Biology and Behavior*, edited by J. McGaugh and S. Kiesler, 11–29. New York: Academic Press.

Watling, Jennifer, Myrtle P. Shock, Guilherme Z. Mongeló, Fernando O. Almeida, Thiago Kater, Paulo E. De Oliveira, and Eduardo G. Neves. 2018. "Direct archaeological evidence for Southwestern Amazonia as an early plant domestication and food production centre." *PLoS One* 13 (7): e0199868.

Wei, Yongbin, Siemon C. de Lange, Lianne H. Scholtens, Kyoko Watanabe, Dirk Jan Ardesch, Philip R. Jansen, Jeanne E. Savage, et al. 2019. "Genetic mapping and evolutionary analysis of human-expanded cognitive networks." *Nature Communications* no. 10 (1): 1–11.

Weibel, Ewald R. 2000. *Symmorphosis: On Form and Function in Shaping Life*. Cambridge, MA: Harvard University Press.

Weiss, Alexander, James E. King, Miho Inoue-Murayama, Tetsuro Matsuzawa, and Andrew J. Oswald. 2012. "Evidence for a midlife crisis in great apes consistent with the U-shape in human well-being." *Proceedings of the National Academy of Sciences* 109 (49): 19949–19952.

Weiss, Kenneth M. 1973. "Demographic models for anthropology." *American Antiquity, Memoirs of the Society for American Antiquity*, edited by H. Martin Wobst, i–ix, 1–186. Cambridge, UK: Cambridge University Press.

Weiss, Kenneth M. 1975. "The application of demographic models to anthropological data." *Human Ecology* 3 (2): 87–103.

Weiss, Kenneth M. 1981. "Evolutionary perspectives on human aging." In *Other Ways of Growing Old*, edited by P. Amoss and S. Harrell, 25–28. Stanford, CA: University Press.

Wentzensen, Ingrid M., Lisa Mirabello, Ruth M. Pfeiffer, and Sharon A. Savage. 2011. "The association of telomere length and cancer: A meta-analysis." *Cancer Epidemiology, Biomarkers and Prevention* 20 (6): 1238–1250.

Wettstein, Gal, and Alice Zulkarnain. 2017. "How much long-term care do adult children provide?" *Issue in Brief*, pp. 17–11. Chestnut Hill, MA: Center for Retirement Research at Boston College.

Whitney, Craig R. 1997. "Jeanne Calment, world's elder, dies at 122." *New York Times*, August 5, 1997, p. 27.

World Health Organization (WHO). 2008. *Older Persons in Emergencies: An Active Ageing Perspective*. Geneva, Switzerland. https://extranet.who.int/agefriendlyworld/wp-content/uploads/2014/06/WHO-Older-Persons-in-Emergencies-An-Active-Ageing-Perspective.pdf.

Wiessner, Polly. 1982. "Risk, reciprocity and social influences on !Kung San economics." In *Politics and History in Band Societies*, edited by E. Leacock and R. B. Lee, 61–84. Cambridge, UK: Cambridge University Press.

Wiessner, Polly W. 2014. "Embers of society: Firelight talk among the Ju/'hoansi Bushmen." *Proceedings of the National Academy of Sciences* no. 111 (39): 14027–14035.

Wilbert, Johannes. 1972. *Survivors of Eldorado: Four Indian Cultures of South America*. New York: Praeger.

Will, Manuel, Mario Krapp, Jay T. Stock, and Andrea Manica. 2021. "Different environmental variables predict body and brain size evolution in Homo." *Nature Communications* 12 (1): 1–12.

Williams, George C. 1957. "Pleiotropy, natural selection and the evolution of senescence." *Evolution* 11:398–411.

Williams, P. D., T. Day, Q. Fletcher, and L. Rowe. 2006. "The shaping of senescence in the wild." *Trends in Ecology and Evolution* 21 (8): 458–463.

Wilmoth, John R. 1995. "The earliest centenarians: A statistical analysis." In *Exceptional Longevity*, edited by B. Jeune and J. W. Vaupel, 109–116. Odense, DK: Odense University Press.

Wilmsen, Edwin N. 1989. *Land Filled with Flies: A Political Economy of the Kalahari*. Chicago: University of Chicago Press.

Winkelman, Michael James. 1990. "Shamans and other magico-religious healers: A cross-cultural study of their origins, nature, and social transformations." *Ethos* 18 (3): 308–352.

Winterhalder, Bruce, William Baillargeon, Francesca Cappelletto, I. Randolph Daniel Jr., and Chris Prescott. 1988. "The population ecology of hunter-gatherers and their prey." *Journal of Anthropological Archaeology* 7 (4): 289–328.

Wiria, Aprilianto E., Firdaus Hamid, Linda J. Wammes, Margaretta A. Prasetyani, Olaf M. Dekkers, Linda May, Maria M. M. Kaisar, et al. 2015. "Infection with soil-transmitted helminths is associated with increased insulin sensitivity." *PLoS One* 10 (6): e0127746.

Wiria, Aprilianto E., Erliyani Sartono, Taniawati Supali, and Maria Yazdanbakhsh. 2014. "Helminth infections, type-2 immune response, and metabolic syndrome." *PLoS Pathogens* 10 (7).

Wood, Brian, Jacob Harris, David Raichlen, Herman Pontzer, Katherine Sayre, Amelia Sancilio, Colette Berbesque, et al. 2021. "Gendered movement ecology and landscape use in Hadza hunter-gatherers." *Nature Human Behaviour* 5 (4): 436–446.

Wood, Brian, and Frank Marlowe. 2013. "Household and kin provisioning by Hadza men." *Human Nature* 24 (3): 280–317.

Wood, Brian M., Jacob D. Negrey, Janine L. Brown, Tobias Deschner, Melissa Emery Thompson, Sholly Gunter, John C. Mitani, et al. 2023. "Demographic and hormonal evidence for menopause in wild chimpanzees." *Science* 382 (6669): eadd5473.

Wood, Brian, David Watts, John Mitani, and Kevin Langergraber. 2017. "Favorable ecological circumstances promote life expectancy in chimpanzees similar to that of human hunter-gatherers." *Journal of Human Evolution* 105:41–56.

Wood, James W., and Peter Smouse. 1982. "A method of analyzing density-dependent vital rates with an application to the Gainj of Papua New Guinea." *American Journal of Physical Anthropology* 58:403–411.

Wrangham, Richard. 2009. *Catching Fire: How Cooking Made Us Human*. New York: Basic Books.

Wrangham, Richard, Michael Wilson, and Martin Muller. 2006. "Comparative rates of violence in chimpanzees and humans." *Primates* 47:14–26.

Wrigley, Edward Anthony, Ros S. Davies, James E. Oeppen, and Roger S. Schofield. 1997. *English Population History from Family Reconstitution 1580–1837*. Cambridge, UK: Cambridge University Press.

Wrigley, Edward Anthony, and Roger S. Schofield. 1989. *The Population History of England 1541–1871*. Cambridge, UK: Cambridge University Press.

Wroblewski, Andrew P., Francesca Amati, Mark A. Smiley, Bret Goodpaster, and Vonda Wright. 2011. "Chronic exercise preserves lean muscle mass in masters athletes." *Physician and Sportsmedicine* 39 (3): 172–178.

Wu, Song, Scott Powers, Wei Zhu, and Yusuf A. Hannun. 2016. "Substantial contribution of extrinsic risk factors to cancer development." *Nature* 529 (7584): 43–47.

Wu, Xinhui, Haixia Yang, Sixian He, Ting Xia, Diang Chen, Yexin Zhou, Jin Liu, M. Liu, and Zhen Sun. 2022. "Adult vaccination as a protective factor for dementia: A meta-analysis and systematic review of population-based observational studies." *Frontiers in Immunology* 13:872542–872542.

Yao, Song, Chi-Chen Hong, Edward A. Ruiz-Narváez, Sharon S. Evans, Qianqian Zhu, Beverly A. Schaefer, Li Yan, et al. 2018. "Genetic ancestry and population differences in levels of inflammatory cytokines in women: Role for evolutionary selection and environmental factors." *PLoS Genetics* 14 (6): e1007368.

Yashin, Anatoli, Alexander S. Begun, Serge I. Boiko, Svetlana V. Ukraintseva, and Jim Oeppen. 2001. "The new trends in survival improvement require a revision of traditional gerontological concepts." *Experimental Gerontology* 37 (1): 157–167.

Yashin, Anatoli, Alexander S. Begun, Serge I. Boiko, Svetlana V. Ukraintseva, and Jim Oeppen. 2002. "New age patterns of survival improvement in Sweden: Do they characterize changes in individual aging?" *Mechanisms of Ageing and Development* 123:637–647.

Yassin, A., K. AlRumaihi, R. Alzubaidi, S. Alkadhi, and A. Al Ansari. 2019. "Testosterone, testosterone therapy and prostate cancer." *Aging Male* 22 (4): 219–227. https://doi.org/10.1080/13685538.2018.1524456.

Yusuf, Salim, Philip Joseph, Sumathy Rangarajan, Shofiqul Islam, Andrew Mente, Perry Hystad, Michael Brauer, et al. 2020. "Modifiable risk factors, cardiovascular disease, and mortality in 155 722 individuals from 21 high-income, middle-income, and low-income countries (PURE): A prospective cohort study." *The Lancet* 395 (10226): 795–808.

Zhao, Yang, Ji-Long Ren, Ming-Yang Wang, Sheng-Ting Zhang, Yu Liu, Min Li, Yi-Bin Cao, et al. 2013. "Codon 104 variation of p53 gene provides adaptive apoptotic responses to extreme environments in mammals of the Tibet plateau." *Proceedings of the National Academy of Sciences* 110 (51): 20639–20644.

Zhao, Zhongwei. 1997. "Long-term mortality patterns in Chinese history: Evidence from a recorded clan population." *Population Studies* 51 (2): 117–127.

Zheng, Hui, Yang Yang, and Kenneth C. Land. 2011. "Heterogeneity in the Strehler-Mildvan general theory of mortality and aging." *Demography* 48 (1): 267–290.

Zheng, Rongshou, Shaoming Wang, Siwei Zhang, Hongmei Zeng, Ru Chen, Kexin Sun, Li Li, Freddie Bray, and Wenqiang Wei. 2023. "Global, regional, and national lifetime probabilities of developing cancer in 2020." *Science Bulletin* 68 (21): 2620–2628.

Zhu, Pingfen, Weiqiang Liu, Xiaoxiao Zhang, Meng Li, Gaoming Liu, Yang Yu, Zihao Li, et al. 2023. "Correlated evolution of social organization and lifespan in mammals." *Nature Communications* 14 (1): 372. https://doi.org/10.1038/s41467-023-35869-7.

Zink, Albert, L. Samuel Wann, Randall C. Thompson, Andreas Keller, Frank Maixner, Adel H. Allam, Caleb E. Finch, et al. 2014. "Genomic correlates of atherosclerosis in ancient humans." *Global Heart* 9 (2): 203–209.

Zipes, Jack. 2003. *The Complete Fairy Tales of the Brothers Grimm All-New Third Edition.* New York: Bantam.

Zuk, Marlene. 2013. *Paleofantasy: What Evolution Really Tells Us about Sex, Diet, and How We Live.* New York: WW Norton.

INDEX

Page numbers in *italics* refer to pictures, graphs, and tables.

Aboriginal Australians: acculturation of, 95, 99, 116; causes of death by age in, *434*; cultural traditions of, *98*; and dancing, *98*; and death in old age, 116; dementia in, 333, 334; diabetes in, 319–21; and health interventions, 319–21; heart disease in, 320–21; heart disease risk factors in, 325; modal lifespan in, *117*, 118, *433*; postreproductive lifespan in, *120*; role of elders in, *98*; study of, 95–96; and use of painting, *98*

accidental deaths, 108–9, 114–15, 434

acculturation: of Aboriginal Australians, 95, 99, 116; of Ache (Paraguay), 129; of Agta (Philippines), 126, 129, 130; and child care, 384; and contemporary hunter-gatherer groups as tainted, 71; effects of at different times, 127; effects of on mortality, 129–30, 450n80; of Herero (Botswana), 129; and infant and child mortality, 104, 107, 125; of Ju/'hoansi (Botswana and Namibia), 129, 450n80; and life expectancy, 125; and modal lifespan, *117*, 118; and perceptions of aging, 382; and population decline, 123; and roles of elders, 247; and study of contemporary hunter-gatherer groups, 70, 75; of Tsimane (Bolivia), 129; and violence, 114; and well-being, subjective, 371; of Yanomamö (Nigeria), 130

Ache (Paraguay): accidental deaths in, 108, 129, *434*; acculturation of, 129; age estimations of, 79, 447n29; causes of death by age in, 108, *434*; child sacrifice in, 108; cultural changes in, 422; depression in, 376; elders in, 262, 266, 267–68, 463n36; and grief, 377; history of, 79; homicide in, 113, 129; and hunting, 169, 175, 177, 211; illness in, 107, 129; infant and child mortality in, 101, *103*, 105, 107, 108, 129–30; life expectancy of,

111; male fertility in, 197; maternal mortality in, 112; modal lifespan in, *433*; mortality rates in, 129; mothers in, 78; and numbers, 15; and pejorative terms, 441n1; population changes in, 126, 166; postreproductive lifespan in, *120*; sharing and cooperation in, 79; social living in, 425; study of, 2, 78–79; survival curve for, *103*; violent deaths in, 107, 108; work hours in, 215

activities of daily living (ADL), 293, *294*

Aeta (Philippines), 123

ageism, 11–12, 36, 263, 380–82, 417

aging: actuarial, 37, 39; as art, 424–25; cultural views on, 264–65; definition of, 37; as disease, 36–37, 57–60; effects of modernization on, 382–83; and environment, 52–54; expectations vs. reality, 401, *402*; fear of, 401; features of in humans, 38, 48–49, 53, 283–84; features of in non-human animals, 31, 50, 165; language of, 17; mechanisms of, 9, 56–57, 58; and mythology, 41, 158, *158*; onset of, 53; of populations, 18–19; rate of, 4, 52, 53, 54, 62; reimagining of, 390–92, 417. *See also* aging, evolution of; aging, physiological; aging, reasons for

aging, biological. *See* senescence

aging, evolution of: and antagonistic pleiotropy, 44–46; and cognitive changes, 304–6; and disposable soma theory, 47–48, 57–58, 455–56n44; and elder contributions, 170, 192–93, 216, 246, 421; and life history, 39, 46; and longevity, 3–4, 7, 186, 208–13; and natural selection, 38–40, 42–44, 47, 53, 57, 60, 199, 304; in nematodes, 55; and "out-with-the-old, in-with-the-new" hypothesis, 40–41, 443n5; and reproduction, 3, 4, 49, 53, 163; and safety factors, 284–86, *285*, 297, 298. *See also* aging, reasons for; evolutionary mismatch